W9-CRB-488

Radiotracer Studies of Interfaces

INTERFACE SCIENCE AND TECHNOLOGY
Series Editor: ARTHUR HUBBARD

In this series:

INTERFACE SCIENCE AND TECHNOLOGY – VOLUME 3

Radiotracer Studies of Interfaces

Edited by

G. Horányi

Institute of Materials and Environmental Chemistry
Chemical Research Center, Hungarian Academy of Sciences
Hungary

2004

ELSEVIER
ACADEMIC
PRESS

Amsterdam – Boston – Heidelberg – London – New York – Oxford – Paris
San Diego – San Francisco – Singapore – Sydney – Tokyo

chem
0134462544

ELSEVIER B.V.
Sara Burgerhartstraat 25
P.O. Box 211, 1000 AE Amsterdam
The Netherlands

ELSEVIER Inc.
525 B Street, Suite 1900
San Diego, CA 92101-4495
USA

ELSEVIER Ltd
The Boulevard, Langford Lane
Kidlington, Oxford OX5 1GB
UK

ELSEVIER Ltd
84 Theobalds Road
London WC1X 8RR
UK

First edition 2004

Library of Congress Cataloging in Publication Data
A catalog record is available from the Library of Congress.

British Library Cataloguing in Publication Data
A catalogue record is available from the British Library.

ISBN: 0-12-088495-x
ISSN (series): 1573-4285

∞ The paper used in this publication meets the requirements of ANSI/NISO Z39.48-1992 (Permanence of Paper).
Printed in The Netherlands.

Working together to grow
libraries in developing countries

www.elsevier.com | www.bookaid.org | www.sabre.org

ELSEVIER   BOOK AID
             International   Sabre Foundation

ac

XP   ccc
     2/2/05

Preface

This collection from prominent scholars offers an authoritative survey on the application of the radiotracer technique for the study of interfacial phenomena. Considering the wealth of this field the reader should not expect that the authors have covered the whole field, however, an attempt was made to present a selection of illustrative examples characteristic for different types of interfaces. The structure of this volume, a part of a series dealing with interfaces, is based on the classical subdivision according to the nature of the interfaces namely:

solid/fluid (solid/gas and solid/liquid)

solid/solid

liquid/fluid (liquid/gas and liquid/liquid)

It was, however, evident that a strict adherence to this subdivision would cause problems and could lead to misunderstanding as generally the different fields of the applications of the radiotracer technique are distinguished according to the branch of science or technology where the methods are used. This is the reason why separate chapters are devoted to the sorption and transport phenomena in biomembranes, to the role of sorption phenomena in contamination and decontamination of nuclear reactors and to environmental problems. It is hoped that owing to the diversity of the topics covered the volume will be useful reading for academic and industrial researchers working on a wide variety of subject areas, for example, catalysis, electrochemistry, colloid science, corrosion science and biochemistry.

This volume is devoted to the memory of George Charles de Hevesy, a Hungarian born chemist, who was awarded the fourth Nobel Prize for research related directly to chemical dynamics in 1943, "for his work on the use of isotopes as tracers in the study of chemical processes".

George de Hevesy was born in Budapest on August 1st 1885, the son of Louis de Hevesy, Court Counsellor, and Eugénie, née Baroness Schosberger. After matriculating at the Gymansium of the Piarist Order in 1903 he studied at Budapest University and Berlin Technical University and he gained his doctor's degree at the University of Freiburg im Breisgau in 1908. He worked for two years as an assistant at the Institute of Physical Chemistry, Technical University of Switzerland. He traveled to England in 1910 to study under Professor Ernest Rutherford at Manchester. Early in 1913 he interrupted these studies to carry out jointly with Frederic Paneth the first radioactive-tracer experiment at the Vienna Institute of Radium Research. During his stay in Vienna he obtained the Venia Legendi in the University of Budapest. In 1914 he was drafted into the Austrian-Hungarian Army. After the end of the war he taught for 6 months at the University of Budapest and left in the spring of 1919 for Copenhagen to discuss his future activities at Niels Bohr's Institute, which was to be erected. In 1920 he settled in Copenhagen, where in 1923 together with Coster he discovered the element hafnium. In 1926 he returned to Freiburg as Professor of Physical Chemistry. In 1930 he was appointed Baker Lecturer at Cornell University, Ithaca. Four years later he again took up his activities at Niels Bohr's Institute. In 1943, following the German occupation of Denmark, Hevesy escaped to Sweden and became professor at Stockholm.

He did the pioneering work in the use of isotopic indicators both in inorganic and life sciences and he was involved in the first clinical use of isotopes. He is considered as the Father of Nuclear Medicine.

He was the author of several important books on radiochemistry. He was awarded the Cannizaro Prize (Academy of Sciences, Rome) in 1929, was elected as a Member of the Hungarian Academy of Sciences in 1945, he was the Copley Medallist (Royal Society, London) in 1949, Faraday Medallist in 1950, Baily Medallist in 1951 and Silvanus Thompson medallist in 1955. In 1959 he received the Ford Foundation's Atoms for Peace Award Medal, in 1961 the Niels Bohr Medal and the Rosenberger Medal of the University of Chicago. Honorary degrees conferred upon Professor de Hevesy include Doctor of Philosophy, Uppsala, Freiburg, and Copenhagen; Doctor of Science, Ghent, Liège, London, and Capetown; and Doctor of Medicine, São Paulo, Rio de Janeiro, Turin, and Freiburg. He was a Fellow of the Royal Society (London), the Swedish Academy of Sciences, Gothenburg Academy, and eleven other scientific academies. He was Honorary Fellow of the Chemical Society (London), the Royal Institution (London), the British Institute of Radiology, the Finnish Chemical Society, the German Bunsen Society, the German Physiological Society, the Chemical Society of Japan, and the American Society of Nuclear Medicine. In addition, he held honorary memberships of many more learned societies.

George de Hevesy died in 1966 in Freiburg. His ashes were returned to Hungary recently and interred with great reverence in Hungarian soil.

The community of Hungarian scientists is not only very proud of Hevesy's achievements, but we are following in his footsteps. This volume could be a proof of this statement. The continuity and existence of the "Hungarian link" is demonstrated by the expertise of the authors of this volume. The legacy of radiotracer studies started by Hevesy and Groh in the basement of a building of the University in Budapest is preserved and developed in many laboratories in Hungary.

G. Horányi
Institute of Materials and Environmental Chemistry,
Chemical Research Center, Hungarian Academy of Sciences

CONTENTS

Chapter 4

Chapter 5

Chapter 6

Chapter 7

Chapter 9

Chapter 10.1

Chapter 10.2

Radiotracer Studies of Interfaces
G. Horányi (editor)

Chapter 1

Historical background

G. Horányi

Institute of Materials and Environmental Chemistry, Chemical Research Center, Hungarian Academy of Sciences, PO Box 17, H-1525, Budapest, Hungary

The birth of the radiotracer technique dates back to the beginning of the twentieth century. The veritable narrative of the story of the birth of the radiotracer technique, the application of isotopic indicators, can be found in Hevesy's famous two volume book of his collected papers. "Adventures in Radioisotope Research" [1]. In the following the story will be narrated by Hevesy himself.

"...The method of isotopic indicators had its ultimate origin in the Institute of Physics at the University of Manchester, which then was under the inspiring leadership of the great physicist, the late Lord Rutherford...."

"...The years I had the privilege to spend in Rutherford's laboratory in Manchester, between 1911 and 1914, witnessed some of the greatest discoveries in the history of physics. I could follow from close quarters the discovery of the atomic nucleus and how Rutherford devised, carried out, and interpreted the results of experiments. All this was done with the greatest ease, without visible effort...."

"...When I was in Manchester, Rutherford was much interested to come into the possession of a strong radium D sample. Large amounts of radium D were stored in the laboratory, but imbedded in huge amounts of lead...."

"...One day I met Rutherford in the basement of the laboratory where the lead chloride was stored. He addressed me by saying: "If you are worth your salt, you separate radium D from all that nuisance of led." Being a young man, I was an optimist and felt sure that I should succeed in my task. Trying during a year all sorts of separation methods and making the greatest efforts, it looked sometimes as if I succeeded, but I soon found out that it was radium E, the disintegration product of radium D, a bismuth isotope, which I separated. The result of my efforts was entire failure. To make the best of this depressing situation, I thought to avail myself of the fact that radium D is inseparable from lead, and to label small amounts of lead by addition of radium D of known activity obtained from tubes in which radium emanation decayed. From such tubes pure radium D can be obtained.

It was the Vienna Institute for Radium Research which owned in those days by far the greatest, amount of radium and, correspondingly, of radium emanation. This fact induced me to interrupt my stay in Manchester and to proceed to Vienna. In the Vienna Institute there were very large amounts of lead chloride, obtained from pitchblende as well, and Paneth, assistant at the Institute, unaware of my efforts at Manchester, made very extensive studies to achieve separation. His great efforts were as abortive as mine. At my suggestion we associated in the application of labelled lead. The first use of this method, early in 1913, was the determination of the solubility in water of sparingly soluble salts such as led sulphide and lead chromate...."

In the paper they published in 1913 [2] a clear formulation of the principles of radiotracer technique can be found.

"...The fourth decay product of radium emanation, RaD, is known to exhibit all the chemical reactions of lead; if RaD is mixed with lead or lead salts it cannot be separated from the lead by any chemical or physical method and if complete mixing of the two substances has taken place then the same concentration ratio is maintained whatever amount of lead is withdrawn from the solution. Sice RaD can be determined in much smaller amounts, owing to its radioactivity, than lead, it may be employed for the qualitative and quantitative estimation of lead to which it as been added; the RaD is an indicator of the lead...."

Remembering those days spent in Vienna Hevesy wrote:

"...When we started with Paneth in the first days of 1913 to apply radium D as a tracer of lead, the word "isotope" as not yet coined. Groups of radioactive substances such as mesothorium and radium, or ionium and thorium, were denoted by Soddy as "chemically inseparable elements"...."

The first radiotracer studies by Hevesy and Paneth constitute an important contribution to the clarification of the problem of isotopic elements. In their paper dealing with the isotope concept [3] they point out the problems about this concept.

"...The theory of isotopic elements was not readily acceptable to chemists and physicists; to the former, because ever since the formulation of the periodic system they had been accustomed to regard the atomic weight as a fundamental property of an element; to the latter, because there was no known instance in which two different elements exhibited the same spectrum and such a hypothesis seemed difficult to unify with the prevailing ideas on the origin of the nature of isotopic elements was simultaneously given considerably more weight by the ideas, developed by E. Rutherford and N. Bohr, on the constitution of the atom, and by the experiments of Moseley on the X-ray spectra of the elements...."

In the same paper they pose the question: "Can isotopic elements replace each other?" and give an unambiguous answer: yes.

(According to an anecdote the first "radiotracer experiment" was carried out by Hevesy in 1911, i.e. two years earlier. At that time he lived at a boarding house in Manchester, England. He suspected that the landlady was recycling previously served meat at dinner, which she indignantly denied. The coming Sunday Hevesy in an unguarded moment added some active deposit to the freshly prepared pie and on the following Wednesday, with the aid of an electroscope he demonstrated to the landlady the presence of the active deposit in the soufflé.)

Beginning from 1913 the next 10-15 years saw a remarkable extension of radiotracer studies. Instead of going into details a list of the titles of the papers published by Hevesy and his co-workers gives a good oversight on the topics concerned:

The velocity of dissolution of molecular layers (1914) [4].

The exchange of atoms between solid and liquid phase (1915) [5].

The intermolecular exchange of atoms of the same kind (1920) [6].

Self-diffusion in solid lead (1921) [7].

Self-diffusion in solid metals (1925) [8].

Radiochemical method of studying the circulation of bismuth in the body (1924) [9].

The application of radioactive indicators in the biology (1926) [10].

In connection with this list may be it is worthy of note that the self-diffusion studies were started in his native city, Budapest [7]. In the following section of his autobiography he mentions this period:

"...For several months after the end of the war it was not possible to leave Hungary. During these months I started with my friend Groh to study self-diffusion in molten and in solid lead, using radium D as an indicator. We fused a radiolead-rod on to the top of an inactive lead-rod, heated this solid system to $200^{\circ}-300^{\circ}$, and determined the dislocation of the radium D atoms. From the extent of dislocation, the rate of self-diffusion of lead was calculated. This early, rough method was improved later during my stay in Copenhagen...."

According to Hevesy the characteristic epochs of the first fifty years of the history of the application of radioactive tracers are as follows.

From 1913 to 1934 only radioactive isotopes found in nature were applied in tracer studies. The fundamental discovery of artificial radioactivity by Frédéric Joliot and Iréne Curie made it possible to label almost every element. From 1934 to 1937 artificial isotopes produced under the action of neutrons emitted by radium-beryllium sources were applied in the production of radioactive tracers. The year 1937 witnessed the opening of a new epoch due to the availability of strongly active samples produced by the cyclotron. After the second world war nuclear reactors produced radioactive isotopes of almost unlimited activity became available, among others ^{14}C and ^{3}H. Samples of minute activity of these radioactive bodies were available prior to that date. ^{14}C discovered in 1940 by Ruben and Kamen became the most frequently applied radioactive isotope in life sciences, instead to the formerly most extensively used ^{32}P.

Although in the atmosphere of the cold war between the 1950s and the 1970s certain fields of nuclear chemistry and physics were under military and national security control, the application of the radiotracer technique in different fields of science and technology saw an unprecedented development. Nowadays the problem of the possible environmental contamination by the nuclear industry and the questions connected with the safe treatment of nuclear waste require the application of the radiotracer technique in the relevant model studies.

Nuclear medicine is now clearly recognized as a new area of medical care and radiopharmaceutical chemistry has become an important branch of pharmacology and biochemistry.

ACKNOWLEDGEMENT

Financial support from the Hungarian Scientific Research Fund is acknowledged (OTKA Grants T 031703 and T 045888).

REFERENCES

[1] G. Hevesy, Adventures in Radioisotope Research, Pergamon Press, New York, 1962.
[2] G. Hevesy and F. Paneth, Z. anorg. Chem., 82 (1913) 322.
[3] G. Hevesy and F. Paneth, Phys. Z. 15 (1914) 797.
[4] G. Hevesy and E. Rona, Z. phys. Chem., 89 (1914) 294.
[5] G. Hevesy, Phys. Z., 15 (1915) 797.
[6] G. Hevesy and L. Zechmeister, Ber. dtsch. chem. Ges. 53 (1920) 410.
[7] J. Groh and G. Hevesy, An. Phys., 65 (1921) 216.
[8] G. Hevesy and A. Obrutsheva, Nature, 115 (1925) 674.
[9] I.A. Christiansen, G. Hevesy and S. Lomholt, C. R. Acad. Sci., Paris 178 (1924) 1324.
[10] G. C. Hevesy, Chem. Soc. J. (1951) 1618.

Radiotracer Studies of Interfaces
G. Horányi (editor)
© 2004 Elsevier Ltd. All rights reserved.

Chapter 2

Advantages of the radiotracer technique

G. Horányi

Institute of Materials and Environmental Chemistry, Chemical Research Center, Hungarian Academy of Sciences, PO Box 17, H-1525, Budapest, Hungary

Despite the demonstration of the value of the tracer technique in early studies, the technique did not come into common use until after World War II when relatively large amounts of cheap radionuclides became available through the use of nuclear reactors.

The main advantages of using radiotracers are as follows:

a) The radiation emitted by radiotracers is generally easy to detect and measure with high precision.

b) The radiation emitted is independent of pressure, temperature, chemical and physical state.

c) Radiotracers do not affect the system and can be used in nondestructive techniques.

d) The radiation intensities measured furnish direct information concerning the amount of the labeled species and no special models are required to draw quantitative conclusions.

The primary assumption for the use of radioactive isotopes is that radioactive isotopes are chemically identical to stable isotopes of the same element, i.e. the substitution of a radioactive isotope for a stable one in a compound does not change the type or strength of the chemical bonds nor does it effect the physical properties of the compound. Although the validity of this assumption from a theoretical point of view is hardly acceptable as in most cases the difference in mass between the various isotopes does cause some changes in the properties of the molecule considered, in most case this isotope effect is rather small and difficult to detect.

Only in the case of systems involving hydrogen-tritium substitution could the isotope effect play an important role.

For heavier elements these effects are neglected, however, it should be taken into consideration that the possible role of isotope effect should be examined depending on the precision and sensitivity of the measurements.

An important assumption in the use of the tracer technique is that the radiation emitted does not affect the chemical and physical properties of the system studied, and the decay product following the radioactive disintegration does not have any influence on the behavior of the system. However, in the case of high radiation intensities secondary radiolytic effects could occur. In order to avoid these effects the tracer experiment should be designed carefully. The radiation intensity should be high enough to provide reliable data but small enough not to produce noticeable radiolytic effects.

In the case of some radionuclides the parent-daughter relationship may cause complications. For instance, if ^{137}Cs (β^-, γ, $t_{1/2}$ 30 y) is used for the study of a process occurring with Cs the radiation measured is originating simultaneously from the parent ^{137}Cs and ^{137}Ba (IT, $t_{1/2}$ 2.5 min) daughter elements. However, considering the low half-life value of the latter isotope, the so-called radioactive equilibrium is reached in 20–25 minutes. (As it is well known that this is not an equilibrium state, we have to do with a steady state.)

In the "equilibrium" the ratio of the radiation intensities corresponding to ^{137m}Ba and ^{137}Cs is constant thus the total intensity measured during the experiment is a true measure of amount of cesium (Cs) alone.

In other cases when the attainment of the radioactive equilibrium requires longer time the discrimination of parent and daughter elements should be carried out by use of energy discrimination techniques.

Finally, in the case of radiotracers with short half-life (in comparison to the time scale of the experiments) the changes caused by the decay of the radioisotope should be taken into consideration.

The radiotracer studies can be carried out with radionuclides both without and with carrier. A sample containing a pure radionuclide is called a carrier-free tracer. Adding to the sample some amount of non-radioactive isotope of the same element, the carrier, we obtain a tracer with carrier as the non-radioactive isotope "carries" the radioactive species.

The concentration of radioactive species in the most carrier-free cases is very low.

Considering the relationship

$$\frac{dN}{dt} = \frac{\ln 2}{t_{1/2}} N$$

where, N is number of the molecules of the radioactive isotope present in the sample considered, $\dfrac{dN}{dt}$ is the rate of the decay, the activity of the sample (the number of atoms disintegrating in 1 sec = 1Bq) and $t_{1/2}$ is the half-life of the isotope (in seconds) the chemical amount of isotopes with various half-life corresponding to 1 MBq radioactivity is shown in Table 1.

Table 1.
Amount of the isotope in the case of 1 MBq radioactivity

half life of the isotopes	chemical amount of the isotope/mol
1 hour	1.6×10^{-16}
1 day	3.8×10^{-15}
1 year	1.4×10^{-12}

Dissolving 1 MBq carrier-free $H_2{}^{35}SO_4$ (half life about 80 days) in one liter water the concentration of the solution with respect to H_2SO_4 will be about 3×10^{-13} mol dm^{-3}. It is important to emphasize that the chemical and physical behavior of the molecules at very low concentrations should differ significantly from that observed at macro concentrations. Therefore, in many cases the tracer experiments are carried out with species containing a macro amount of carrier. Nevertheless carrier-free radionuclides are used to study phenomena occurring at trace concentrations. Some studies of this kind were carried out at the very beginning of the application of the tracer technique.

To this group belong the study of the adsorption of radioactive species at the walls of vessels and containers, the investigation of the formation of radio colloids and equilibrium precipitation and crystallization studies at low concentrations.

The electrochemistry of species present in trace concentrations was also an interesting field for a long period.

Another field is the elaboration of tracer separation methods with aim of separation and isolation of carrier-free radioisotopes.

The main fields where radiotracers (mostly with carrier) are used can be summarized as follows:

1) Analytical chemistry (radiometric analysis, isotope dilution analysis, activation analysis etc).

2) General physical and colloid chemistry (determination of reaction paths study of exchange processes, determination of equilibrium constants, studies of surface processes and phenomena occurring in solids).

3) Biology and life sciences.

The use of labeled compounds in the life sciences is very extensive. (Radiotracer study of the interaction of living species with the environment, study of uptake of various components; biochemical analysis: autoradiography, radioimmunoassay, DNA analysis emission computer tomography, radiation therapy.)

4) Industrial applications. (Level and thickness measurement flow measurement, leak detection, residence time studies etc.)

In accordance with its title the present volume is devoted to give a survey on the application of radiotracer technique for the investigation of phenomena occurring at interfaces in general.

Considering the wealth of the experimental evidences and theoretical considerations reported in the literature only some illustrative examples will be shown for solid/fluid and liquid/fluid interfaces.

Separate chapters deal with sorption and transport phenomena in biomembranes and with some practical aspects, laying the stress on the problems connected with nuclear reactors and environmental issues. A chapter is devoted to characteristic features of the experimental technique and to the problems of instrumentation.

It is self-evident that only radiotracer studies carried out in heterogeneous systems will be discussed so no attention will be paid to phenomena taking place in homogeneous phases.

The main stress is laid on studies devised to clarify the behavior of the interfacial layer of two adjoining phases.

The radiotracer technique is extremely suitable for detecting the presence (adsorption) or absence of a given species in the interfacial layer and for giving direct information concerning the dynamic behavior of adsorbed species related to the question of the reversibility of the adsorption. The reversibility of adsorption implies that desorption of adsorbed species takes place at a measurable rate. The rate of desorption can be easily estimated by the exchange of adsorbed labeled species with nonlabeled ones added to the solution phase or vice versa. In the case of irreversibility of the adsorption, the absence of any exchange clearly demonstrates the situation.

The clarification of these questions can be considered as the first phenomenological level of understanding and interpretation of interfacial processes.

This first step should be taken in almost all radiotracer studies, before devising techniques enabling us to gain quantitative data for the description of the interface studied. A more detailed discussion of these questions can be found in Chapter 4 in connection with the radiotracer study of phenomena occurring at electrode surfaces.

ACKNOWLEDGEMENT

Financial support from the Hungarian Scientific Research Fund is acknowledged (OTKA Grants T 031703 and T 045888).

Radiotracer Studies of Interfaces
G. Horányi (editor)
© 2004 Elsevier Ltd. All rights reserved.

Chapter 3

Adsorption and catalytic reactions at solid/gas interfaces

Z. Schay

Institute of Isotope and Surface Chemistry, Chemical Research Center, Hungarian Academy of Sciences, P.O. Box 77, H-1525 Budapest, Hungary

1. INTRODUCTION

The extensive use of radioisotopes in heterogeneous catalysis goes back to the sixties of the previous century. At that time the main research area was the elucidation of the mechanism of the reactions using ^{14}C and ^{3}H labeled molecules. Campbell and Thomson [1] and Thomson [2] have reviewed in 1975 and in 1980 already the use of radioactivity in surface science and its application to characterization and observation of working catalysts. Later the research area expanded to study the development of the active catalyst and to measure simultaneously and independently adsorption and catalysis on the working catalyst. Recently, the majority of publications deals with the use of ^{35}S in hydrodesulfurization and positron emitters as ^{11}C, ^{13}N and ^{15}O in operando studies of the working catalysts. It has been recognized that catalysts never remain unchanged during catalysis and the reactant(s) and the catalysts form a constantly changing system in which the so called active sites, which are the sites where the catalytic cycle runs, develop during the catalytic reaction itself. In many cases the number of the active sites seems to be only a small fraction of the possible sites and so-called spectator species are present on the surface of the catalyst.

Some advantages of using radioisotopes against stable ones are the extremely low detection limit of the label, the relative simple and cheep instrumentation, the possibility to measure not only the gas phase but directly the surface of the working catalyst, and thus to measure directly the mass balance, to locate the label along the catalyst bed under real catalytic conditions by using positron emitters. Some drawbacks are the general concern in the public to use radioactivity and the increasingly strict regulations, it is also difficult to determine the position of the label inside the molecule, and very fast processes are difficult to follow because of the random nature of radioactive decay.

There are some special advantages and disadvantages in radiation protection, detection and quantification when soft β^- emitters as ^{14}C or ^3H are used. In one side there is no need for additional shielding of the normal laboratory glass equipment and the risk of contamination and incorporation is low. On the other side the self-absorption in the solid has to be considered when the radioactivity on the catalyst is measured as discussed in an early study with ^{14}C labeled compounds [3]. To minimize self absorption the specific radioactivity of the reaction products in gas phase was determined by burning to CO_2, precipitating as $BaCO_3$ and measured in the form of a thin layer using Geiger-Müller counter with thin mica window.

In this chapter the use of radioisotopes in heterogeneous catalysis including adsorption during or before catalysis will be outlined. Some typical applications will be selected to show the reader the possibilities and capabilities of using radioisotopes in heterogeneous catalysis. It is not the goal of this chapter to review all the literature in this subject. Application of radioisotopes in adsorption processes not directly relevant to heterogeneous catalysis such as air filters or adsorbers, ion–exchangers, migration in soil, etc. as well as the study of exoelectrons are excluded.

2. EQUIPMENT

In the first studies using mainly ^{14}C labeled hydrocarbons, Geiger-Müller detectors with mica window were used. Two basic approaches were applied to measure the radioactivity. In the beginning only the products of the catalytic reaction were analyzed via transforming them into $BaCO_3$ as outlined in the introduction of this chapter. Later, the radioactivity on the catalyst was measured *in situ*. A typical arrangement after Webb et al. [4] for the latter is shown in Fig. 1.

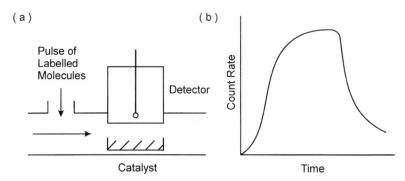

Fig. 1. (a) Geiger-Müller detector for measurement of adsorption.
(b) Detector signal of a pulse of labelled compound.
Reproduced from [4].

In the first version, a thin layer of the catalyst was placed into a boat and moved magnetically in and out of sight of the detector. Later on, two detectors were used one in sight of the catalyst the other one detecting the gas phase with the same geometry. The arrangement was used to study the selective hydrogenation of ethyne in excess ethene near to room temperature. Typical count rates for a pulse of labeled hydrocarbon are also displayed in Fig 1(b). The maximum of the curve gives the sum of reversible and irreversible adsorption, from the slopes the rates of adsorption and desorption can be calculated and the residual radioactivity is proportional to the irreversible adsorbed amount of the labeled compound. Some authors used proportional counter instead of Geiger-Müller one, which resulted better counting efficiency, especially when it was used in the flow through mode connected to a gas chromatograph. It was also typical to collect the fractions after GC separation and count in a liquid scintillation spectrometer to get absolute activities. Davis et al. [5] used a rotatable surface barrier detector attached to an ultrahigh vacuum machine to study the chemisorption of ^{14}C-benzene on Pt(111) crystal face.

More sophisticated detectors were constructed using europium doped CaF_2 window as scintillator, which converts β^- particles from the radioactive decay into light. The light is conducted to cooled photomultipliers by cooled glass rods using optical silicone to create contacts between the scintillator, the glass rod and the photomultiplier as shown in Fig. 2. [6].

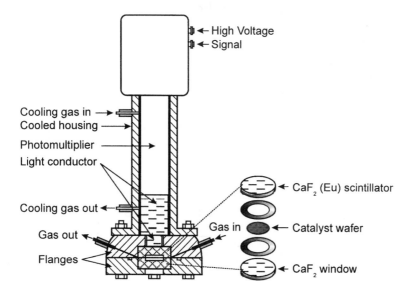

Fig. 2. Reactor and scintillation detector for in situ measurement of adsorbed radioactive species. Reproduced from [6].

The cell is similar to that used for *in situ* transmission FT-IR studies and the catalyst sample is in the form of a self-supporting wafer. The gas flow is distributed on both sides of the wafer. The cell can be heated to 700-800 K, when using graphoil gaskets. Instead of Geiger-Müller or proportional counters, nowadays Eu doped CaF_2 is used in radio gas chromatography, too.

The spread of the use of PET (positron emission tomography) initiated the application of the short half-time β^+ radioisotopes in operando studies. By using a special arrangement of an array of detectors shown in Fig. 3. [7] the position of the label in the reactor under real catalytic conditions can be followed with high spatial and time resolution. This is due to the annihilation process in which two γ-photons of the same energy of 511 keV are radiated in opposite directions. This energy is high enough to cross the wall of a catalytic reactor as well as the furnace around it, and can easily be detected in a distance by using conventional γ-detectors. The position of the detectors sensing the two photons in coincidence defines the line (in the case of a quasi one dimensional catalyst bed, the position) on which the annihilation occurred. Because of the short half-life of the positron emitters used in catalysis (^{11}C 20.4 min, ^{13}N 9.96 min ^{15}O 2.07 min) the catalytic reactor has to be close to a cyclotron. This is a real constraint in the application of this technique.

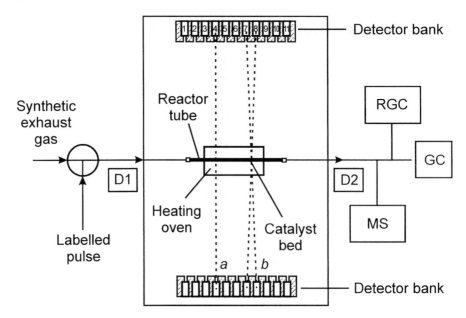

Fig. 3. Schematic drawing of the experimental setup to measure positron emission in operando studies. Reproduced from [7].

3. DETERMINATION OF ACTIVE CENTERS

In heterogeneous catalysis it is essential to characterize the sites on which the catalytic cycle runs. This includes the determination of the site or active center density on the surface. The reader is reminded that usually only a small fraction of the surface of the catalyst participates in the catalytic cycle itself. A large fraction of the surface is occupied by so called spectator species which are strongly bounded and do not undergo the desired catalytic transformation. The use of tracers, especially radioactive isotopes, makes it easy to distinguish between the two forms. In one approach, labeled reagent is added to the reaction mixture when the system is in steady state conditions and the time dependence of the build up of radioactivity on the catalyst and in the products is measured. In the other one the reverse sequence is applied, namely steady state is reached using the tracer and its disappearance is followed after a switch for unlabeled reactants. In special reactions stepwise poisoning of the catalyst can be used to estimate the number of the active sites from the catalytic activity versus the amount of poison added.

3.1. Polymerization catalysts

In the polymerization of ethene and propene the use of Ziegler- or Phillips-type heterogeneous catalysts is dominating. A basic understanding of the mechanism of the systems requires information on the number of active sites under real catalytic conditions. Tait et al. [8-10], Fink et al. [11,12] as well as some other recent papers [13,14] propose the use of ^{14}CO radio-labeling technique to determine the number of growing polymer chains on the catalysts. The polymerization reaction is quenched by addition of a well-determined amount of ^{14}CO and left in contact for 10-30 min before isolation of the polymer by alcohol. The polymer is decontaminated with ethanol acidified with HCl, washed, filtered and dried to constant weight. To eliminate the effect of the particle size the samples are ground in liquid nitrogen before counting in scintillation gel. Another possibility to measure the radioactivity of the polymer is burning the polymer in oxygen and absorbing the liberated CO_2 in a basic solution as 2-aminoethanol/methanol and measuring by liquid scintillation technique. Under ideal conditions the number of the active center equals the number of ^{14}CO. The proposed mechanism for radiolabeling of the polymer involves co-ordination and insertion of ^{14}CO to the active metal site and quenching (decontaminating) with ROH/H^+:

$$^{14}CO$$
$$\downarrow$$
$$L_xMe - CH_2 - CH_2 \sim P + {}^{14}CO \rightleftharpoons L_xMe - CH_2CH_2 \sim P$$

$$^{14}CO$$
$$\downarrow$$
$$L_xMe - CH_2 - CH_2 \sim P \longrightarrow L_xMe - {}^{14}CO - CH_2 - CH_2 \sim P$$

$$L_xMe - {}^{14}CO - CH_2 - CH_2 \sim P \xrightarrow{ROH/H^+} H{}^{14}C - CH_2 - CH_2 \sim P$$
$$\underset{O}{\overset{\parallel}{}}$$

The reliability of the method depends on the stoichiometry of the interaction of CO with the active transition metal - growing polymer bonds. No multiple insertion reactions or co-polymerization reactions should take place but the insertion reaction should be complete during the contact time. It is typical that the ^{14}CO uptake increases with contact time and pressure as reported by Tait et al. in Fig. 4. [8].

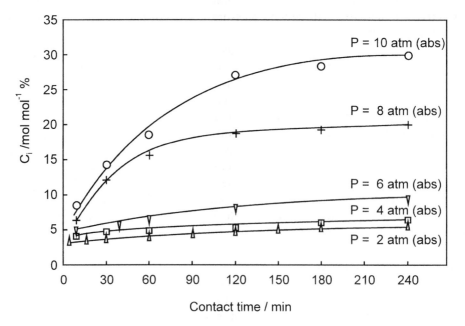

Fig. 4. Specific radioactivity of the polymer as a function of contact time at varying pressures at 80 $^{\circ}$C. Reproduced from [8].

There is no consensus of this phenomenon but the results are consistent with the involvement of monomer in side reaction, which incorporate additional [14]CO into the polymer chains resulting in an overestimation of the number of the active sites. This becomes more important at higher polymerization pressures and longer contact times and justifies the use of short contact time. On the other hand very short contact times may result in incomplete insertion of [14]CO into the transient metal - polymer chain resulting in an underestimation of the number of active sites. A careful selection of the amount and contact time of [14]CO is essential to obtain realistic values of active centers using this method.

Tritium labeled methanol can also be used for the determination of the number of active centers [9]. The main limitation of this method is that CH_3O^3H reacts with nonpropagative or inactive metal-polymer bonds and an extrapolation to zero polymerization time is needed.

A dual labeling approach using a combination of the [14]CO labeling and ^{3}H quenching [13] can harmonize the apparent discrepancies of the individual methods and under optimized conditions can provide comparable data.

3.1.1. Examples

In the polymerization of ethene by supported chromium-SiO_2 Phillips-type catalyst it was established [8] that only about 3.2-7.3% Cr is active; the fraction of active Cr increases with ethene pressure but is constant throughout a given polymerization run.

In the polymerization of ethene by $TiCl_4/MgH_2$-supported Ziegler-type catalyst [12,13] also a maximum of only 14% of the titanium is active and the dramatic decrease in the polymerization activity with increasing Ti content is a consequence of the decrease of the number of active centers. It was found that in ethene homo- and co-polymerization an $AlEt_3$/Ti mole ratio of about 50 results in maximum activity and maximum number of active centers. Low Ti-content in the $TiCl_4/MgH_2$ catalyst results in the formation of active centers for homo-polymerization, while high Ti content favors co-polymerization.

In the polymerization of propene on the complex $TiCl_4$-diisobutyl phthalate-$MgCl_2$-$AlEt_3$-phenyltriethoxysilane catalyst [14] the number of active centers decreases but the propagation rate constant increases in the presence of hydrogen due to steric reasons.

3.2. Occupancy principle in catalysis

Norval and Thomson [15] applied the occupancy principle derived by Orr and Gillespie [16] for the determination of adsorption during catalysis using ^{14}C-chlorobenzene hydrogenation on silica supported palladium catalyst as an example. The basic idea used in occupancy principle is to divide the system into interacting compartments and by monitoring the time course of the count rate of a labeled pulse passing across the compartments under steady state conditions the capacity (the size of the active pool) of the compartments can be determined. The occupancy principle states that the ratio of occupancy to capacity is the same for all parts of the system and equals the reciprocal of the entry flow. From this the population in each compartment, in catalysis the number of molecules adsorbed reversible on the catalyst, can be determined. Irreversibly adsorbed or spectator species are not detected and the pool of the active molecules on the catalyst may be used as measure of the number of active centers.

The experimental set-up was similar to the one shown in Fig.1. An additional chamber of the same geometry was applied to measure the elution from the gas phase without catalyst. It was concluded that in the temperature range of 193 to 448 K the support acted as a reservoir for chlorobenzene to react on the palladium particles. A large number of chlorobenzene molecules were involved in adsorption and exchange without forming benzene.

Norval et al. [4] applied this principle to determine the number of adsorbed ethene molecules on Ir/Al_2O_3 catalysts under hydrogenation reaction. ^{14}C labeled pulses of ethene in hydrogen were passed over the catalyst and the radioactivity of the effluent gas was monitored continuously, as described above. The adsorbed amount corresponded to a turnover number of 3.6 molecules per second per active site. It was concluded that only 20% of the exposed metal surface took part in the hydrogenation.

4. DEVELOPMENT OF ACTIVE CATALYST

In the previous section the use of tracers for counting the number of active centers is outlined. There is an other important question in heterogeneous catalysis namely to understand how do the active centers develop during the different steps of preparation and activation of the catalyst. In some cases the use of radioisotopes offers unique possibilities as discussed in the followings.

4.1. Acid sites

Chlorination of alumina results in enhancement of Brönsted or Lewis acidity and is an important step in regeneration of noble metal based naphtha reforming catalysts. Thomson et al. [17] studied the chlorination of alumina with ^{36}Cl and ^{14}C labeled CCl_4 as well as ^{36}Cl labeled HCl. $^{14}CCl_4$ adsorbed weekly on calcined γ-alumina at room temperature. On increasing the temperature of the adsorption, the ^{14}C radioactivity sharply decreased and no radioactivity was observed at 500 K. Using ^{36}Cl labeled CCl_4 the radioactivity sharply decreased over the range 293-473 K, a case similar to $^{14}CCl_4$, but increased again over the range 473-503 K indicating a breakdown of the C-Cl bond. Only Cl was retained by the surface and the carbon part left it as $OCCl_2$. Part of the ^{36}Cl activity could be removed by isotope exchange using HCl. In contrast, chlorination with $H^{36}Cl$ resulted in surface chlorine species which could be nearly completely removed by isotope exchange using HCl. Obviously there are two types of chlorine, a labile and a stable one which can be assigned to Brönsted and Lewis acid sites, respectively. Brönsted sites containing labile chlorine form by replacement of terminal hydroxyl groups by terminal chlorine, whereas Lewis acid sites form as unsaturated $AlCl_2$ in which both chlorine bridge neighboring aluminum atoms on and directly below the surface.

4.2. Reduction of catalysts

Paál and Thomson [18,19] studied the effect of hydrogen left on platinum black catalyst after reduction with ^3H. In these studies tritium occupied subsurface positions and could be removed by exchange with inactive hydrogen pulses. Surprisingly, the first inactive hydrogen pulse injected into helium flow at room temperature did not contain any tritium. Subsequent pulses contained more and more tritium indicating a slow migration of tritium to the surface. The metal and the tritium together formed the active catalyst for hydrocarbon reaction as surface tritium was incorporated into hydrocarbons reacted on it.

4.3. Hydrodesulfurization catalysts

The increasing demand for low sulfur containing fuels revitalized the use of ^{35}S in the development of so called deep-hydrodesulfurization (HDS) catalysts. In these studies radiotracers are used to follow the formation of the active catalyst from the oxide to the final sulfide form as well as to study the mobility of sulfur in sulfide form as the catalytic activity is supposed to be connected with it via formation of coordinatively unsaturated sites (CUSs) and vacancies.

In recent papers Kogan et al. [20-22] proposed a method to evaluate the relative number of CUSs and surface SH groups in $CoMo/Al_2O_3$ catalysts. The calcined $CoMo/Al_2O_3$ catalysts were sulfidized in an autoclave by using elemental sulfur-^{35}S and 6MP H_2 at 360 °C. 100 mg sulfided samples were transferred into a microreactor and treated in He flow at 360 °C for 30 min to remove adsorbed sulfur compounds, followed by hydrogenation for 7-8 min to activate the catalyst by partial reduction and creating vacancies. Any increase in the time of hydrogenation did not result in additional removal of ^{35}S. After this pretreatments the catalysts were kept in hydrogen flow and pulsed by 1 µl pulses of unlabeled tiophene. The radioactivity of the products was measured by radio-gas chromatography. In each case H_2S was the only product containing ^{35}S thiophene remained always unlabeled. The molar radioactivity of H_2S produced by each thiophene pulse was plotted as a function of total H_2S formed. The experimental curves were approximated by a sum of exponentials of $A_i exp(-\lambda_i x)$ type, where x stands for the amount of H_2S formed, and each exponential describes one type of active site. The number of exponentials to fit the experimental curve gives the number of different active sites, A_i relates to the number and λ_i to the mobility of S in the same type of sites, respectively. If more than one exponential is needed to fit the experimental curve, it means that there are different sites for HDS reaction and they differ in the mobility of their SH groups. For details see [20]. Based on this characterization of active sites Kogan et al. found that unpromoted Mo/Al_2O_3 catalysts have only one type of active sites and on increasing the Mo content in the range of 2-12 wt. % A_i increases nearly linearly but λ_i remains practically constant. On promotion by cobalt, an other type of active site forms, differing in reactivity of SH groups. The two sites were named as "rapid" and "slow" sulfur and related to Mo and Co sites, respectively. Increasing the amount of Co, there is an increase in the number of "slow" and a decrease in the "rapid" sites, respectively. The same time P_i, the productivity defined as the ratio of the number of converted thiophene molecules to the number of the SH groups of the site of a given λ_i, dramatically increases for the "rapid" sites. It was concluded that for HDS of light molecules e.g. a gasoline fraction, the catalyst has to have the highest activity of the "rapid" sites, whereas for heavy oils the "slow" sites are efficient. Poisoning by N-containing compounds results in a decrease in the productivities of all types of sites and the decrease is practically independent of the type of the poison. The mechanism of the poisoning is related to a decrease in the mobility of surface SH groups and the reduction of vacancies.

Massoth et al. [23] studied the sulfur uptake and exchange in a flow recirculation reactor on combinations of Mo, W, Ni, Co, Pd and Pt, supported on alumina using ^{35}S labeled H_2S. By monitoring the gas phase radioactivity by a Eu doped CaF_2 scintillation detector they concluded that the amount of exchangeable sulfur correlates well with thiophene HDS activity measured in separate experiments. In a theoretical analysis of the exchange experiments, equations were derived in which exchange equilibrium, adsorption equilibrium, and gas phase mole balance were considered for the calculation of the amount of sulfur exchange in the sulfided catalysts.

Dobrovolszky et al. [24-26] studied the initial sulfur uptake of the oxidized catalysts by pulsing it with ^{35}S labeled H_2S in H_2. Using alternating pulses of H_2S as sulfiding agent and tiophene or cyclohexanol as reactants the development of catalytic activity and sulfur uptake and release were monitored. The thiophene conversion was inversely proportional with the amount of irreversibly hold sulfur. Pulses of cyclohexanol stopped sulfur uptake. Pulsing labeled H_2S into a continuous stream of unlabeled H_2S/H_2 [26] indicated the permanent interaction between the sulfided NiW and NiMo catalyst and gas phase H_2S.

The group of Kabe [27-34] studied HDS of dibenzothiophene (DBT) not only on Mo based catalysts but used Ru, Cs, Cr, Ni, Pt and Co, too. In these studies ^{35}S labeled DBT and H_2S were used and the radioactivity was measured by liquid scintillation technique. H_2S was absorbed in a basic scintillation solution and samples of the liquid products and were collected for measurement of their radioactivity. The sulfur uptake of the catalysts was calculated from the mass balance. Typical sequence of the measurements was to stabilise the sulfided catalysts in HDS of unlabeled 1 wt % DBT in decalin, followed by a purge with decalin and admitting ^{35}S labeled DBT as shown in Fig. 5. The amount of labile sulfur was calculated from the area of labeled H_2S released on introduction of unlabeled DBT after purging with decalin. This experiments also proved that labile sulfur couldn't be removed by purging alone. As mentioned before [20] the labile sulfur is not uniform and depending on the concentration of the organosulfuric compound only some part but not all participates in the ultra deep HDS [34], the fraction of which decreases on decreasing sulfur content of the feed. A summary of the most recent results is in press [35]. Analyzing the tracer experiments on Cr/Al_2O_3 catalysts [28] it was found that the cause of the increase in DBT activity with increasing Cr content was not the increase in the number but the change of the nature of the active sites. On the other hand the increase in catalytic activity with the addition of Co or Ni to Mo/TiO_2 increased the number of active sites [30] without changing their nature. Obviously each system has its characteristic feature with respect to the formation of the active centers.

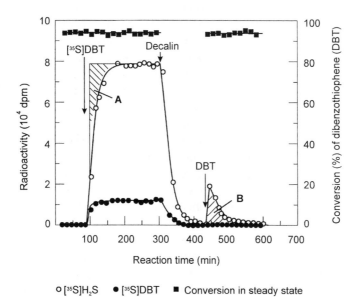

Fig. 5. Changes in the radioactivities of unreacted ^{35}S]DBT (dibenzoltiophene) and formed ^{35}S]H_2S with reaction time at 360 °C over sulphide/Mo/TiO2 catalyst.
  A ^{35}S remained on the catalyst
  B ^{35}S released
  Reproduced from [29].

5. TRACER STUDIES TO DETERMINE REACTION MECHANISM

The use of radioactive tracers in studying the mechanism of reactions has the advantage over the use of stable ones that due to the extreme high sensitivity of the detection of the label, products of side reactions or reaction routes of minor contribution to the main one can easily be detected. The analysis of the experiments is similar to that of SSITKA (steady state isotopic transient kinetic analysis) or TAP (transient analysis of products). In the experiments either pulses or a continuous feed of the labeled molecule is applied in steady state conditions. An other possibility is to dose the catalyst before the actual reaction with labeled molecules and follow the appearance of the label when using unlabeled reactants (or in reverse sequence) or purge the system (in inert or reactive flow) and follow the removal of the label.

5.1. Mechanism of hydrodesulfurization

In 4.3. the formation of the active HDS catalyst was outlined. In most of these studies the role of the mobile sulfur is also discussed. The general conclusion is that this type of sulfide sulfur is part of the active center of the reaction but do not participate in the reaction itself. Their role is to govern the formation of vacancies for the adsorption of the reactant [e.g.20]. A so called "forcing out" mechanism operates according to which the sulfurorganic compound is adsorbed on the active center, characterized by a balance between SH groups and vacancies. The C-S bond breaks at the expense of dissociatively adsorbed hydrogen not belonging to any SH group and the sulfur part forms a new SH group. The balance between the vacancies and SH group is disturbed and a metastable center forms. This stabilizes by forcing out any of the SH groups with reaction of hydrogen to form gas phase H_2S creating simultaneously a new vacancy and closing the catalytic cycle. This mechanism is additionally supported by the fact that under HDS conditions gas phase H_2S do not exchange with the active sulfided catalyst as proved with ^{3}H-labeled H_2S [36]. A computer simulation of the regeneration of the CUS in combination with ^{35}S experiments is described by Dumeignil et al. in [35]. Additional reactions such as isomerization, hydrogenation etc. are also important but are outside of the scope of the present chapter.

The role of adsorbed carbon overlayers in thiophene HDS was studied by Bussell and Somorjai [37] over Mo(100) single-crystal surface using ^{14}C labeling. The carbon overlayers were produced by thermal decomposition of ^{14}C labeled C_2H_4 at 1100K. The radioactivity was measured in situ in the UHV chamber with a Si(Li) surface barrier detector. They have shown that carbon from ethene remains on the Mo(100) surface during HDS and the activity of the clean Mo(100) is the same as that of carbided one.

5.2. Selective hydrogenation

The polymerization of ethene or propene requires ultra pure feedstock, especially the removal of traces ethyne or higher acetylenes as well of dienes. This initiated extensive research to remove this compounds in low concentration by selective hydrogenation. Besides the industrial importance there was a theoretical interest in this reactions as to find out why was it possible to remove by hydrogenation practically all of ethyne in large excess to ethene without considerable conversion of ethene itself into ethane. The use of ^{14}C labeled compounds was extremely successful in elucidating the mechanism of the reactions. The pioneering work of Al-Ammar and Webb [38-40] was followed by the group of Weiss [41-46] who used not only ^{14}C labeling but introduced the double labeling technique [44] using deuterium and ^{14}C to elucidate simultaneously the routes of carbon and hydrogen in the course of the reaction. The typical catalyst for this reaction is Pd/Al_2O_3 type with low palladium content. In some formulations silver or other modifiers are added to reduce the unselective hydrogenation of ethene and suppress the formation of oligomers. For the same purpose in some applications CO is added to the feed. As the reaction runs typically at or around room temperature it was easy to measure not only the radioactivity of the products but also the radioactivity remaining on the catalyst surface.

The general assumption is that the selectivity is determined by thermodynamic factors, which control the surface coverages of ethyne and ethene together with mechanistic ones, which determine the hydrogenation of ethene without intermediate desorption from the surface. The selectivity strongly depends on the experimental conditions, especially on the concentration of ethyne and hydrogen. When ethyne concentration is low, hydrogenation of gas phase ethene becomes significant. There is a side reaction in which C_4 and higher hydrocarbons are formed. ^{14}C tracers proved that ethyne is the source of this polymerization type reaction. Margitfalvi et al. [41] proved by using $^{14}C_2H_2$ that there is a direct route for ethane formation from ethyne via a concerted mechanism. This route is dominant at low ethyne partial pressure. Under this conditions ^{14}C labeled ethene was converted to ethane, too, however on increasing the partial pressure of ethyne, this route completely ceased. On fresh catalyst the main reaction route of ethyne is formation of ethene [42]. As time on stream increases and C_{4+} oligomers accumulate on the catalyst, hydrogenation of ethene becomes significant without changing the reaction routes of ethyne. Addition of traces of CO to the reactants [43] immediately blocks the direct formation of ethane and also the hydrogenation of gas phase ethene. Addition of copper to the Pd/Al_2O_3 catalyst results in similar changes in selectivity as addition of CO to the gas phase.

Using ^{14}C tracers, Webb and co-workers [38,39,47] studied the hydrogenation of ethyne on other supported metal catalysts, too. The general feature of the reaction was the same as for Pd/Al$_2$O$_3$ one, namely ethyne reacts mainly to form ethene but the direct route for ethane formation is also operative. There is a build up of olygomers on the catalyst surface, too. The ratio of the different reaction routes depends on the type of the catalyst and the partial pressure of the reactants.

5.3. Fischer-Tropsch synthesis

In the middle of the 20th century there was an increased interest in improvement of the Fischer-Tropsch (FT) process due to the dramatic increase in the price of crude. A large number of publication as well as some books dealt with different aspects of the reaction including studies of the initialization and chain growth mechanism. ^{14}C labeled hydrocarbons, alcohols and other oxygenates were used [48-54]. It has been found that adding various oxygen containing labeled compounds to the synthesis gas, the label appears in the products of the FT reaction on a reduced iron catalyst. The incorporation of primary alcohols was very efficient and yielded products with constant radioactivity per mole for hydrocarbons of higher carbon number. It was proved that growth of chains occurred at the carbon atom to which the alcoholic OH group was attached. The mechanism generally accepted today, namely the formation of CH$_2$ units via carbide on the surface and they polymerization, were not supported by early tracer experiments [3]. In this experiments Kummer et al. carbided the catalyst by decomposition of ^{14}CO and the appearance of the label was followed in the products using unlabeled CO + H$_2$ feed. They concluded that no more than 8-30% of the methane formed originates from the labeled carbide even at very low conversion of the carbide. In this early publication from 1948, the experiments were planned in a very careful way and the possible pitfalls were discussed, too, including the possibility of a rapid isotope exchange in the carbide. The failure to observe the importance of the so called carbide mechanism is due to the rapid transformation of the catalytically active surface carbide, formed only in the reaction under real catalytic conditions, into inactive bulk and surface carbides if not all the reactants are present. This example underlines that extreme care and caution is essential in planning and interpreting tracer experiments as the authors stressed it, too.

Blyholder and Emmett [52] used ketene labeled on the CH$_2$ group to distinguish between possible reaction routes. According to their assumptions if ketene acts as a chain initiator in a manner similar to alcohols the molar radioactivity of the products would be independent of the carbon number. If the ketene dissociates into CO and CH$_2$ but the latter does not participate in the chain growth then the products would be practically not radioactive and the hydrogenation of the CH$_2$ species to methane results in radioactive product. If the polymerization of CH$_2$ groups is the main route of the synthesis than the molar radioactivity of the products would be proportional to the carbon number.

The tracer experiments carried out on iron catalyst at one atmosphere resulted in nearly constant molar radioactivity of the products. The slight increase in molar radioactivity with increasing carbon number was basically considered as experimental error, but the possibility of a very slight build in of CH_2 groups was not excluded. A chain building mechanism was proposed in which the growing chain is bound to the surface by a metal-oxygen bond as well as a carbon-metal bond.

Tau et al. [55,56] reinvestigated recently the product distribution of the FT reaction using an industrial C-73 "doubly-promoted" catalyst under real industrial conditions in a stirred autoclave reactor. There was a complete product analysis up to C_{40}. ^{14}C - ethanol and pentanol labeled in the carbinol position and ^{14}C-ethene were added to the synthesis gas. They found evidence for two chain growth mechanisms. One of them involves incorporation of labeled alcohol via oxygen containing surface intermediate while the other reaction pathway operates via an oxygen-free one and produces only alkanes. The two pathways are completely independent. Adding ^{14}C-ethene to the synthesis gas proved that the primary role of ethene is to initiate chain growth, which accounts for more than 85% of the total incorporation.

Our current knowledge of the mechanism of the Fisher-Tropsch synthesis is summarized by Davis in [57]. It is concluded that the so-called oxygenate mechanism dominates over Fe based catalysts but the surface carbide one does over Co based ones.

5.4. Dehydrocyclization of alkanes and hydrogenation of aromatics

5.4.1. Dehydrocyclization

Dehydrocyclization is an essential reaction in naphtha reforming therefore an enormous number of papers was published in this field. In a recent review by Davis [58] the present knowledge of dehydrocyclization mechanism has been summarized. Radioactive isotopes are used in two ways. In the first one a special position of the reactant is labeled and the distribution and if possible the position of the label in the products is determined. A possible mechanism consistent with all the results is proposed. In the other approach, the so-called kinetic isotope method [59], a compound supposed to be an intermediate, is labeled and added to the reactants (or inversely an inactive intermediate is added to the labeled reactant). Again, the distribution of the label is determined and a possible mechanism is proposed. By this method only the existence of a reaction pathway can be proved but it does not give any information of the contribution of this pathway to the overall transformation. In other words the kinetic isotope method may detect a minor or insignificant reaction pathway. The consumption of the original form of the key component means the limit beyond which the method cannot be applied. Both methods have been widely utilized not only with radioactive but also with stable isotopes, as well.

A typical example for the first method is a study by Sárkány [60] of aromatization of [3-^{14}C] methylpentane over Ni catalysts. It was concluded that benzene formed both by intramolecular rearrangements and by CH_4 addition-abstraction mechanism as shown in Fig. 6. The hydrogen partial pressure plays an important role in controlling the ratio among the possible reaction routes as it governs the amount of the carbonaceous deposit formed in the reaction which in turn changes the apparent ensemble size of Ni.

a) C$_5$ cyclization-ring enlargement

b) Vinyl shift-1,6-dehidrocyclization

c) CH$_x$ addition-abstraction

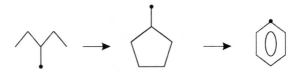

Fig. 6. Routes of benzene formation from [3-^{14}C]methylpentane.

A typical example for the second method is the work of Paál and Tétényi [61,62] on dehydrocyclization of hexenes and n-hexane on platinum catalysts. In this experiments a mixture of [1-^{14}C] hexene-1 and inactive hexatriene-1,3,5 or [1-^{14}C]n-hexane and inactive n-hexene-1 were converted. As the label appeared in the hexatriene and in the hexene fraction, respectively, it was concluded that the reaction sequence hexane => hexene => hexadiene => hexatriene => cyclohexadiene => benzene takes actually place. In the dehydrogenation of six-membered cyclic hydrocarbons [63] no detectable transfer of radioactivity has been found from n-hexane or n-hexene to cyclohexane, indicating that the probabilities of the dehydrocyclization via cyclohexane intermediate is negligible.

Kilner et al. [64] used a combination of radioactive ^{14}C and stable ^{13}C labeling on converting $[1\text{-}^{14}C\text{-}2\text{-}^{13}C]$-ethyl-cyclopentane with chromia-alunima catalyst. The position of both labels in toluene, the product of the dehydrocyclization, was determined and compared with the predictions of different pathways. The distribution of the labels was consistent with a bifunctional mechanism in which ethyl-cyclopentene, formed by dehydrogenation at the metallic site, underwent ring expansion at the acidic sites of the support and was dehydrogenated on the metallic site to form toluene. The other pathway, in which the first step is hydrogenolysis of the ring to produce labeled n-heptane and 3-methyl-hexane followed by cyclization made little contribution. The determination of the position of ^{13}C label is usually made by mass spectrometry making use of the fragmentation of the molecule in the ion source, by microwave spectroscopy or NMR. The determination of the position of the ^{14}C label needs stepwise chemical degradation of the molecule by special reactions, which split the original molecule only at one position in each step.

Dehydrocyclization of various ^{14}C labeled hydrocarbons on different catalysts [58, 65-72] revealed that there is no general mechanism for dehydrogenation. The pathway depends on the reactant, hydrogen pressure, the catalyst and also the time on stream. The latter is rather unexpected, but is mainly due to changes in the structure and carbon coverage of the catalyst in the initial period. Pines and co-workers [66,68] observed an unexpectedly low methyl label of about 18% for the first samples collected from the conversion of $[1\text{-}^{14}C]$-n-heptane or $[1\text{-}^{14}C]$-n-octane. This is close to the value, which results from a complete scrambling of the label to each position of toluene and xylene, respectively. Later the ratio of the methyl label increased to about the value, expected for a direct six-carbon ring closure.

The vast majority of the experiments with ^{14}C labeled hydrocarbons on mono-functional catalysts, using nonacidic supports, is consistent with more than 80% of the aromatics being formed by a direct six-carbon ring closure mechanism. The contribution of other reactions as chain lengthening, ring expansion, isomerization, CH_x addition-abstraction, bond-shift is usually marginal under real industrial conditions of high temperature and hydrogen pressure, provided that nonacidic support is used. At low temperatures, below about 300 °C, a five-carbon ring forms in the first step of the reaction. It can undergo secondary reactions as hydrogenolysis, ring enlargement, etc., resulting in isomers, aromatics or even coke. The reader has to be reminded that on monofunctional catalysts isomerization occurs much faster than the formation of aromatics and the above conclusions refer to the formation of aromatics only. The isomers are formed through C_5-ring formation and subsequent hydrogenolysis with a faster rate than the formation of aromatics.

On bifunctional platinum catalysts with acidic supports, the selectivity for aromatics production from n-octane is considerably higher than on monofunctional ones. There are two parallel pathways. The first one is C_6-ring closure followed by dehydrogenation to C_8 aromatics, both taking place on the platinum sites. The second one involves C_8 isomerization by both sites in three steps (olefins – isomerization – hydrogenation), dehydrogenation of the isomers on platinum, C_5 ring formation followed by ring expansion on the acidic sites and hydrogenation on platinum. This pathway is more rapid than the cyclization on the platinum site.

It can be concluded that dehydrogenation is a complex reaction. To find the detailed mechanism one has to be extremely cautious in interpreting experimental data, including tracer experiments, because of the potential for many kinetic disguises [58].

5.4.2. Hydrogenation of aromatics

In the seventies Debrentsev et al. [73,74] studied the hydrogenation of benzene on a series metal catalyst to clear up the discrepancy between the predictions of Balandin's multiplet theory and the observation of cycloolefins among the products of benzene hydrogenation. The multiplet theory predicts a one step hydrogenation of benzene via so called sextet mechanism, without unsaturated intermediates. Using the kinetic isotope method, the authors added inactive cyclohexadiene to the ^{14}C labeled benzene and measured the radioactivity of all products. Based on a careful analysis of the distribution of the label in the products the authors concluded that both stepwise and direct hydrogenation of benzene did actually occur. Because of the uncertainty in the determination of specific radioactivities as well as considering the limitations of the method, the author have not given any ratio for the contribution of the two pathways to the overall reaction rate.

5.5. Other hydrocarbon reactions
5.5.1. Hydrogen transfer

Hydrogen transfer steps are essential in hydrogenation, hydrogenolysis, hydrocracking, cyclization etc. reactions. The activation of molecular hydrogen has been extensively studied by H_2-D_2 exchange reaction over all metals of catalytic importance. In this paragraph the application of isotopic exchange labeling for the study of hydrogen transfer step between hydrogen donor and acceptor hydrocarbon species will be outlined. Parravano [75] studied the hydrogen transfer between benzene and other C_6 hydrocarbons by measuring the rate of redistribution of ^{14}C tracer between the hydrocarbons in thermodynamic equilibrium with the catalyst surface. From these data the rate coefficient for the hydrogen transfer reaction can be calculated which gives the rate of the hydrogen transfer step between hydrocarbon and catalyst surface. In [75],

Parravano applied the idea to the ^{14}C-benzene – cyclohexane pair; the reaction was carried out at different benzene/cyclohexane partial pressures over Pt, Pd, Ir, Rh and Ru supported on alumina or silica with or without adding hydrogen. The effect of the metal particle size was also studied. It was possible to calculate the number of adsorbed hydrogen atoms added to benzene in the rate-controlling step. On Pt/Al$_2$O$_3$ catalyst at 117 °C up to 50 nm Pt particle size one hydrogen atom was added in the rate-controlling step while 5-6 hydrogen atoms on larger Pt particles. As expected, the number of hydrogen atoms added increased with increasing temperature. The rate constants per unit of metal surface area, calculated from the rate coefficients, were practically independent of the particle size.

5.5.2 Alkyl transfer and homologation

Parravano [76] investigated the redistribution of ^{14}C label in binary mixtures of benzene – toluene, benzene – ethylbenzene, toluene – xylene, toluene – ethylbenzene and cyclohexane – methylcyclohexane over alumina and magnesia supported Pt, Ir, Ru, and Au catalysts in a flow system. The experimental conditions were chosen to avoid the formation of products other than the labeled and unlabeled starting compounds. Alkyl transfer occurred under mild catalytic conditions for each pair and each catalyst. The kinetic effects were interpreted as a result of the competitive and reactive adsorption of the corresponding hydrocarbon mixture.

Paál and co-workers [77,78] studied in a pulse-microcatalytic apparatus the homologation of [1-^{14}C]-n-hexane and [3-^{14}C]-methylpentane over Pt black. By comparing the measured and calculated specific radioactivities they could differentiate between the two extreme mechanisms discussed in the literature before, namely methylene transfer or complete breakdown of the alkane into C$_1$ units followed by insertion of one of the C$_1$ units into the unreacted alkane. They concluded that the complete breakdown of the parent alkane is the dominating pathway.

5.5.3. Hydrogenolysis, cracking, isomerization and dehydrogenation

Guczi et al. [79] studied the hydrogenolysis of ethane over nickel powder catalyst in the presence of deuterium or tritium. The specific radioactivity of methane formed was practically constant and about twice of that of tritium added. The specific radioactivity of ethane increased linearly in time. The results were interpreted by assuming equilibrium exchange in ethane and a negligible exchange in gas phase methane. The authors concluded that the initial step for hydrogenolysis is the dehydrogenation of ethane to form 1,2-diadsorbed species followed by the rate-determining step of C-C bond rupture. The CH$_x$ fragments formed, may undergo further exchange before desorbing in the form of labeled methane.

Hightower and Emmett [80] studied the catalytic cracking of n-hexadecane over a commercial silica-alumina catalyst. Small amounts of ^{14}C labeled primary products as ethene, propene, propane, pentene, benzene, toluene, n-heptane, and n-decane were mixed with a stream of helium saturated with n-hexadecane to investigate the importance of some secondary reactions. About 60 components and the coke remaining in the catalyst were identified and analyzed for radioactivity in each tracer experiment. With the exceptions of propene and pentene the other labeled hydrocarbons underwent less than 10% conversion. About 40% of propene was hydrogenated and 25% participated in nonaromatic alkylation. For pentene cracking, aromatization, coke formation, and isomerization were also significant. Most of the paraffins above C_2 were formed by hydrogen transfer to the corresponding olefins.

Hightower et al. [81,82] studied the double bond isomerization in butenes over alumina and silica-alumina catalysts using [1-^{14}C]-n-butene-1 – cis-butene-2 and [1-^{14}C]-cis-butene-2 – n-butene-1 reactant mixtures. Butenes were the only reaction products with no more than 1% of carbonaceous deposit formed on the surface of the catalysts. In the tracer experiments the isomerization of the labeled and unlabeled butenes were measured simultaneously as two separate and independent reactions. The product distributions were calculated as a function of time. Assuming that the surface reactions are rate controlling and using a fitting procedure, the rate constants for the so called triangle mechanism, in which there is a direct conversion of each butene into the two other ones, were determined. It has to be noted that there is an equivalent description of the isomerization, called as Y mechanism, in which the reaction proceeds *via* a common intermediate. For the deeper evaluation of the results the authors used a special time dependent term, α_a^i, defined as the ratio of the specific radioactivity of isomer i to the special radioactivity of the initially labeled isomer at a given time in the course of the reaction. This parameter is very sensitive to changes in the rate constants and reflects minor changes in the mechanism. The time dependence of the α value was converted to it's dependence on the conversion of the unlabeled isomer. This α vs. conversion plots were used to identify which of the two mechanisms operates on a given catalyst. Over silica-alumina catalysts the Y mechanism operates, the common intermediate being a carbonium ion. Over alumina the reaction proceeds through different surface complexes and on different sites.

Paál et al. [83] used also radiotracers to distinguish between the triangular and consecutive pathway for the formation of cyclohexanone and phenol from cyclohexanol over Ni powder catalyst. Using ^{14}C-cylohexanol – inactive cyclohexanone reactant mixture the authors concluded that the consecutive pathway cyclohexanole => cyclohehanone => phenol is the dominant one over the direct formation of phenol, as the specific radioactivities of cyclohexanone and phenol were about the same and both were lower than that of cyclohexanole.

Davis [84] studied the dehydrogenation of methyl-[1-^{14}C]-cylohexane during naphtha reforming over acidic and nonacidic Pt/Al$_2$O$_3$ catalysts. He found that under real reforming conditions about 90% of the methylcyclohexane was dehydrogenated to toluene, about 4% was demethylated to benzene and the balance was converted to C$_1$-C$_5$ gases, liquid paraffins and naphthenes. He concluded that dehydrogenation and demethylation occur on the metallic function, while the conversion of the primarily formed aromatics to paraffins needs metal-acid sites.

6. ADSORPTION, CHEMISORPTION

In these studies the adsorbed amount is measured directly using detector arrangements shown in Figs. 1, 2. In some cases the adsorbed amount is determined from the mass balance or by measuring the desorbed amounts by elution techniques. The application of positron emitters will be discussed in a separate section. As in the previous sections, some typical applications will be presented only.

Tétényi and Babernics [85] investigated the chemisorption of ^{14}C-benzene on nickel, platinum, and copper powders. Labeled benzene was adsorbed and the chemisorbed quantity was determined by the specific radioactivity of the solution obtained after the labeled compound was eluted by unlabeled benzene or hydrogen. The eluate was condensed in unlabeled benzene – cyclohexane mixture, which was later separated by gas-chromatography and the radioactivity in each peak eluting from the gas-chromatograph was measured. From the specific activity of cyclohexane and benzene the authors made conclusions on the mechanism of benzene desorption. The weekly chemisorbed benzene was removed by benzene at 150 °C in the form of 5% cyclohexane and 95% benzene. Additional hydrogenation at 150 °C removed strongly chemisorbed benzene in the form of 85% cyclohexane and 15% benzene. The ratio of the two forms was 2:1. Additional hydrogenation at 380 °C removed only traces of cyclohexane. The authors found dissociative adsorption of benzene on nickel and platinum, while there was no adsorption on copper.

Davis et al. [5] studied the chemisorption and rehydrogenation of ^{14}C-benzene in submonolayer amounts on Pt(111) single crystal surface using in situ in the ultrahigh vacuum chamber a surface barrier detector optimized for low energy β^- radiation. Irreversible, dissociative chemisorption was observed in the whole temperature range of 45–350 °C. The maximum surface coverage at 350 °C was about 2 carbon atoms per one surface platinum atom. As discussed in [85], the authors also proved that benzene chemisorbed around room temperature can completely be removed by hydrogen. On the other hand, the radioactivity cannot be completely removed by hydrogenation when benzene was chemisorbed at 300 °C. The rehydrogenation of this strongly bound benzene (or carbon deposit) proceeded in two distinct stages. As there was no product analysis reported, no conclusions about the form of this strongly hold radioactivity were given. A side product of the tracer work was the absolute calibration of the Auger C/Pt signal of the AES part of the machine against the ^{14}C counts. From the latter, the absolute amount of carbon on the surface could be calculated, as the detector response was calibrated against a ^{14}C-polymethylmethacrylate sheet placed in the position of the Pt single crystal.

Schay et al. [86] studied the adsorption and removal of ^{35}S on spent and regenerated industrial Pt/Al_2O_3 and $Pt-Re/Al_2O_3$ naphtha reforming catalysts by direct measurement of the amount of the radioactivity retained by the catalysts after sulfidation. It has to be reminded that in the industrial practice some sulfur is added to the feed during start up, when a Pt-Re bimetallic catalyst is used, to avoid excessive cracking. The authors observed that the two catalysts retained about the same amounts of sulfur, but only 35% and 19% of sulfur was removed by hydrogen at 470 °C in 2 hours from the Pt and the Pt-Re catalysts, respectively. After repeated oxidative regeneration cycles the catalysts still contained about 39% and 65% of the original amount of sulfur, respectively, indicating that the bimetallic catalyst retains sulfur very strongly. About 20% of the initial sulfur was located on the support, the remaining 80% on the metals. In the hydrogenolysis of n-hexane, used as a test reaction, the product distribution over the sulfur containing catalysts was similar to that of the carbon formed at ageing. Similar results were reported by Pönitzsch et al. [87] by measuring the desorbed amounts of $H^{35}S$ absorbed in KOH. This indicates that both approaches, namely the direct counting of the catalysts and the measurement of the desorbed species, can be used in adsorption studies.

In a series of publications with the main title "Radiochemical studies of chemisorption and catalysis" Webb and co-workers [88-91] studied the adsorption and co-adsorption of ^{14}C-ethene, ^{14}C-ethyne, ^{14}C-propene, ^{3}H, and ^{14}C-CO over Ni, Rh, Pd, Ir, and Pt catalysts supported on alumina and silica using in situ detection of radioactivity on the catalysts by Geiger-Müller detector as shown in Fig. 1. In these studies a possible correlation was looked for between the retention, adsorption strength, mobility of the adsorbed species, displacement of the adsorbed species by an other one etc. and the selectivity for ethyne hydrogenation. As a general tendency a large fraction of the chemisorbed ethene and propene was retained by the catalysts as an unreactive strongly adsorbed species. Using tritium, it was proved that not only the metal but the support also participates in the hydrogenation and hydrogen migration between the support and the metal is of importance. The same time, migration of the hydrocarbon species from the metal to the support was also proved. Chemisorption of CO reduced or completely blocked the chemisorption of hydrocarbons on the metals; only ethyne could displace about 40% of CO.

7. POSITRON EMITTERS IN CATALYSIS

The use of β^+ emitters offers some advantages over the use of other radioisotopes as it was outlined in section 2. of this chapter. The main advantage is that they can be used in operando studies, it means under real catalytic conditions, to determine the position and the amount of the label along the catalyst bed in the time scale of about one second by using an array of detectors opposite to each other as shown in Fig. 3. The arrangement is a simplified version of the PET camera used in nuclear medicine as in catalysis there is no need for the location of the label in all three spatial dimensions; it is sufficient to locate it in one dimension along the axis of the reactor. An additional advantage is they high specific activity of 10^{20} Bq mol^{-1}, resulting in a high sensitivity of the detection in the picomole range. The disadvantage is the short half-life of the isotopes of catalytic interest, which makes the quantitative evaluation difficult and does not allow the study of slow processes in the catalyst bed. Because of the short half-life the catalytic reactor has to be close to the production site of the radioisotope, it means close to a cyclotron. This limits the application of the method to a few laboratories where a cyclotron and catalytic research meet, like the State University of Ghent in Belgium, the Eindhoven University of Technology in The Netherlands and the Paul Scherrer Institute in Switzerland.

The nonmedical applications of a PET camera were already reviewed in 1991 [92]. The techniques for studying the dynamics of processes with time scales ranging from milliseconds to days were outlined on examples of industrial applications. In 1992 Jonkers et al. [7] published a short paper in Nature about the use of positron emitters as tracers in catalysis. In 1993 Miranda et al. [93] have discussed the perspectives of positron annihilation spectroscopy (PAS) in heterogeneous catalysis. In this technique the lifetime of the positrons is measured, which is sensitive to electron states of inner surfaces such as pores. ^{22}Na, ^{57}Ni, and ^{58}Co built into the catalysts are the most commonly used sources of catalytic interest in PAS. Acid sites, spreading or segregation of catalytic components, defects and ionic vacancies can be studied by PAS. In some other review type publications [94-96] the application of special variations of the use of positron emitting tracers are described. Positron emission particle tracking (PEPT) is capable of measuring the flow pattern inside a fluidized-bed reactor by tracking a labeled particle. Positron emission profiling (PEP), the one dimension variant of PET, gives the time dependence of the concentration profile of the label along the catalyst bed and these data are used to set up mathematical models of the kinetics of the reaction. The PEP data are collected in the form of a three dimensional function (x,y,z) as the corresponding values of location, radioactivity or concentration, and time. The data are visualized as a two dimensional time-location map called as "reaction image", where the intensity is represented by colors or by graduation in black and white as shown in Fig. 7. In other representations the concentration vs. location at constant time or the concentration vs. location at a constant position, as parameters, respectively, are plotted.

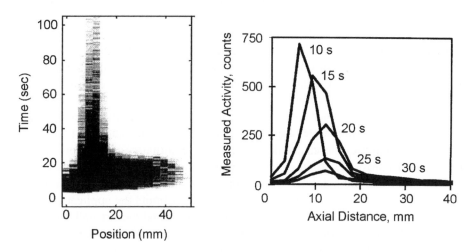

Fig. 7. (a) PEP spectrum showing n-hexane hydrogenolysis, (b) Constant time cross-sections of   (a) on freshly reduced Pt/H-mordenite; gray graduation represent measured radioactivity. Reprinted from [94].

7.1. Environmental catalysis

Vonkeman et al. [97,98] studied the oxidation of ^{11}CO on Pt-Ceria-Alumina and Pt-Rh-Ceria-Alumina catalysts used in automotive exhaust gas purification. In these experiments small pulses of the labeled molecules were dosed into the feed, without disturbing the steady state over the catalyst, and the time dependence of the distribution of the label along the axis of the catalyst bed was monitored with resolutions of 1.2 s and 8 mm, respectively. Besides ^{11}CO, they used $^{11}CO_2$, $C^{15}O$, $C^{15}OO$, ^{15}OO, and ^{13}N, too. By ^{13}NN the dispersion in the catalyst bed, by ^{15}OO the irreversible adsorption of oxygen at noble metal surfaces, by $^{11}CO_2$ and $C^{15}OO$ the exchange of oxygen atoms with ceria were studied. Product identification of the labeled compounds at the exit of the reactor by radio gas chromatography revealed that about 24% of the ^{15}OO molecules leave the reactor unchanged with a time dependence corresponding to ^{13}NN. The remaining part shows a long tailing and is converted to CO_2 via reaction with CO indicating an irreversible adsorption of O_2. Experiments with CO_2 labeled either in the C or in the O positions, revealed that the oxygen atoms remain much longer in the catalyst bed than the carbon ones. The same was observed for CO, too, indicating the dissociation of the molecules. To simulate the CO oxidation by O_2, a mathematical model was constructed which contained 10 equations. Kinetic reaction parameters based on the reaction images of the different compounds labeled in different positions were derived for the model. The values were in reasonable agreement with independent data found in the literature. Adding 1400 ppm unlabeled NO to the reaction mixture at 100 °C a decrease in both the amount and desorption rate of CO was observed.

Baltensperger et al. [99] studied the selective catalytic reduction of ^{13}NO by NH_3 over vanadia-titania catalyst at extremely low NO concentrations of 5×10^{-9} ppm. They measured the amount of labeled compounds at the exit of the catalytic reactor by trapping first NO and NO_2 by a Co_2O_3 trap followed by a zeolite trap cooled with liquid nitrogen for N_2. Different pretreatments as oxidation in air, reduction in hydrogen and partial reduction with ammonia, were given to the catalyst prior to the kinetic measurements in which ^{13}NO was mixed into a helium stream containing 15% O_2 but no NH_3. Only the catalyst pretreated with ammonia, showed conversion of ^{13}NO to ^{13}N^{14}N. The reaction rate at 92 °C was in reasonable agreement with the ones extrapolated from 1000 ppm NO concentration by assuming first order for NO and zero order for NH_3 in the reaction rate. The same holds for the activation energy. The low NO concentration allowed to investigate the kinetics after pretreatment with various gases without adding these gases during the catalytic experiment.

In other experiments [99] the author used thermochromatography to determine adsorption enthalpies on various oxides. In these experiments a linear temperature profile between 70 and -193 °C was produced along a column filled with the samples. The carrier gas containing the ^{13}N labeled adsorbate in extremely low concentration was passed through the column for 20 min. The column was removed and the axial distribution of the radioactivity in the column was measured with a collimated NaI detector. The temperature of the adsorption was determined from the position of the radioactivity and adsorption enthalpies for NO and NO_2 at extremely low, practically zero coverage were calculated.

Sobczyk et al. [100] studied the low temperature ammonia oxidation over Pt-alumina catalyst by adding small pulses of ^{13}NH$_3$ or ^{15}OO into 2% NH_3 and 1.5% O_2 containing He stream when the catalyst reached steady state. Based on the PEP images the authors concluded that at 75 °C most of NH_3 reversibly adsorbs on alumina and only a small amount is converted to nitrogen species, which poison the platinum. At 200 °C most of the NH_3 is converted to N_2 and N_2O and no detectable ^{13}N activity remains on the catalyst. The ^{15}O PEP images showed that most of the oxygen leaves the catalyst very fast but about 10% is irreversibly adsorbed and poisons the catalyst. Preadsorbed ^{15}O was not removed by the ammonia-oxygen flow or by hydrogen and poisoned the catalyst. Oxygen spillover to the alumina was small. A drawback of the application of alumina support is that below 120 °C it preferentially adsorbs ammonia and therefore the platinum is oxidized. In the curse of the reaction the reaction takes place in excess of oxygen at the platinum resulting in high local temperatures, which enhance the oxide formation resulting in the deactivation of the catalyst. Only above 250 °C could a stable activity be reached because of the removal of adsorbed oxygen and NH_x species.

7.2. Hydrocarbon reactions

The surface coverage of H-mordenite by n-hexane was measured at temperatures typical for hydroisomerization as a function of n-hexane partial pressure by using PEP for monitoring the progress of small pulses of [^{11}C-1]-n-hexane in hydrogen/n-hexane mixture as a function of time and location within the reactor [101]. In some experiments 2-[^{11}C]-methylpentane was also used. The profile of the injected pulse before the catalyst was measured by a NaI scintillation detector. The surface coverage of n-hexane was near to 1 at 150 °C, and decreased to 0.5 at 230 °C and 8.4 torr. Adsorption enthalpies at zero coverage were calculated from the retention of pulses of 10^{-6} moles ^{11}C-hexane in pure hydrogen. They obtained 66.7 and 68.2 kJ mol^{-1} for n-hexane and 2-methylpentane, respectively. From the images recorded at various flow rates and temperatures, diffusion constants were calculated [102] by fitting the activity-position-time surface to a model based on the time-dependent mass balance equations for biporous packed-bed adsorber. The calculations showed that for H-ZSM-5 diffusion in the micropores has a small contribution to the overall mass transport of hexane, while it has a large contribution in smaller pore size zeolites as H-ferrierite and H-ZSM-22. PEP images of pulses of n-hexane into a hydrogen stream over a freshly reduced Pt-H-mordenite catalyst at 150 °C [102] showed only diffusion of n-hexane through the catalyst bed. At higher temperatures of 190, 210, and 230 °C, respectively, hydrogenolyses of n-hexane at the front of the reactor was observed. The light hydrocarbons together with C_7 and C_8 hydrocarbons were rapidly eluted and a carbonaceous overlayer formed on the catalyst, which could be removed as methane by temperature programmed hydrogenation to 400 °C.

The labeled hexane was produced in a two-step pentene homologation reaction over vanadium promoted Ru-silica catalyst [103]. First ^{11}CO is pulsed into helium stream over the catalyst at 350 °C to form reactive $^{11}C_1$ surface species. The temperature is rapidly reduced to 110 °C and 1-pentene is pulsed over the catalyst. The gas flow is switched for hydrogen to hydrogenate the surface species. The hydrocarbons leaving the reactor are trapped in a liquid-nitrogen trap. By rapid heating of the trap the hydrocarbons are injected into a radio gas-chromatograph to completely separate all C_{3+} products. The hydrocarbon needed for further experiments described in [101,102], is trapped and is injected into the catalytic reactor by rapid heating.

LIST OF ABBREVIATIONS

AES   Auger Electron Spectroscopy
CUS   Coordinatively Unsaturated Sites
DBT   Dibenzothiophene
FT   Fischer-Tropsch
FT-IR   Fourier Transform Infra-Red
GC   Gas Chromatography
HDS   Hydrodesulfurization
MS   Mass Spectrometry
PAS   Positron Annihilation Spectroscopy
PEP   Positron Emission Profiling
PEPT   Positron Emission Particle Tracking
PET   Positron Emission Tomography
RGC   Radio Gas Chromatography
SSITKA Steady State Isotopic Transient Kinetic Analysis
TAP   Transient Analysis of Products
UHV   Ultra High Vacuum

REFERENCES

[1] K.C. Campbell and S.J. Thomson, Progr. Surface and Membrane Sci., 9 (1975) 163.
[2] S.J. Thomson, in Characterisation of Catalysts, Eds.: J.M. Thomas and R.M. Lambert, John Wiley and Sons (1980) 214.
[3] J.T. Kummer, T.W. DeWitt and P.H. Emmett, J. Am. Chem. Soc., 70 (1948) 3632.
[4] S.V. Norval, S.J. Thomson and G. Webb, Appl. Surf. Sci., 4 (1980) 51.
[5] S.M. Davis, B.E. Gordon, M. Press and G.A. Somorjai, J. Vac. Sci. Technol., 19 (1981) 231.
[6] U. Schroder, L. Cider and N.H. Schoon, Stud. Surf. Sci. Catal., 75 (1993) 643.
[7] G. Jonkers, K.A. Vonkeman, S.W.A. Vanderwal and R.A. vanSanten, Nature, 355 (1992) 63.
[8] S.M. Wang, P.J.T. Tait and C.E. Marsden, J. Mol. Catal., 65 (1991) 237.
[9] M.M. Marques, P.J.T. Tait, J. Mejzlik and A.R. Dias, J. Polym. Sci. Part A., 36 (1998) 573.
[10] P.J.T. Tait, G.H. Zohuri and A.M. Kells, Macromolecular Symp., 89 (1995) 125.
[11] I.A. Jaber and G. Fink, J. Mol. Catal. A., 98 (1995) 135.
[12] I.A. Jaber and G. Fink, Makromol. Chem. Phys., 190 (1989) 2426.
[13] S.N. Gan, P.S.T. Loi, S.C. Ng and D.R. Burfield, Stud. Surf. Sci. Catal., 89 (1994) 91.
[14] G.D. Bukatov, V.S. Goncharov, V.A. Zakharov, V.K. Dudchenko and S.A. Sergeev, Kinet. Catal., 35 (1994) 358.
[15] S.V. Norval and S.J. Thomson, J. Chem. Soc., Faraday T. I., 75 (1979) 1798.
[16] J.S. Orr and F.C. Gillespie, Science, 162 (1968) 138.
[17] J. Thomson, G. Webb and J.M. Winfield, J. Mol. Catal., 67 (1991) 117.
[18] Z. Paál, S.J. Thomson, Radiochem. Radioanal. Lett., 12 (1972) 1.
[19] Z. Paál, S.J. Thomson, J. Catal., 30 (1973) 96.
[20] V.M. Kogan, N.N.Rozhdestvenskaya and I.K. Korshevets, Appl. Catal. A., 234 (2002) 207.
[21] V.M. Kogan, Appl. Catal. A., 237 (2002) 161.
[22] V.M. Kogan, R.G. Gaziev, S.W. Lee and N.N. Rozhdestvenskaya, Appl. Catal. A., 251 (2003) 187.
[23] F.E. Massoth, T. Koltai and P. Tétényi, J. Catal., 203 (2001) 33.
[24] M. Dobrovolszky, P. Tétényi and Z. Paál, Chem. Eng. Commun., 83 (1989) 1.
[25] M. Dobrovolszky, Z. Paál and P. Tétényi, Catal. Today, 9 (1991) 113.
[26] M. Dobrovolszky, T. Koltai, Z. Paál and P. Tétényi, Appl. Catal. A., 166 (1998) 65.
[27] A. Ishihara, J. Lee, F. Dumeignil, R. Higashi, A. Wang, E.W. Qian and T. Kabe, J. Catal., 217 (2003) 59.
[28] F. Dumeignil, H. Amano, D. Wang, E.W. Qian, A. Ishihara and T. Kabe, Appl. Catal. A., 249 (2003) 255.
[29] D. Wang, W. Qian, A. Ishihara and T. Kabe, Appl. Catal. A., 224 (2002) 191.
[30] D. Wang, X. Li, E.W. Qian, A. Ishihara and T. Kabe, Appl. Catal. A., 238 (2003) 109.
[31] D. Wang, X. Li, W. Qian, A. Ishihara and T. Kabe, J. JPN Petrol. Inst., 45 (2002) 39.
[32] W. Qian, T. Kawano, A. Funato, A. Ishihara and T. Kabe, Phys. Chem. Chem. Phys., 3 (2001) 261.
[33] D. Wang, W. Qian, A. Ishihara and T. Kabe, J. Catal., 209 (2002) 266.
[34] E.W. Qian, Y. Hachiya, K. Hirabayashi, A. Ishihara, T. Kabe, K. Hayasaka, S. Hatanaka and H. Okazaki, Appl. Catal. A., 244 (2003) 283.

[35] F. Dumeignil, J.-F. Paul, E.W. Qian, A. Ishihara, E. Payen and T. Kabe, Res. Chem. Intermediat., in press

[36] G.V. Isagulyants, A.A. Greish and V.M. Kogan, 9th Int. Congr. On Catal., Eds.: M.J. Phillips and M. Ternan, The Chem. Inst. Of Canada, Ottawa, 1988 35.

[37] M.E. Bussel and G.A. Somorjai, J. Catal., 106 (1987) 93.

[38] A.S. Al-Ammar and G. Webb, J. Chem. Soc., Faraday T., I., 74 (1978) 195.

[39] A.S. Al-Ammar and G. Webb, J. Chem. Soc., Faraday T., I., 74 (1978) 657.

[40] A.S. Al-Ammar and G. Webb, J. Chem. Soc., Faraday T., I., 75 (1979) 1900.

[41] J. Margitfalvi, L. Guczi and A.H. Weiss, React. Kinet. Catal. L., 15/4/ (1980) 475.

[42] L. Guczi, Z. Schay, A.H. Weiss, V.Nair and S. LeViness, React. Kinet. Catal. L., 27/1/ (1985) 147.

[43] S. LeViness, V. Nair, A.H. Weiss, Z. Schay and L. Guczi, J. Mol. Catal., 25 (1984) 131.

[44] L. Guczi, R.B. LaPierre, A.H. Weiss and E. Biron, J. Catal., 60 (1979) 83.

[45] J. Margitfalvi, L. Guczi and A.H. Weiss, J. Catal., 72 (1981) 185.

[46] Z. Schay, A. Sárkány, L. Guczi, A.H. Weiss and V. Nair, Proc. Vth Intern. Symp. Heterogeneous Catalysis, Varna, Part I., (1983) 315.

[47] E.A. Arafa and G. Webb, Catal. Today, 17 (1993) 411.

[48] J.T. Kummer, T.W. DeWitt and P.H. Emmett, J. Am. Chem. Soc., 70 (1948) 3632.

[49] J.T. Kummer, H.H. Podgurski, W.B. Spencer and P.H. Emmett, J. Am. Chem. Soc., 73 (1951) 564.

[50] J.T. Kummer and P.H. Emmett, J. Am. Chem. Soc., 75 (1953) 5177.

[51] W.K. Hall, R.J. Kokes and P.H. Emmett, J. Am. Chem. Soc., 79 (1957) 2983.

[52] G. Blyholder and P.H. Emmett, J. Phys. Chem., 63 (1959) 962.

[53] G. Blyholder and P.H. Emmett, J. Phys. Chem., 64 (1960) 470.

[54] W.K. Hall, R.J. Kokes and P.H. Emmett, J. Am. Chem. Soc., 82 (1960) 1027.

[55] L-M. Tau, H. Dabbagh, S. Bao and B.H. Davis, Catal. Lett., 7 (1990) 127.

[56] L-M. Tau, H.A. Dabbagh, B. Chawla and B.H. Davis, Catal. Lett., 7 (1990) 141.

[57] B.H. Davis, Fuel Processing Technol., 71 (2001) 157.

[58] B.H. Davis, Catal. Today, 53 (1999) 443.

[59] M. Haissinsky, Nuclear Chemistry and its Applications, Addison-Wesley (1964) 656.

[60] A. Sárkány, J. Catal., 105 (1987) 65.

[61] Z. Paál and P. Tétényi, Acta Chim. Acad. Sci. Hung., 58 /1/ (1968) 105.

[62] Z. Paál and P. Tétényi, Acta Chim. Acad. Sci. Hung., 54/2/ (1967) 175.

[63] Z. Paál and P. Tétényi, Acta Chim. Acad. Sci. Hung., 55/3/ (1968) 273.

[64] A.E.H. Kilner, H.S. Truner and R.J. Warne, Radioisotope Conference, 1954, Proc. 2nd Conf. Oxford, 19-23 July 1954, vol. 3, Physical Sci., Industrial Applications, Butterworths, London, 1954, p. 23.

[65] A. Sárkány, J. Catal., 89 (1984) 14.

[66] H. Pines and C.T. Chen, 2nd Congr. Int. Catal. Paris, 196 (1961) 367.

[67] J.J. Mitchell, J. Am. Chem. Soc., 80 (1958) 5848.

[68] H. Pines and C.T. Chen, J. Org. Chem., 26 (1961) 1057.

[69] B.H. Davis, J. Catal., 29 (1973) 395.

[70] S.M. Csicsery and R.L. Burnett, J. Catal., 8 (1967) 74.

[71] C.-S. Huang, D.E. Sparks, H.A. Dabbagh and B.H. Davis, J. Catal., 134 (1992) 269.

[72] F.R. Cannings, A. Fisher, J.F. Ford, P.D. Holmes and R.S. Smith, Chem. Ind. (1960) 228.

[73] Yu.I. Derbentsev, Z. Paál and P. Tétényi, Acta Chim. Acad. Sci. Hung., 70/4/ (1971) 369.

[74] Yu.I. Derbentsev, Z. Paál and P. Tétényi, Z. Phys. Chem. Neue Folge, 80 (1972) 51.

[75] G. Parravano, J. Catal., 16 (1970) 1.
[76] G. Parravano, J. Catal., 24 (1972) 233.
[77] M.A. Dobrovolszky, Z. Paál and P. Tétényi, Acta. Chim. Acad. Sci. Hung., 119 (1985) 95.
[78] Z. Paál, M. Dobrovolszky and P. Tétényi, J. Chem. Soc., Faraday T. I, 80 (1984) 3037.
[79] L. Guczi, B.S. Gudkov and P. Tétényi, J. Catal., 24 (1972) 187.
[80] J.W. Hihtower and P.H. Emmett, J. Am. Chem. Soc., 87 (1965) 939.
[81] J.W. Hightower, H.R. Gerberich and W.K. Hall, J. Catal., 7 (1967) 57.
[82] J.W. Hightower and W.K. Hall, J. Phys. Chem., 71 (1967) 1014.
[83] Z. Paál, A. Péter and P. Tétényi, Z. Phys. Chem. Neue Folge, 91 (1974) 54.
[84] B.H. Davis, J. Catal., 29 (1973) 395.
[85] P. Tétényi and L. Babernics, J. Catal., 8 (1967) 215.
[86] Z. Schay, K. Matusek and L. Guczi, Appl. Catal., 10 (1984) 173.
[87] P. Tétényi, M. Dobrovolszky and Z. Paál, Appl. Catal. A., 86 (1992) 115.
[88] D. Cormarck, S.J. Thomson and G. Webb, J. Catal., 5 (1966) 224.
[89] J.A. Altham and G. Webb, J. Catal., 18 (1970) 133.
[90] J.U. Reid, S.J. Thomson and G. Webb, J. Catal., 29 (1973) 421.
[91] J.U. Reid, S.J. Thomson and G. Webb, J. Catal., 30 (1973) 372.
[92] M.R. Hawkesworth, D.J. Parker, P. Fowles, J.F. Crilly, N.L. Jefferies and G. Jonkers, Nucl. Inst. And Methods in Phys. Res. Section A, 310 (1991) 423.
[93] R. Miranda, R. Ochoa and W.-F. Huang, J. Mol. Catal., 78 (1993) 67.
[94] B.G. Anderson, R.A. vanSanten and L.J. Ijzendoorn, Appl. Catal. A., 160 (1997) 125.
[95] B.G. Anderson, R.A. vanSanten and A.M. deJong, Topics in Catal., 8 (1999) 125.
[96] G. Jonkers, Handbook of Heterogeneous Catalysis, Eds.: G. Ertl, H. Knözinger and J. Weitkamp, Wiley-VCH, Vol. 3. (1997) 1023.
[97] K.A. Vonkeman, G. Jonkers and R.A. vanSanten, Stud. Surf. Sci. Catal., Vol. 71 (1991) 239.
[98] K.A. Vonkeman, G. Jonkers, S.W.A. Vanderwl and R.A. vanSanten, Ber. Binsenmges, J. Phys. Chem., 97/3/ (1993) 333.
[99] U. Baltensperger, M. Ammann, U.K. Bochert, B. Eichler, H.W. Gaggeler, D.T. Jost, J.A. Kovacs, A. Turler, U.W. Scherer, J. Phys. Chem., 97 (1993) 12325.
[100] D.P. Sobczyk, E.J.M. Hensen, A.M. deJong and R.A. vanSanten, Topics in Catal., 23 (2003) 109.
[101] R.A. vanSanten, B.G. Anderson, R.H. Cunningham, A.V.G. Mangnus, J. vanGrondelle and L.J. vanIjzendoorn, 11th Int. Congr. On Catal., Stud. Surf. Sci. Catal., 101 (1996) 791.
[102] B.G. Anderson, N.J. Noordhoek, D. Schuring, F.M.M.M. deGauw, A.M. deJong, M.J.A. deVoigt and R.A. vanSanten, Catal. Lett., 56 (1998) 137.
[103] R.H. Cunningham, A.V.G. Mangnus, J. vanGrondelle and R.A. vanSanten, J. Mol. Catal. A., 107 (1996) 153.

Radiotracer Studies of Interfaces
G. Horányi (editor)

Chapter 4

Studies of electrified solid/liquid interfaces

G. Horányi

Institute of Materials and Environmental Chemistry, Chemical Research Center, Hungarian Academy of Sciences, PO Box 17, H-1525, Budapest, Hungary

The solid/liquid interface can be considered as a version of the solid/fluid interfaces and in this sense – disregarding some practical problems – no differences in the application of the tracer technique could be expected in the case of solid/gas and solid/liquid systems.

This approach is acceptable in the case of systems containing apolar solvents and solutes, however, in the case of polar solvents and electrically charged solutes (ions) the formation of electrical double layer in the interphase is unavoidable. Owing to this situation the behavior of the system cannot be interpreted without the consideration of electrochemical parameters i.e. the system should be treated in terms of electrochemistry. In the literature there is a clear distinction for these cases even in the nomenclature using the terms electrosorption instead of adsorption and electrocatalysis instead of catalysis indicating that the interfacial processes are influenced by the electric field of the double layer formed in the interphase. The formation of double layer at the solid/liquid interface is evident in the case of conducting (metals) and semiconducting (semiconductor) solid phases, even in the case of isolators a double layer could be formed via the interaction of the surface of the solid phase with a charged component of the solution phase. One of the best-known examples for this process is the protonation of such non-conducting oxides as Al_2O_3 or Fe_2O_3. In accordance with these considerations, in this chapter radiotracer studies of electrified interfaces will be presented on the basis of the following distinctions:

1) Applications of the radiotracer technique in electrochemistry of
   a) conducting
   b) non-conducting system (mainly in the case of oxides)
2) Corrosion studies.

Colloid systems with solid/liquid interfaces will be discussed separately in Chapter 5.

1. APPLICATION OF THE RADIOTRACER TECHNIQUE IN ELECTROCHEMISTRY

The application of radiotracer method in electrochemistry dates back to the pioneering works by Hevesy in 1914. The aim of these studies was to demonstrate that isotopic elements can replace each other in both the electrodeposition and equilibrium processes (Nernst law). Nevertheless Joliot's fundamental work in 1930 is considered by electrochemists as a landmark in the application of radiochemical methods in electrochemistry.

Beginning from the 1950-s simultaneous formation of working groups could be observed. For long period Frumkin's Institut and Karpov Institute in Moscow, the groups directed by Schwabe in Leipzig and Dresden and Bockris' group in Pennsylvania were the centers of radiotracer studies. The historical background is well documented by the reviews published during the last thirty years [1-13]

During the last 40 years new centers were formed in Hungary [14-32], Poland [33-38], United States [39-50], Lithuania [51-55]. Together with the new groups, the application of radiotracer methods has continued in Bockris' laboratory [56-60] and the late Professor Kazarinov's laboratory [61-64].

Time to time works based on radiotracer technique appear from various laboratories [65-67]

The electrochemical phenomena studied by radiotracer techniques range from equilibrium adsorption of ions and neutral species to corrosion processes involving formation of surface layers (films) and deposition and dissolution of metals, etc. The variety of fields involved in these studies, and the development of the experimental technique, have been well demonstrated in the reviews cited above[1-13].

In the following three main fields will be discussed:
a) Dissolution and deposition of metals
b) Electrosorption phenomena at metal/solution interfaces.
c) Adsorption at oxide/solution interfaces.

Before going into details, it is very important to make some general comments how radiochemical methods can be used for the interpretation and understanding of phenomena occurring at electrodes.

1.1. General remarks concerning the application of radiochemical methods for the interpretation and understanding of phenomena occurring at electrodes

The radiochemical methods can be used to answer a great variety of questions appearing in the course of the investigation of electrochemical systems.

The answer and the method to be used for answering depend on the level of understanding and interpretation aimed at by the question posed. In the following we consider three various levels of understanding and interpretation.

1.1.1. Phenomenological level

a) Simple questions to be answered by yes or no.

The radiotracer technique is extremely suitable to answer question connected with the presence or absence of a given species on the electrode surface or to answer the question concerning the mobility or immobility of adsorbed species i.e. to make a distinction between reversible and irreversible adsorption. Thus simple radiochemical experiments play very important role in answering fundamental questions. Some examples will be discussed briefly.

i) The problem of co-adsorption or induced adsorption.

It is an important question whether the electrosorption of species in a given system is accompanied with the simultaneous adsorption of other molecules or not. Radiotracer method offers a unique possibility of answering this question by labeling the species considered as possible partners in co-adsorption. The important, and nowadays widely known, fact that the underpotential deposition of metals is accompanied very often by the co-adsorption (induced adsorption of anions) was demonstrated first time by radiotracer technique labeling the anions.

ii) Another fundamental question is the reversibility of the adsorption. The reversibility of adsorption implies that desorption of adsorbed species takes place with a measurable rate. The rate of desorption can be easily estimated by the exchange of adsorbed labeled species with non-labeled ones added the solution phase or vice versa. In the case of irreversibility of the adsorption the absence of any exchange clearly demonstrates the situation.

b) Determination of simple relationships.

i) Potential and concentration dependence of the adsorption.

ii) Relationship between the rate of electrode processes and the apparent coverage of reacting species.

In most cases radiotracer methods furnish reliable direct data for the determination of isotherms and adsorption vs. potential relationships.

In some cases the sensitivity of the method could be very high depending on the specific activity of the isotopes used. It is important that for the calculation of the adsorption values no specific assumption are required and the separation of the signal coming from the adsorbed layer from that coming from the solution phase is relatively easy in comparison with other techniques.

The radiotracer method allows to determine directly the coverage vs. current or potential relationship under steady state condition (potentiostatic or galvanostatic) in the course of electrode processes even in the absence of adsorption equilibrium with respect to the reacting species i.e. without the knowledge of the adsorption isotherms.

1.1.2. Overall chemical level (What happens)

At this level the chemical events, the chemical composition of products appearing in the solution phase should be clarified in order to give an interpretation in terms of plausible chemical transformations without analyzing the mechanism of processes going on the surface of the electrode. Various combination of electrochemical and radiochemical methods are applied to investigate the "real" transformations in "real" systems (polycrystalline (rough) surfaces, polymers, structural materials etc.).

1.1.3. Physical-chemical level (Why and how it happens)

For the mechanistic interpretation of the phenomena occurring with the adsorbed species a deeper insight is required than in the case of overall chemical level. Experiments should be carried out at well-defined surfaces (single crystal faces), consequently the radiotracer technique should be adjusted to this requirement.

At the same time the coupling of radiotracer methods with other methods of investigation is also a characteristic feature of these studies. (Coupling with spectroscopic methods, and application of radiotracer technique in conjunction with the combination of ultra-high-vacuum (UHV) surface science techniques and electrochemical methods.)

1.2. Radiotracer studies of dissolution and deposition of metals

The application of radiotracer methods for the study of dissolution is surveyed in the monograph by Kazarinov and Andreev [3].

The principle of these methods is based on the labeling of a component of the metal phase by one of its radioactive isotopes and calculating the dissolution rate of the metal specimen by measuring either the increase in radiation coming from the solution phase, or the decrease in radiation coming from the solid phase.

The main steps characterizing radioactive tracer methods used for investigation of dissolution processes of metals and alloys can be summarized as follows:

i) Introduction of the radioisotope into the specimen. This task can be achieved at least by three methods: (a) through melting (b) by electrolytic deposition of the radioactive metal and (c) by subjecting the metal specimen to neutron irradiation in a nuclear reactor.

ii) Measurement of the changes in radiation intensity caused by the dissolution process.

As mentioned above this step can be performed in two ways.

a) Determination of the radioisotope concentration in electrolyte solution either through "sampling" or continuous measurement of the radiation intensity of the solution phase.

b) Determination of radiation intensity coming from the metal sample. This method was used in the case of amalgams. The dissolution process of solid electrodes can be studied by any radiotracer method used for the investigation of adsorption phenomena. (The principle of these methods will be discussed later.)

An interesting version is the application of β backscattering for the study of the electrochemical formation and dissolution of thin metal layers.

The backscattering of β radiation is a well-known and widely studied phenomenon in the field of nuclear physics β backscattering serves as a basis for various analytical methods. This phenomenon is often used to measure the thickness of thin layers formed on various surfaces. In some cases changes in layer composition can be determined by measuring the intensity of the backscattered radiation.

The phenomenon of backscattering follows from the very nature of processes occurring in the case of interaction of β-particles with a material. A significant part of the β-radiation is absorbed, however, a great number of β-particles are scattered in various directions. Those particles that are scattered in a direction about $180°$ (in comparison to the direction of the incident radiation) are the so-called backscattered particles forming the "backscattered radiation". The ratio of the intensity of the backscattered radiation to the incident radiation depends strongly on the thickness, density and atomic number of the scattering material.

Surprisingly, for a long period, β backscattering has not been applied to the in-situ study of electrode processes leading to the formation or dissolution of layers on a support and it has not been used to study transformations resulting in changes in the composition of layers covering electrode surfaces.

In Refs. 68,69 it was shown that, under appropriately chosen experimental conditions, β backscattering could be used to follow the electrochemical formation or dissolution of thin metal layers in a cell originally developed for radiotracer adsorption studies. ^{99}Tc was used as the β-radiation source ($E = 0.29$ MeV).

Layers formed by electrodeposition from solutions containing Bi^+, Ag^+, Cu^{2+} and ReO_4^- ions were studied. The supporting electrolytes were 1 mol dm^{-3} H_2SO_4 and $HClO_4$ solutions.

Only a brief survey can be given here of some of the main features of β backscattering.

Two arrangements of the β source can be considered: (i) internal source; (ii) external source.

In the case of an internal source, the electrode surface or a layer deposited on it (support) is the source of β radiation and the backscattering by a layer formed on the support (scatterer) is measured. This arrangement is shown in Scheme. 1.

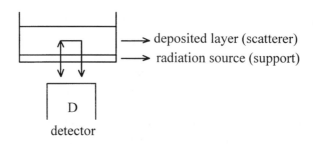

deposited layer (scatterer)
radiation source (support)

D

detector

Scheme 1.

In this case the measured radiation intensity (I_T) consists of the intensity I of the direct radiation and the intensity I_b of the backscattered radiation:

$$I_T = I + I_b \tag{1}$$

In the case of an external source the situation is quite different and there are several options depending on the position of the source. Two possible arrangements of the β source are shown in Scheme 2.

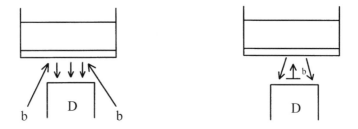

(The external source is indicated by b.)

Scheme 2.

In this case the measured radiation intensity consists of the intensity I_b^s of the radiation backscattered by the support and the contribution ΔI_b to the intensity of backscattered radiation caused by the layer formed on the support:

$$I_T = I_b^s + \Delta I_b \qquad (2)$$

In the present discussion the characteristic features of backscattering will be considered in connection with the application of an internal β source using a ^{99}Tc layer to this end.

The first step is the preparation of the radiation source by the electrodeposition of a ^{99}Tc layer on the gold plated plastic foil forming the bottom of the cell.

The formation of this layer that is same time is the radiation-source, can be monitored by measuring the increase in radiation intensity during the electrodeposition of the layer.

Electrodeposition from solutions of Cu^{2+}, Ag^+ and Bi^{3+} ions was studied. It follows from the discussion of the fundamental phenomena [68] that the higher the atomic number of the scatterer the higher will be the sensitivity of the method.

Curves a, b and c in Fig. 1 show how the intensity of the backscattered, radiation changes with the thickness of the deposited layer (calculated from the charge used in the deposition process).

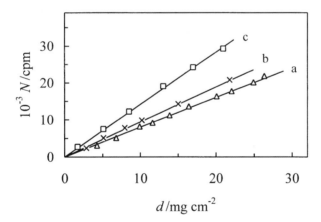

Fig. 1. The dependence of the intensity of the backscattered radiation on the thickness d of the deposited layer: (a) Cu; (b) Ag; (c) Bi.

It can be shown that not only the electrodeposition but also the dissolution of a layer can be easily be monitored by measuring the changes in the radiation intensity.

The results obtained from a voltradiometric study of the dissolution of a Bi layer are shown in Fig. 2.

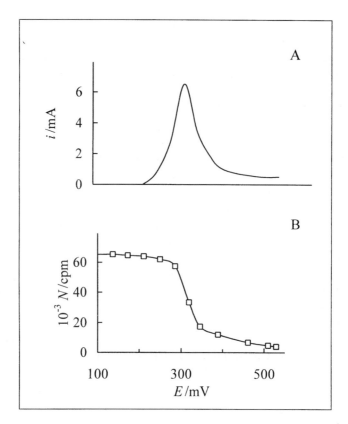

Fig. 2. Dissolution of a Bi layer during an anodic sweep (sweep rate, 0.6 mV s^{-1}) (a) current
vs. potential relationship (on RHE scale) and (b) the simultaneous change in the
intensity of the backscattered radiation.

The potentialities of the method using external sources with the application
of five different β-emitters (^{36}Cl, ^{60}Co, ^{99}Tc, ^{134}Cs and ^{204}Tl) were also
investigated [69].

The experimental results reported in [69] demonstrated that β-
backscattering, using an external β-source, can be a useful tool for the in-situ
study of electrodeposition and dissolution of thin layers. The most important
advantage of this method over other methods using internal β-sources is that
only minimal handling of radioactive solutions is required. The sealed sources
can be used several times and the treatment of radioactive waste can be avoided.
The other advantage of the external radiation source over the internal source is
that a larger number of substances can be used for electrochemical studies. Thus
it offers a significant extension of the applicability of this method. Following

appropriate calibration the backscattered radiation intensity values can be unambiguously transformed into layer thickness values.

Study of the adsorption of radiolabeled anions in the course of deposition and/or dissolution of metals may contribute to the mechanistic interpretation of the phenomena observed.

Two illustrative examples could be mentioned.

In a recent work [70] combined EQCM and voltammetric studies in conjunction with radiotracer adsorption investigations were carried out in order to obtain direct information on the interfacial behavior of the Cu^{2+}–Cu system first of all to clarify the role and participation of Cu^+ ions, formed by the interaction of cupric ions and copper metal, in the interfacial processes. The appearance and accumulation of Cu^+ ions in the solution phase were clearly demonstrated through radiotracer experiments using ^{36}Cl labeled Cl^- ions.

According to the mechanistic picture suggested in Ref. 70 the interaction of Cu^{2+} ions with copper surfaces leading to the formation of Cu^+ species involves adsorbed anions, namely adsorbed labeled Cl^- ions. Depending on the concentration of chloride ions, some surface sites on copper are occupied by adsorbed chloride species. The interaction of adsorbed chloride ions with cupric ions can be given by the following equation: Cu^{2+} (solution) + 2 Cl^- (ads.)+ e^- = CuCl (surface) + Cl^-(solution) while cuprous ion should appear in accordance with CuCl(surface) = Cu^+(solution) + Cl^-(solution) and CuCl(surface) + Cl^-(solution) = $CuCl_2^-$(solution). (The assumption of the reaction Cu^{2+}(sol) + Cl^-(ads) + e^- = CuCl(surface) should be rejected as the transfer of Cu^{2+} from the solution phase to the surface should be connected with the arrival of two electrons to the interface.)

Another interesting example is the study of the specific adsorption of radiolabeled anions in the course of deposition and dissolution of Cd. [24, 71].

In a recent communication [24], the specific adsorption of ^{36}Cl-labeled Cl^- ions and ^{35}S-labeled HSO_4^- ions were studied in 1 mol dm^{-3} $HClO_4$ supporting electrolyte in the presence of Cd^{2+} ions at a polycrystalline gold support over a wide potential range involving the potential regions corresponding to electrodeposition, alloy formation, underpotential deposition of Cd species and an adatom-free surface. The distinct sections found in the potential dependence of the adsorption of labeled anions were ascribed to the changes in the state of the electrode surface, the deposition/dissolution of the bulk Cd phase and the slow formation /elimination of a Cd/Au alloy.

The alloy formation in the course of underpotential deposition (upd) of Cd on gold from $HClO_4$ supporting electrolyte was studied by coupled voltammetric and electrochemical microbalance techniques [71]. The experimental results were discussed in the light of data obtained from the radiotracer study of induced adsorption of anions. It has been demonstrated that the alloy formation is well reflected in the results of EQCM measurements and the changes in the induced adsorption of anions.

1. 3. Experimental methods for electrosorption studies

1.3.1. Cell types

Various methods have been developed for in situ radiotracer adsorption studies depending on the requirements of the problems to be studied. The methods applied and the main factors leading to the choice of the most appropriate method are discussed in detail in review articles [2-6,60]. In Chapter 9 a survey on the experimental technique can be found. In the present Chapter only the main features of the most important methods will be considered.

In the case of the in situ studies the central problem is how to separate the signal (radiation) to be measured from the background radiation, and how to attain the optimal ratio of these quantities.

From this point of view methods can be divided into two main groups:

a. Radiation of the solution background is governed and minimized by self-absorption of the radiation i.e. by the attenuation of the radiation intensity by the radioactive medium itself (thin-foil method).

b. Background radiation intensity is minimized by mechanical means (thin-gap method).

a. The thin-foil method

In the case of the thin-foil method, the detector "sees" simultaneously both components of the radiation coming through a thin foil [metal or metal-plated (gold) plastic film] forming the bottom of the cell. The adsorbent is either the foil itself or a thin layer deposited on the bottom of the cell serving as a mechanical support and electric conductor (if the foil is metal-plated).

The situation can be visualized by the scheme shown in Scheme 3.

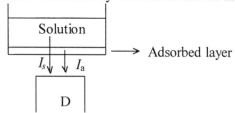

Scheme. 3. The principle of the thin foil method.

The radiation measured (I_T) is

$$I_T = I_s + I_a \qquad (3)$$

(where I_s and I_a are the intensities of the radiation coming from the solution phase and from the adsorbed layer, respectively). Using isotopes emitting soft β^--radiation (radiations characterized by high mass absorption coefficient (μ) (^{14}C, ^{35}S, and ^{36}Cl) the self-absorption of the radiation in the solution phase is so high that the thickness of the solution layer effective in the measured solution background radiation is very low. (Similar phenomena could be observed for isotopes emitting low energy X-ray.)

In these cases, on the basis of a simplified model, the solution background radiation is equal to

$$I_s = \alpha I_0 q \int_0^\infty c\exp[-\mu x \rho]dx = \alpha I_0 q \frac{c}{\mu\rho} \tag{4}$$

while the radiation coming from the adsorbed species is

$$I_a = \alpha I_0 q \gamma \Gamma \tag{5}$$

where α is a proportionality factor, I_0 is the specific activity of the labeled species present in concentration c in the solution phase, q is the geometric surface area of the electrode, γ is its roughness factor, μ is the mass absorption coefficient of the radiation, ρ is the density of the solution phase, Γ is the surface concentration of adsorbed species, x is the coordinate measuring the distance from the electrode surface in the solution phase. The integration goes from 0 to the infinity. However, the "infinity", for instance, in the case of the soft β^--radiation emitted by ^{14}C is less than 10^{-2} cm, as the radiation is completely absorbed in a solution layer of this thickness.

On the basis of equations 3, 4 and 5 Γ can be determined easily:

$$\Gamma = \frac{I_a}{I_s} \cdot \frac{c}{\mu\rho\gamma} = \frac{I_T - I_s}{I_s} \frac{c}{\mu\rho\gamma} = \left(\frac{I_T}{I_s} - 1\right)\frac{c}{\mu\rho\gamma} \tag{6}$$

It follows from the preceding simple equations that the solution background radiation is proportional to the concentration of the labeled species and reliable results can be expected only if $I_T/I_s > 1$. I_a is proportional to the roughness factor (Eq. 5).

Various technical versions of the foil method have been elaborated [2-4,72]. For details of the technical description, we refer to Chapter 9.

b. The thin-layer or gap method

In cases where radiation coming from the solution background would be too high in comparison with that originating from the adsorbed layer (smooth surfaces, γ-radiation), some kind of mechanical means should be applied to reduce the role of the background radiation.

Mechanical control of the background radiation can be achieved in two different ways: fixed rigid reduction of the solution layer thickness (classical thin-layer or gap method), and flexible mechanical or temporary reduction of the thickness of the solution layer (electrode-lowering technique). In the first case, specially designed cells are used where only a very thin solution layer flows between the electrode and the detector, as shown in Scheme 4.

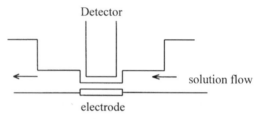

Scheme 4. Classical version of thin-layer or gap method.

This method is not often used, as it has several drawbacks, such as complexity of the equipment (a pump for solution circulation), the problem of determining exact electrochemical parameters owing to the presence of a thin solution layer separating the auxiliary and reference electrodes from the main electrode, etc.

These problems are avoided by the application of a flexible version, the so-called electrode-lowering technique. In this technique, the solution gap between the electrode and detector is minimized temporarily only during the time of measurement of the adsorption. The electrode is positioned in two positions, as shown by the scheme in Scheme 5.

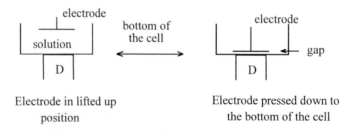

Scheme 5. The electrode lowering technique.

In the "lifted-up" position, attainment of the adsorption and electrochemical equilibrium (or steady state) proceeds without disturbance; the detector measures only the solution background radiation. In pressed-down position, the intensity measured comes from the species adsorbed on the electrode surface and from the solution layer present in the gap between the bottom of the cell and the electrode. The thickness of this gap depends on the mechanical state of the electrode surface, the stability of the bottom of the cell etc.

In the first version of this method, developed by Kazarinov [3], the bottom of the cell was a plastic foil i.e., a material with not very much mechanical stability. Wieckowski [73] eliminated this drawback by using a glass scintillator. He embedded in a ceramic disc to form the bottom of the cell. Owing to this construction, the gap thickness attained in the electrode's pressed-down position could be about $1-2 \times 10^{-4}$ cm. This thickness is one order of magnitude lower than the limiting thickness value determined by the self-absorption of β^--radiation of ^{14}C in the case of the foil method. The cell developed by Wieckowski [73] is shown in Chapter 9. Detailed analyses of the calculation of surface concentrations from radiation intensities can be found there.

In the case of isotopes emitting β^--radiation instead of Eq. 4, in the first approximation, the following relationship could be used for the calculation of Γ:

$$\Gamma = \frac{I_p}{I_b} \frac{c}{\mu \rho f_b (\exp - \mu \rho d)} \tag{7}$$

where

$I_p =$ the counting rate when the electrode is pressed down against the detector surface

$I_b =$ the counting rate when the electrode is far from the detector

$d =$ the thickness of the solution layer between the electrode and the detector

$f_b =$ the backscattering factor (for the material of the electrode)

A more detailed discussion see in Chapter 9.

1.3.2. Direct and indirect methods

As to the role of the labeled species in the radiotracer study of adsorption phenomena, two different versions of the method may be distinguished. In the first, the direct method, the species to be studied is labeled and the radiation measured gives direct information on the adsorption of this species. However, this method cannot be used in several cases owing to technical restrictions related to the very nature of the radiotracer method (the available concentration range is limited; no distinction can be made between the adsorption of the labeled compound studied and that of a product formed from it; the number of commercially available labeled compounds is restricted.)

Considering all these problems the use of the so-called indirect radiotracer methods was suggested. Instead of labeling the species to be studied another adequately chosen labeled species (indicator species) is added to the system and the adsorption of this component is followed by the usual radiotracer measuring technique. Evidently, the sorption of the indicator species should be in relation with that of the species to be studied. The nature of this link could be different in different systems. For instance, in some cases competitive adsorption with the labeled species while in other cases induced adsorption of the labeled species may provide information on the adsorption behavior of a given molecule. The principle of the study in the former case can be demonstrated by the scheme presented in Scheme 6.

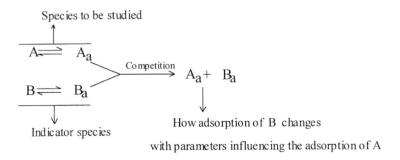

Scheme 6. The principle of the indirect radiotracer technique.

1.3.3. State of the specimen studied.

The nature (metals, alloys, oxides, modified electrodes, etc.) and the structure (powdered or compact, polycrystalline or single crystal materials, smooth or rough surfaces) of the specimen to be studied are important factors determining the approach to the sorption phenomena and the radiotracer methods used for their study.

The main groups are as follows:

a. Compact metals and electrodeposited metal layers.

For a long period of time the adsorption phenomena had been studied only on polycrystalline surfaces. Both smooth and rough surfaces with high real surface area were considered, using preferably the electrode-lowering technique in the former and the foil method in the latter cases.

For many years only platinized platinum electrode, as representative of electrodes with high real surface area, has been in the foreground of studies. During the last decades investigations have been extended to other metals as well, using copperized [74-76], silverized [77,78], nickelized [79], rhodized [14,15,80], and aurized [75,81], electrodes with high real surface area (roughness factor values above 20). The behavior of electrodeposited Tc and Re layers was studied in detail [82,83]. This extension of adsorption studies involved the elaboration of methods for the preparation of the electrodes and measuring of the roughness factors.

The investigation of electrosorption phenomena at morphologically uniform and ordered single-crystal, i.e. well-defined, electrodes is a new development since the end of the 1980s [4-6,73]. The version of electrode-lowering technique elaborated by Wieckowski and colleagues was created to this end [4-6,73] in conjunction with other methods for the study of single-crystal surfaces. At any rate, the method gives excellent results for polycrystalline smooth electrodes as well.

b. Powdered adsorbents.

The principle of the radiotracer method for the in-situ study of adsorption phenomena on powdered metal samples and the experimental technique were reported more than 30 years ago. The use of powdered samples ensures higher reliability of tracer adsorption measurements as compared to those carried out with smooth surfaces of compact metal samples. It is, however, a drawback of the use of powdered materials that very often their structure cannot be unambiguously defined and determined.

The very principle of the method is the measurement of radiation intensity originating from species adsorbed on a powdered layer sprinkled on a thin gold plated plastic foil that serves simultaneously as the window for radiation measurement.

A series of papers [84-87,] were published presenting results obtained with stainless steel samples used as construction material in nuclear reactors. Later the method was extended to other metals [88-90] and various oxides as well [91-97].

In the case of thin layers using soft β^--emitter for labeling, the radiation intensity measured for a given specific surface area of the powder should be proportional to the amount of the powder sprinkled on the bottom of the measuring cell. However, with increasing thickness the radiation intensity tends to a limiting value in consequence of self-absorption of the soft β radiation.

The determination of Γ values from radiation intensity data measured in the limiting case requires three independent measurements [11].

i) Measurement of radiation intensity (I_1) coming from the solution phase in the presence of the labeled species but in the absence of powder.

ii) Measurement of the total limiting radiation intensity originating from the powder layer following the adsorption of the labeled species (I_2).

iii) Measurement of the radiation intensity (I_3) without the adsorption of the labeled species on the powder surface (or by addition of non-labeled species to the solution phase i.e. by decreasing the specific activity to a very low value).

A very important advantage of the powder method is that the slow dissolution of the powder (for instance, metals or oxides in acidic medium) does not exert any influence on the count rate if the amount of the powder is in the region of the limiting radiation intensity.

All these results verify that the method can furnish reliable results (for the determination of the character of pH and concentration dependence of the adsorption on various non-conducting or conducting powders).

However, in the case of a specimen available in form of compact metal a serious drawback is that the application of the method should be preceded by the preparation of a powder from the compact metal.

c. Polymer films.

A great variety of methods has been used for the investigation of polymer films on electrodes and one of these is the radiotracer method [61,98-108] which enables us to obtain information on the ionic charge transport in electroactive polymer films labeling co- and/or counter ions as well as on the formation and destruction of the films formed from labeled monomers. The main problems were [104] embedding of electrolytes during the formation of polymer films, study of the motion of counter- and co-ions in the course of electrochemical measurements, and study of the exchange processes. For a more detailed discussion of the above problems, we refer the reader to Ref. 109.

The results of radiotracer polymer film studies could be of importance in corrosion science as well if we take into consideration that conducting polymers can form a corrosion protective coating for metals especially for steel or anodized aluminum.

Important informations on polymer adsorption can be found in works by Schlenoff dealing with the application of radiochemical methods for the investigation of interfaces involving polymer species [50, 110-112].

1.4. Adsorption of ions
1.4.1. General

As it is known [113] two fundamental types of the adsorption of ions at electrified interphases should be distinguished:

i) Non-specific adsorption (positive or negative) that occurs when only long-range coulombic interactions are in play.

ii) Specific adsorption when short-range interactions between the ions and the surface become predominant.

Theoretically the specific adsorption can be detected by the deviation of the system from the behavior expected on the basis of the double layer theory (ion free layer and a diffuse layer obeying Gouy-Chapman theory). However, this detection requires sometime troublesome measurements and the evaluation of the data obtained should be based on model assumptions.

Radiotracer technique offers a unique possibility of demonstrating the occurrence of the specific adsorption of an ion by labeling it and studying its adsorption in the presence of a great excess of other ions, electrolytes (supporting electrolyte). For these studies at least a difference of one-two orders of magnitude in concentrations should be considered. Under such conditions it is evident that in the case of non-specific adsorption, determined by the coulombic interactions, no significant adsorption of the labeled species, present in low concentration, could be observed. In contrast to this, the observation of a measurable adsorption can be considered as a proof of the occurrence of the specific adsorption.

For instance, if in the presence of 1 mol dm^{-3} HClO$_4$ supporting electrolyte and 1×10^{-4} mol dm^{-3} ^{35}S labeled H$_2$SO$_4$ a measurable increase in the count-rate (that corresponds to the adsorption of the labeled species on the electrode) can be observed following a potential shift there could be no doubt that a specific adsorption took place. Thus the potential and concentration dependence of this specific adsorption can be systematically measured, however, it should be taken into consideration that these relationships reflect the result of adsorption competition with other species present in the system.

1.4.2. Anion adsorption

The radiotracer studies of anion adsorption can be divided into two groups depending on the nature of the electrode surface

a) adsorption on "bare" or "clean" surfaces

b) adsorption and sorption on surfaces modified by adatoms, organic molecules, polymer films etc.

A significant part of these studies was carried out with "rough" electrodes characterized by high real surface area.

During the last twenty years as an extension of the work with platinized electrodes [16,18,20,22,58,59,114] the potential and the concentration dependence of the adsorption of radiolabeled Cl$^-$, HSO$_4^-$ (SO$_4^{2-}$) and in some cases H$_2$PO$_4^-$ ions was studied at, copperized [74,75], nickelized [79], silverized [77,78], aurized [75,81], ruthenized [115] and rhodized [14,15,80,116-118] electrodes with high real surface area in the presence of a great excess of HClO$_4$ supporting electrolyte i.e. the specific adsorption of these ions was studied.

On the basis of comparison of the results of these studies the following main general conclusions can be drawn:

i) The potential dependence of the adsorption, the Γ vs. E curve, has one or more inflexion points, thus the corresponding dΓ/dE vs. E curves contain one or more peaks.

ii) At potentials where the dissolution of the metals (e.g. Ag, Cu, Ni) or the formation of oxide layer (e.g. Rh, Ru, Au) takes place the anion adsorption attains a high level i.e. the metal surface is covered by adsorbed anions.

iii) The relative adsorption strength of the anions in increasing sense is as follows: $ClO_4^- \langle HSO_4^- \left(SO_4^{2-}\right) \langle H_2PO_4^- \langle Cl^-$.

Using labeled sulfate ions it was demonstrated that the main features characterizing the adsorption of sulfate ions on smooth polycrystalline copper [119], gold [120], platinum [121] and rhodium [122] do not differ from those observed with copperized, aurized, platinized and rhodized electrodes.

The radiotracer study of the interrelation of anion specific adsorption and underpotential deposition (upd) of metal ions, contributed significantly to the clarification of processes occurring in the course of formation of metal adatom submonolayers and monolayers.

For a long period the adsorption of hydrogen on Pt, Rh, Pd electrodes at more positive potential values than the hydrogen equilibrium potential, formation of a monolayer of adsorbed H atoms, was considered in the literature as some specific and interesting phenomenon characteristic for some noble metals and hydrogen.

However, beginning from the sixties of the last century it became quite evident that the adsorption of metal ions on foreign metal substrates is often accompanied by significant charge transfer commensurable with the charge involved in hydrogen adsorption at noble metal electrodes. These phenomena were termed as underpotential deposition characterized by the general equation

$$M_s^{n+} + ne^- = M_{ad} \tag{8}$$

The study of the upd of metal ions belongs to the group of adsorption studies in the presence of a great excess of supporting electrolyte as in most cases studied the concentration of metal ions is chosen to be very low in comparison with that of the ions of the supporting electrolyte.

The simultaneous occurrence of the specific adsorption of anions and the formation of adatoms on a surface could result in significant difficulties in the calculation of the charge and mass balance required for the reliable interpretation of the overall adsorption phenomena. A number of situations should be distinguished in such cases however we consider only two limiting cases.

A) The upd takes place at a surface (in a potential range) where the role of the specific adsorption of anions can be neglected. In such cases the mass and charge balance can be relatively easily treated.

B) The upd takes place on a surface (in a potential range) where the specific adsorption of the anions present in the system is significant. This group of phenomena can be divided in to two subgroups.

B1) Adatoms displace adsorbed anions i.e. as the coverage with respect to adatom increases the opposite happens with adsorbed anions.

B2) Specific adsorption of the anions takes place as a result of interaction with adatoms. The extent of anion adsorption is proportional to the coverage with respect to the adatoms. Even if the adatom displaces the anions adsorbed on the support metal this effect could be compensated or overcompensated by the extent of adsorption induced by the adatoms. In the latter case an increase in the overall anion adsorption can be observed and this increase can be called as induced anion adsorption.

Radiotracer adsorption studies furnish rapid and reliable information on these phenomena using labeled anions in low concentration in the presence of a great excess of supporting electrolyte containing the metal ions forming the adatoms present also in a low concentration [15,17,18,121,123-125].

As an illustrative example the effect of Cd, Cu and Ag adatoms on Au support on the adsorption of labeled HSO_4^- (SO_4^{2-}) ions is shown in Fig. 3.

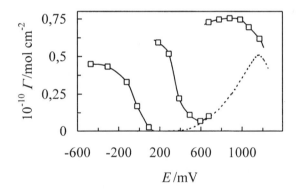

Fig. 3. Effect of the underpotential deposition of Cd^{2+} (1), Cu^{2+} (2), and Ag^+ (3) on the potential dependence of the adsorption of HSO_4^- ions on polycrystalline gold. Curve 4 (---) corresponds to the potential dependence without any addition: $c_{H_2SO_4} = 4 \times 10^{-4}$ mol dm^{-3}, $c_{Me} = 8 \times 10^{-4}$ mol dm^{-3} in 1 mol dm^{-3} HClO$_4$.

The coupled voltammetric and radiometric study of the simultaneous adsorption of hydrogen and anions (HSO_4^- and Cl^-) on a platinized electrode [126] clearly demonstrated that anion specific adsorption and hydrogen adsorption could not be separated from each other. (Smooth polished polycrystalline or disordered well-defined Pt and Rh crystal faces behave similarly to platinized surfaces.)

While the adsorption of Cl^-, HSO_4^- and $H_2PO_4^-$ ions is more or less reversible process depending on the nature of the electrode studied there are anion adsorption processes, which should be considered as strong chemisorption. This type of behavior was found in the case of the adsorption of CN^- ions on platinum electrode [127] and for S^{2-} and HS^- ions on platinum [128] and silver electrodes [78].

More or less strongly adsorbed anions induce cation adsorption. This problem will be discussed in connection with cation adsorption.

It should, however, be noted that in strongly alkaline media the extent of specific adsorption of SO_4^{2-}, Cl^- and PO_4^{3-} ions, is very low on noble metal electrodes. The absence of specific adsorption of these species in an alkaline medium may presumably be ascribed to the relatively strong adsorption of OH^- ions, which prevents the adsorption of other negatively charged species.

No direct evidence is available for the specific adsorption of OH^- ions, however, the unexpected potential dependence of the adsorption of the alkaline earth metal ions (Ca^{2+}, Sr^{2+}, Ba^{2+}) in alkaline medium (observed using radiotracer methods [129-131]) may be considered as an indirect proof for the occurrence of OH^- adsorption. It has been shown that the adsorption of ^{45}Ca labeled Ca^{2+} ions at Pt [130] and Au [109] electrodes in the presence of 0.1-0.15 mol dm^{-3} NaOH supporting electrolytes increases with increasing positive potentials (on the RHE scale), even below potentials where the formation of an oxide layer should be taken into consideration. This process was interpreted as cation adsorption induced by electrosorbed anions (OH^- ions). An increase of Na^+ concentration results in some decrease of Ca^{2+} adsorption indicating the occurrence of an "ion-exchange" process.

The results presented and discussed above were obtained with polycrystalline electrodes. As mentioned previously, the field of radiochemistry of well-defined surfaces was opened by Wieckowski. Most of the achievements on this field are connected with the investigation of HSO_4^- (SO_4^{2-}) adsorption on various crystal faces of Pt, Rh and Au and on the same surfaces after underpotential deposition of adatoms (formation of an ultra-thin layer).

These works were reviewed during the last years in several publications [5-12] thus only some of them should be mentioned here.

The comparison of the results on adsorption of HSO_4^- (SO_4^{2-}) ions obtained on Pt(100), Pt(111), Pt(110) and Pt(poly) electrodes [8,132,133,134] clearly demonstrates that adsorption of sulfate anion on the Pt(hkl) electrodes is strongly surface structure specific. It follows from the analysis of the adsorption data that the extent of adsorption (Γ) vs. potential plot obtained for the polycrystalline platinum electrode is a result of a cumulative adsorption of sulfate on the single crystal facets (111), (100) and (110) of the polycrystalline surface, however, the contribution from ordered (110) plane to the overall behavior is quite small.

Another interesting comparison was made between the adsorption behavior of HSO_4^- ions on Pt(111) and Rh(111) surfaces.

It has been stated that although there is some resemblance between the voltammetry of Rh(111) and Pt(111) in perchloric acid electrolyte, the addition of labeled sulfuric acid to the perchloric acid supporting electrolyte yields distinctively different voltammograms. The difference has been proposed to result from stronger adsorption of anions on rhodium than on platinum, as well as from the fact that the high energy-hydrogen is formed on Pt(111) but not on Rh(111).

Finally, the extensive and comparative studies of the enhancement of HSO_4^- adsorption caused by the underpotential deposition of metal ions (Cd^{2+}, Cu^{2+}, Ag^+) on Pt(111) and polycrystalline Pt surfaces should be mentioned [121,135].

In Refs. 35 and 136 adsorption of sulfate ions was studied on three basal planes and a polycrystalline surface of silver in 0.1 mol dm^{-3} $HClO_4$ by radiometric and electrochemical methods. On all four surfaces adsorption was found to be reversible with respect to the potential and bulk solution concentration of sulfate.

The most important findings are as follows:

i) Radiometrically determined adsorption isotherms show a clear diference in sulfate adsorption on individual planes.

ii) Adsorption of the anion on the basal planes of silver decreases in the sequence: Ag(111) greater than or equal to Ag(110)>Ag(100).

iii) The polycrystalline silver electrode exhibits intermediate adsorption properties between the most and the least active basal planes.

A similar study with monocrystalline copper electrodes was reported in Ref. 37. It was found that the limiting surface concentration of sulfate ions decreases in the sequence: Cu(111)> Cu(110) > Cu(poly) > Cu(100). The highest value of surface concentration of sulfate for the (111) plane was accounted for by the match between the tetrahedral structure of the anion and the trigonal distribution pattern of the surface atoms of copper on this plane.

1.4.3. Cation adsorption

In comparison with the anions relatively less attention was paid to the cation adsorption in direct radiotracer studies, at least during the last ten years.

A radiotracer technique combined with cyclic voltammetry was used for the investigation of underpotential deposition of ^{204}Tl labeled thallium onto

(111), (110) and (100) silver crystalfaces from perchloric acid solutions [38]. Two modes of thallium adsorption were found: a weak adsorption without charge transfer and a strong adsorption with charge transfer.

The underpotential deposition of labeled thallium on a polycrystalline gold electrode in alkaline solution was studied in Ref. 137. It was found that at potentials close to that of the Tl/Tl(I) equilibrium a close packed monolayer of Tl adatoms is formed.

In connection with the study of the underpotential deposition of ^{65}Zn labeled Zn^{2+} ions on a polycrystaline gold substrate a novel method for the evaluation of the surface excess of X-ray emitting species was elaborated [138,139].

Induced cation adsorption was found to be a tool for the study of the adsorption of some anions and organic species [140].

As mentioned in connection with OH^- adsorption the unexpected potential dependence of the adsorption of Ca^{2+} ions on Pt and Au electrodes in alkaline medium was explained by the assumption that this behavior should be ascribed to the electrosorption of OH^- ions i.e. the observed Γ vs. E relationship reflects that of OH^- adsorption which induce the cation adsorption.

Although this interpretation seemed to be acceptable, further direct evidence was required to verify the concept of induced cation adsorption under the experimental conditions considered.

In order to solve this problem some fundamental requirements should be fulfilled:

i) an anion with higher adsorbability than that of OH^- ions should be found;

ii) one must know or must determine the potential dependence of the adsorption of this species;

iii) a study of the adsorption of Ca^{2+} ions should be carried out in the presence of adsorbed anions.

The choice of the CN^- ion as model species for the study of the induced adsorption of labeled Ca^{2+} ions seemed to be well founded in light of radiotracer [141,142] and other studies [143-147]. Really, it was shown [142] that the strong chemisorption of CN^- ions results in a significant increase of the Ca^{2+} adsorption in an alkaline medium leading to a potential dependence of ^{45}Ca-labeled Ca^{2+} adsorption that is characteristic for the potential dependence of the adsorption of CN^- ions. This result confirms the assumption that the unexpected potential dependence of alkaline earth metal ions (increasing adsorption with increasing potential) in alkaline medium is a consequence, of their induced adsorption via adsorbed OH^- species.

As a continuation of this work experiments where carried out in acid medium in the presence of Cl^-, $H_2PO_4^-$, $HSO_4^-(SO_4^{2-})$ anions studying the adsorption of ^{45}Ca labeled Ca^{2+} ions. The results obtained indicated that the cation adsorption induced by the specific adsorption of anions is a general feature [148].

Another interesting aspect of induced cation adsorption is that observed in the case of $-SO_3H$ and $-COOH$ groups anchored to the surface of a platinum electrode via strongly chemisorbed organic species [149]. It was found that the chemisorption of such compounds as benzene sulfonic and maleic acids enhance the adsorption of ^{45}Ca labeled Ca^{2+} ions.

It was assumed that the organic molecules are anchored to the surface via the interaction of the aromatic ring and the double bond with the surface leading to a strong chemisorption while the $-SO_3H$ and $-COOH$ groups are involved in ion-exchange processes.

$$-COOH \leftrightarrow COO^- + H^+$$
$$-SO_3H \leftrightarrow SO_3^- + H^+$$
$$-COO^- + M^+ \leftrightarrow -COOM$$
$$-SO_3^- + M^+ \leftrightarrow COOM$$

1.4.4. The link between the relative surface excesses determined by thermodynamic means and the surface excess concentration determined by radiotracer method

The problem: what is the link between the relative surface excesses determined by thermodynamic means and the surface excess concentration determined by radiotracer method is of great theoretical importance.

There are, a number of investigations where the results of radiotracer adsorption studies are compared with those of measurements of thermodynamic nature.

To answer the question posed above we should consider the fundamental relationships. In the simplest case the relative surface excess of a component i (Γ_i') referred to the solvent (water) is given by the following relationship

$$\Gamma_i' = \frac{1}{A}\left(n_i - n_w \frac{x_i}{x_w} \right) \tag{9}$$

where n_i and n_w are the amount of component i and water, in the whole system. x_i and x_w are the corresponding mole fractions. A is the area of the interface.

The radiotracer method allows us to make a distinction between the amount of species i in the solution phase n_i^s, and the amount of i that could be considered as accumulated in the interface n_i^a. Thermodynamically it means, that we are using a reference system, the composition and volume of which coincides with that of the homogeneous bulk phase.

Thus

$$n_i = n_i^a + n_i^s \tag{10}$$

It is evident that together with n_i^a some amount of solvent should be connected (the adsorption of other species is also possible), therefore

$$n_w = n_w^a + n_w^s \tag{11}$$

Considering that

$$\frac{x_i}{x_w} = \frac{n_i^s}{n_w^s} \tag{12}$$

we obtain:

$$\Gamma_i' = \frac{1}{A}\left(n_i^a - n_w^a \frac{x_i}{x_w} \right) = \Gamma_i - \Gamma_w \frac{x_i}{x_w} \tag{13}$$

which is an equivalent form of Eq. (9) containing the surface excess concentrations Γ_i and Γ_w. (Eqs (9) and (13) well demonstrate that the value of the relative surface excess does not depend on the choice of the reference system.)

Assuming, for instance, a monolayer adsorption model, the total surface area A can be expressed as:

$$\sum_{i \ne w} A_i n_i^a + A_w n_w^a = A \tag{14}$$

where A_i and A_w are the area occupied by one mole species "i" and water, respectively.

From this equation

$$n_w^a = \frac{A - \sum\limits_{i \ne w} A_i n_i^a}{A_w} \tag{15}$$

and thus

$$\Gamma_i' = \frac{1}{A}\left(n_i^a - \frac{A - \sum\limits_{i \neq w} A_i n_i^a}{A_w} \frac{x_i}{x_w} \right) \tag{16}$$

In the case if the concentration of species i in the solution is low, $x_i \ll x_w$, we can write:

$$\Gamma_i' \approx \frac{1}{A} n_i^a \tag{17}$$

On the other hand, according to Eq. (10), the surface excess concentrations Γ_i can be defined by the equation:

$$\Gamma_i = \frac{1}{A}\left(n_i - n_i^s \right) \tag{18}$$

where n_i is the amount of component i in the whole system, and n_i^s is the amount of component i in the reference system (the bulk solution phase). Using Eqs. (10),(13) and (18) the difference between Γ_i and Γ_i' can be expressed as

$$\Delta\Gamma_i = \Gamma_i' - \Gamma_i = -\frac{1}{A}\frac{x_i}{x_w} n_w^a = -\frac{1}{A}\frac{x_i}{x_w}\left(n_w - n_w^s \right) = -\frac{x_i}{x_w}\Gamma_w \tag{19}$$

According to Eq. (19) the smaller the absolute value of Γ_w, the smaller the absolute value of the difference between surface excess concentrations and (relative) surface excesses. Since the extent of the reference phase can be selected arbitrarily, and the value of $\Delta\Gamma_i$ depends directly on this choice, the latter is ill defined from thermodynamic point of view. However, by introducing some nonthermodynamic assumptions, the value of $\Delta\Gamma_i$ can be estimated for several special cases, e.g. in the cases, if the above mentioned monolayer model can be considered as a reliably description of the interphase, the value of Γ_w is small, and the concentration of the labeled species i in the solution (x_i) is also low, therefore $\Delta\Gamma_i$ is negligible, and Eq.(17) is always true.

The above results show that in the cases discussed above the relative surface excess practically coincides with the surface excess concentration determined by radiotracer method at low concentration of the labeled species.

Similar result could be obtained in the case of two or three monolayer thickness of the surface layer.

It is, however, well known that in the case of charged species (ions) the thickness of the adsorption layer, the diffuse part of the double layer, strongly depends on the total ion concentration, on the ionic strength. Consequently, if the labeled ions are present in a low concentration and the total ionic concentration is also rather low, the monolayer model cannot be used without restrictions. It seems to be interesting to analyze the meaning of the results in these cases.

Since in radiotracer experiments the determination of the adsorbed amount of the labeled species is aimed, besides the normal radioactive background (dark count rate), the true solution background is also measured. One of the most frequently applied methods is the following:

1) The potential of the electrode is adjusted to a value, where no adsorption of the labeled species takes place, and the counting rate is measured. The difference of this value and the dark count rate is accepted as the counting rate of the background (N_{bg}).

2) The potential is then adjusted to the desired value, and measurement of the counting rate is carried out. The difference of this value and the dark count rate is accepted as the total counting rate (N_{tot}).

Since N_{bg} remains constant during the measurements, from

$$N_{tot} = N_{bg} + N_{ad} \tag{20}$$

the count rate of the adsorbed species (N_{ad}) can easily be obtained. In the case of thin foil method and isotopes emitting soft β radiation the surface concentration Γ_i of the labeled species is calculated from the formula:

$$\Gamma_i = \frac{N_{ad}}{N_{bg}} \frac{c}{\mu R}, \tag{21}$$

where c is the concentration of the labeled species, μ is the linear coefficient of absorption in the solution phase, and R is the roughness factor. In the case of electrode lowering technique (19) is modified:

$$\Gamma_i = \frac{N_{ad}}{N_{bg}} \frac{c}{\mu R f_b \exp(-\mu x)}, \tag{22}$$

where f_b is the backscattering factor, and x represents the distance between the electrode face and the detector.

On the basis of the above considerations it is clear, that in this way the surface excess concentration of the labeled species can be determined. (Similarly to the case discussed earlier, from the point of view of thermodynamics it means, that the composition and volume of the reference system is the same as that of the bulk phase of the real system.)

However, in the case of diluted solutions, the difference between the (relative) surface excesses Γ_i' and surface excess concentrations Γ_i can be considerable. It may cause serious problems especially in such cases, if data originating from radiotracer measurements are compared directly with data calculated on the basis of thermodynamical equations, (like e.g. the electrocapillary equation), since the latter ones are true (relative) surface excesses.

For example, the characteristic thickness of the diffuse layer calculated for the aqueous solution of a strong, 1:1 electrolyte at 25 °C is about 1 nm, if the concentration of the electrolyte is 0.1 M, and 3 nm at 0.01 M. Of course, the "whole" thickness of the layer, namely the thickness, where the spatial change of the concentrations becomes negligible is considerably greater. If we still assume, that the concentration of a 1:1 electrolyte is 0.1 M, and the thickness of the diffuse layer is only 1 nm, we can estimate the lower limit of the error. Let the surface charge at a given potential be 5 $\mu C/cm^2$ = $5 \cdot 10^{-2}$ C/m^2, originating completely from the excess amount of cations with an ionic charge of +1 and a molar mass of 100 g/mol, then in the 1 nm thick layer we have some

$$\frac{5 \cdot 10^{-2} \frac{C}{m^2}}{10^5 \frac{C}{mol}} \cdot 100 \frac{g}{mol} \cdot 1 m^2 = 5 \cdot 10^{-5} g \, (\text{or } 5 \cdot 10^{-7} \text{ mol}) \text{ ions per m}^2 \text{ surface. (This is, of}$$

course, only a rough approximation, since anions are also present in the diffuse layer.)

If the average density of this layer is close to 1000 kg/m^3 = 10^6 g/m^3 (this assumption can be, of course, also questioned, since the structure of the surface layer differ from that of the bulk phase) the total mass of a 10^{-9} m^3 segment is 10^{-3} g. It means, that besides the $5 \cdot 10^{-5}$ g ions, there is about 9.5·10^{-4} g (that is roughly $5 \cdot 10^{-5}$ mol) water also present. Since $x_i/x_w \approx 0.1/55.5 = 1.8 \cdot 10^{-3}$, the

value of $n_i^a - n_w^a \frac{x_i}{x_w}$ can be estimated as $5 \cdot 10^{-7} - 5 \cdot 10^{-5} \times 1.8 \cdot 10^{-3} = 5 \cdot 10^{-7} - 0.9 \cdot 10^{-7}$

mol.

Since the above calculations thermodynamically correspond to the choice of a reference system, the composition and volume of which coincides with that of the homogeneous bulk phase, according to the final result the difference of Γ_i' and Γ_i is more than 15 % in this case.

On the basis of the above considerations it is evident, that the situation may be very similar in the case of radiotracer experiments with labeled cations carried out according to steps (1) and (2). In practice, when the solution background is determined at a certain potential where "no adsorption of the labeled species occurs", the concentration of the cations in the solution side of the surface layer is assumed to be negligible. (It should be mentioned here, that the expression "no adsorption of the labeled species occurs" is therefore not really correct from the point of view of thermodynamics, since it apparently excludes the possibility of $\Gamma_i' < 0$.) The surface concentration of the cations at a potential, where adsorption of them takes place, can be calculated practically as a difference of amounts of cations in the double layer at the two potentials, divided by the surface area. In principle the thickness of the diffuse part of the double layer depends on the potential, but the change may be not very significant.

From Eq. (19) it is clear, that in principle the problems concerning the difference between surface excess concentrations and (relative) surface excesses can be circumvented by calculating the counting rate for a reference system of the same composition as the bulk solution phase, and in which the amount of water is equal with the total number of moles of water in the system under test (it is important to note here, that the volume of the above "reference phase" is necessarily not equal with the volume of the solution phase including the diffuse double layer), i.e., in principle it is possible to determine the relative surface excess of the labeled component by radiotracer experiments also in such cases, where the monolayer model cannot be used.

Unfortunately, the accurate calculation of this counting rate is rather a theoretical possibility, since for such calculations the complete characteristics of the measuring system including parameters depending on geometrical arrangements, background activities, counting efficiencies, etc. would be necessary.

On the other hand, the realization of conditions required for an experimental determination of the corresponding solution background seems to be practically impossible, since apart from the huge experimental problems with e.g. the determination of rather great volumes with an accuracy in the nm^3 range, there is no guarantee, that during the subsequent experiments signals originating from exactly equivalent (but not equal) volumes are detected. (The effect of such a difference may be considerable, since the ratio of the counts of the solution background and that of the adsorbed amount is expected to be relatively high in the cases discussed here.) Nevertheless, the use of the same solution background (determined e.g. at a potential close to the p.z.c.) in a series of experiments is practically equivalent with a constant $n_w^a \dfrac{x_i}{x_w}$ term and a variable n_i^a term in Eq. (13).

Consequently, if the relevance of the monolayer (multilayer) model is questionable, the exact physical meaning of the experimental quantities may remain uncertain. Since in these cases it is a rather difficult task to find the link between the results of thermodynamic experiments and values determined by radiotracer method, a detailed analysis of the actual conditions is always necessary.

In contrast to the above systems, in the presence of a great excess of supporting electrolyte (used in most experiments reported in the literature) only the specific (monolayer) adsorption of the labeled minority component is measured, thus the surface excess determined by the radiotracer method can be always used as a good approximate data in thermodynamic calculations.

1. 5. Adsorption of organic species

1.5.1. General

In all adsorption studies one of the most important question pertaining to the treatment of the experimental data is the existence of the adsorption equilibrium i.e. whether it can be assumed that the adsorption process is reversible or not. As mentioned in the Chapter 2 and in a previous section of this chapter the radiotracer methods offer a possibility of studying the dynamics of adsorption. The mobility of labeled species adsorbed on the surface of the electrode can be easily detected by their exchange with non-labeled molecules added to the solution phase.

This type of experiments were carried out several times with various organic species adsorbed on noble metal (first of all platinum) electrodes and it was found that very often the exchange rate is practically zero thus the assumption of adsorption equilibrium in these cases is very questionable. Considering this problem the adsorption behavior of organic species was divided into two categories according to the mobility of their adsorbed molecules probed by the rate of exchange between labeled adsorbed and unlabeled dissolved species [150].

The first group is characterized by significant mobility of the adsorbed molecules.

It includes some saturated carboxylic acids (no formic acid), urea etc.

The second group is characterized by the practical immobility of adsorbed species and it contains aromatic compounds, unsaturated compounds, alcohols, acetone, methanol, formaldehyde, formic acid etc. [2].

Despite the evidences calling the attention to a very cautions treatment of adsorption data in terms of adsorption isotherms there are attempts in the literature to do so without any distinction.

A very interesting confrontation of views in the literature [39,56] enlightens this situation very clearly. In ref. 56 studies on the electrosorption of organic compounds on polycrystalline platinum electrodes were carried out in 0.01 mol dm^{-3} HCl solution. In-situ techniques, radiotracer measurements, ellipsometry and Fourier transform infrared spectroscopy, were used in the measurements. Good agreement in results was obtained between these techniques. The electrosorption process was found to be slow, but a bell-shaped coverage (θ) versus potential (V) curve expected for reversible adsorption processes was generally obtained. The concentration dependence of the adsorption was treated in terms of adsorption isotherm. For the interpretation of the phenomena a water competition adsorption model, based on the statistical mechanics approach, was proposed. Both lateral interactions between adsorbed species and the effect of heterogeneity of the electrode surface were taken into account

Considering the great variety of the organic species studied (n-butyl alcohol, n-valeric-acid, n-valeraldehyde, n-butylamine, n-valeronitrile, 1-butylmercaptan, n-valeroylchloride, phenol, benzoic-acid, benzaldehyde, aniline, benzonitrile, thiophenol, benzoylchloride, 1-naphthol, 1-naphthoic acid, 1-naphthaldehyde, 1.naphthamine and 1-naphthoylchloride) and the previous experiences about the lack of mobility of adsorbed alcohols and aromatic compounds and their high reactivity at a Pt surface [2-4] the general validity of these assumptions seemed to be questionable. This view was clearly expressed by Wieckowski in a comment [39] on ref. 56:

"...(1) since adsorption reversibility was not documented the application of isotherms for adsorption characterization was not justified; ....." even the assumption of equilibrium: "... is contrary to what is found in the literature about adsorption of the studied compounds on platinum [4,150]. In fact, if not prevented by competing surface contaminants or morphological surface constraints it goes to completion. This leads to essentially irreversible adsorption, or an immobile product [150]."

On the other hand, in ref. [56] the role of the competitive adsorption between water and anions is completely ignored. Chloride ions adsorb relatively strongly on platinum [151-153] thus the surface of the electrode, preceding the adsorption of organic species, should be at least partly covered by adsorbed Cl$^-$ ions. Experimental evidences show [2], that adsorbed Cl$^-$ ions are displaced by strongly, irreversibly, adsorbing organic molecules i.e. the role of this exchange reaction cannot be neglected.

Considering this controversy and other similar problems in the literature it is a fundamental task, before treating the adsorption phenomena in terms of equilibrium, to study the mobility of the adsorbed species to furnish reliable information concerning the reversibility (or irreversibility) of the adsorption.

The radiotracer technique, as mentioned, offers the simplest way to fulfill this requirement and the results of such studies should be taken into consideration in the case of any effort aiming at the interpretation of the apparent concentration and potential dependence of adsorption.

1.5.2. Adsorption of aliphatic compounds

The investigation of the electrosorption and electrooxidation of CH_3OH and the possible intermediates of its oxidation, HCHO, HCOOH, CO, has been a central task for fuel cell-oriented electrocatalytic research for several decades. Despite this long history there is a continuous interest in these problems presumably for the very reason that still there are many open questions in connection with the inexpensive practical commercial realization of fuel cells. Surveys on the relevant radiotracer studies were presented in the eighties [2,3]. The main conclusion of these studies was that the adsorption of CH_3OH, CH_2O, COOH, CO leads to the formation of the same strongly chemisorbed CO, CHO species. In the second half of the eighties and in the nineties spectroscopic methods came into the foreground for these studies thus nowadays investigations based on the application of radiotracer technique are very sporadic [8,154] in comparison with the previous years. (See Ref. 67 and literature cited therein.)

In Ref. 67 the adsorption, exchange, and oxidation of formaldehyde have been studied on gold at various potentials, temperatures, concentrations, and pH values by voltammetry and the radiotracer thin-gap method and compartmental analysis.

Studies on the adsorption of urea should be also mentioned. Wieckowski et al. reported a complete experimental and theoretical description of urea adsorption on Pt(100) using voltammetry, radiochemistry, low energy electron diffraction and quantitative Auger-electron spectroscopy [155,156].

The adsorption of thiourea (TU) on copper and silver single crystal electrodes was studied recently using ^{14}C labeled TU [157,158] and perchloric acid supporting electrolyte. A reversible adsorption and crystal plane dependent surface concentrations were found for all crystal faces of both metals.

The radiotracer study of the adsorption behavior of saturated aliphatic mono- and dicarboxylic acids at platinum electrodes has been the subject of several studies [2-4,159]

The behavior of the first member of the homologue series under consideration (formic and oxalic acids) differs from that characteristic for the series. On the other hand, experimental results show that there is no significant difference in the behavior of such simple mono- and dicarboxylic acids as acetic, propionic, malonic, succinic, adipic and pimelic acids.

The adsorption behavior of these species at a platinum electrode at their low concentrations (10^{-4}-10^{-3} mol dm^{-3}) can be characterized by two main features: (i) occurrence of reversible adsorption processes: reversibility with respect to potential and concentration changes; (ii) the potential dependence of the adsorption of these compounds differs significantly from that observed for most organic species containing CH_2–OH, –CO– groups, double bonds or aromatic nucleus. In acidic medium a continuous increase of the adsorption occurs by increasing the potential from 0 to 800 mV (on RHE scale), a feature which is very similar to that observed for the specific adsorption of some anions as Cl^-, HSO_4^- and $H_2PO_4^-$ ("anion-like" behavior).

This behavior can be explained by the absence of a strong irreversible chemisorption and by the solvent (H_2O) and solute (organic acid) adsorption competition (at the surface of the electrode or in the second ad-layer [160]).

Adsorption of acetic acid, being a representative of the reversible adsorption of organic species, was in the focus of interest during several years. An exhaustive analysis of the structural aspects of the adsorption of acetic acid was given by Wieckowski [160]. In this analysis it was shown how the direct quantitative radiochemical information on adsorbate concentration could be used in conjunction with data obtained from infrared spectroscopy to provide an exhausted surface analysis of adsorbate in the solid/liquid interface. On one hand, the radiotracer measurements provide a valuable check on the interpretation of the spectral data on the other the spectral information reveals the nature of the surface coordination and orientation of the adsorbed species.

A survey on the results obtained from radiotracer study of the adsorption of aliphatic primary amino compounds at platinum electrodes can be found in Ref. 161. Adsorption properties of

  i) mono- and diamines (methyl amine, butyl amine, ethylenediamine),

  ii) amino acids (glycine, alanine, γ-amino-butyric acid and aspartic acid),

  iii) other simple amino compounds (taurine, ethanol amine) were studied and compared.

The role of the various functional groups in the overall adsorption behavior was considered and discussed. One of the main conclusion of these studies is that the overall adsorption behavior of a molecule is a resultant of the effects exerted by the different functional groups.

As a continuation of this work the adsorption of:

serine ($HOCH_2$–$CH(NH_2)$–COOH) at platinum electrode [162],

butylamine, ethylenediamine, methionine and ethanol amine at a smooth gold electrode [163], a simple tetra peptide (Arg–Lys–Asp–Vol) at smooth gold and platinized platinum electrodes [164],

a tripeptide (Arg–Lys–Asp) at a smooth gold electrode [165]
was studied.

It is of interest to note that no adsorption of simple amino compounds was found in acid medium at a gold electrode [163] with the exception of methionine. The significant adsorption found in the latter case can be ascribed to the presence of the $-SCH_3$ group.

A potential and concentration dependent adsorption was found in alkaline medium for all the species studied. Two types of adsorption were distinguished. During the first stage of the adsorption loosely adsorbed species are formed mainly, but with increasing adsorption times and concentrations the role of a strongly chemisorbed species becomes much more pronounced.

It is suggested that the formation of strongly chemisorbed species is connected with oxidative transformations and the apparent potential dependence of the adsorption may be explained in terms of oxidative chemisorption and reductive desorption processes.

The adsorption behavior of the peptides studied is roughly determined by the functional groups of the amino acid components. In acid medium only loosely adsorbed species can be observed on gold while a strong chemisorption occurs at platinized platinum. (In agreement with the phenomena observed for amino acids.) In alkaline medium strong chemisorption takes place at both metals.

In Ref. 166 an example for the radiotracer study of protein adsorption can be found. The adsorption of two model proteins human serum albumin (HSA) and immunoglobulin (IgG), on a gold electrode surface was investigated using ^{125}I radiolabeling and cyclic voltammery (CV). ^{125}I radiolabeling was used to determine the extent of protein adsorption, while CV was used to ascertain the effect of the adsorbed protein layer on the electron transfer between the gold electrode and an electroactive moiety in solution, namely, $K_3Fe(CN)_6$. The adsorbed amounts of HSA and IgG showed approximately monolayer coverage. The amount of adsorbed protein increased when a positive potential shift was applied to the electrode, while the application of a negative potential shift resulted in a decrease. When the solution pH varied to alter the charge on the protein, the adsorption trends appeared to follow electrostatic interaction namely, greater adsorption when the electrode and the protein possessed opposite charge and vice versa. The adsorbed protein layer had the effect of blocking the electron transfer.

The oxidation of glucose has been studied very intensively during the last years with the aim of developing sensors for the detection of glucose and to create implantable bio-fuel-cells.

Various methods were used for these studies, however, radiotracer studies constitute a specific approach to the problems [167,168].

The electrosorption of ^{14}C labeled glucose at a platinized platinum electrode can be followed directly and information can be obtained on the coverage with respect to the labeled species. The behavior of adsorbed species formed at different potentials can be studied by electrochemical polarization and via exchange processes with non-labeled species. On the other hand, the study of the adsorption of labeled Cl^- and HSO_4^- ions, in the presence of glucose, gives information on adsorption of the latter species through the competitive adsorption with the former ones. In addition, the effect of the anion adsorption on the oxidation process can be studied, as well.

1.5.3. Adsorption of aromatic compounds

The investigation of the adsorption of aromatic compounds on noble metal electrodes was one of the earliest studies carried out by radiotracer technique in the sixties [2-4]. Adsorption of benzene, benzene sulfonic acid, phenol, benzoic acid and other derivatives on platinum was studied during the last decades [2-4]. Such studies are continuing using, for instance, benzoic acid as a model compound.

Adsorption of ^{14}C labeled benzoic acid on a polycrystalline gold electrode was studied by Zelenay et al. [169]. It was found that a strong adsorption occurs in a wide potential range, however, the exchange of adsorbed labeled molecules with non-labeled species added to the solution phase attests the occurrence of dynamic equilibrium between the adsorbed and solution species.

On the basis of adsorption data and model calculations it was assumed that the adsorbed molecules could be present in two different orientations. Flat (parallel to the surface) orientation dominates at low potentials while at high potentials the vertical (perpendicular to the surface) orientation is predominant. Similar study was carried out with polycrystalline silver electrode [36].

Strong irreversible chemisorption of 2-naphtoic acid was observed at platinized electrode in both acid and alkaline media [170]. It was shown that the elimination of chemisorbed species could be achieved by reduction and oxidation consequently, the apparent potential dependence of the adsorption should be explained by the occurrence of these processes.

The radiotracer study of the adsorption of pyridine on a smooth gold electrode [171] in conjunction with other methods is another important contribution to former investigations carried out with platinum and gold electrodes aiming at the determination of the orientation of adsorbed species.

A study [172] of the adsorption of a derivative of pyridine, nicotinic acid, on platinized electrode clearly demonstrated that the adsorption behavior of nicotinic acid is very similar to its isocyclic aromatic counterpart, benzoic acid, as a strong irreversible chemisorption occurs at potentials where no reduction or oxidation of the chemisorbed species takes place.

The concentration and potential dependence of the adsorption of several aromatic molecules on platinized electrodes was determined by Bockris et al. [56].

1.5.4. Study of the behavior of polymer film electrodes, and adsorption of polymers

The various applications of radiotracer technique for the study of the behavior of some polymer film electrodes are summarized in Refs. 104, 109.

Results obtained by a combined electrochemical-radiotracer technique on poly/vinyl ferrocene (PVF), tetracyanoquinodimethane polyester (PTCNQ), polypyrrole (PP) and polyaniline (PANI) are presented.

Only some illustrative examples will be mentioned here.

In the case PTCNQ film the "break in " phenomenon and the motion of counter-ions during the transformation of redox sites was followed by use of ^{45}Ca labeled Ca^{2+} ions. The measurement of the potential dependence of the equilibrium Ca^{2+} sorption in PTCNQ film revealed the existence of two different states corresponding to partners involved in dimerization equilibrium.

On the other hand, by labeling the co-ions (for instance, ^{35}S labeled SO_4^{2-}) the motion of these ion can be followed as well.

Another interesting example is when the formation and the state of the film is studied labeling the monomer forming the film. This procedure was followed in the case of PANI films using ^{14}C labeled aniline for the preparation (electrochemical polymerization) of the film.

By labeling the monomer the following information can be obtained:

i) correlation between the amount of monomer involved in the film and the charge involved in the corresponding redox transformations,

ii) the charge involved in the formation of the film,

iii) changes in the stability of the film, the possible over oxidation and decomposition of the film.

Thus on the basis of these informations a scheme was proposed for the interpretation of the phenomena occurring in the course of the formation of PANI-film by anodic oxidation of the aniline monomer.

In a series of publications [107,108] studies of the formation of poly(o-phenylenediamine) (PPD) films and their behavior were carried out using quartzcrystal microbalance and radiometric techniques.

The sorption of $H_2PO_4^-$, HSO_4^- and Cl^- ions was studied and compared by the radiotracer technique using ^{35}S, ^{32}P and ^{36}Cl isotopes for labeling. It was found that the anion sorption in both oxidized and reduced films depends on pH, i.e. on the protonation of the film; however, the protonation and anion sorption is not an essential condition for the occurrence of the redox process.

Results furnished by differential voltradiometric measurements advocated in the favor of the assumption that the reduction of film results in the formation of three or four types of center with different behavior with respect to the protonation and anion sorption. It was demonstrated that PPD films are very selective for the sorption of $HSO_4^- (SO_4^{2-})$ ions in the presence of a great excess of other anions.

In Ref. 50 the kinetics of the adsorption of radiolabeled poly(styrene sulfonate) at a positively-charged surface were determined using a surface-charged plastic scintillator as substrate. In situ measuremets indicated diffusion-limited adsorption. It was concluded from the absence of self-exchange of unlabeled for labeled polymer that the sorption is a thermodynamically irreversible process, at least for highly charged polymers. Charge compensation of sulfonate was determined by displacement of labeled calcium ($^{45}Ca^{2+}$) probe ions. It was demonstrated that the use of radiometric techniques facilitates studies of polyelectrolyte multilayer formation.

Chemisorption via metal–sulfur interactions has proven a convenient means for attaching species to the surface of metals such as gold. Typically, a clean metal is immersed in a dilute solution of thiols or disulfides to produce well-defined, organized structures, selfassembled monolayers (SAMs), at the metal/liquid interface. Nowadays these systems are in the focus of interest.

The radiotracer technique found its application in the investigation of SAMs using alkanethiols bearing radiolabeled (^{35}S) head groups [173]. Labeled octadecanethiol monolayers formed on a variety of substrates (Au, Pt, Cu and Ag) were studied to determine coverage, thermal- and photostability and surface roughness. Problems connected with the enhancement of the stability of SAMs mode from thiols were discussed in detail.

1.6. Adsorption studies at oxide/solution interfaces

The application of radiotracer technique for the study of the adsorption phenomena occurring at oxide/solution interphase was prompted on the basis of the following argumentation:

a) We have little direct in situ information on the specific adsorption of charged species (ions) at the oxide–aqueous solution interface in strong acid medium despite the fact that the role of surface chemistry of metal (hydr)oxide–solution interface is well documented.

b) Most of the methods used in colloid chemistry for the investigation at oxide-solution interface are far from being able to follow in situ the adsorption phenomena and to study the effect of various parameters on the adsorption without disturbing the original experimental conditions.

The radiotracer technique elaborated for the study of electrode surfaces, presented in the previous sections, offers a unique possibility of solving this problem using metal oxides in the form of a powder in contact with the solution phase containing various labeled anions.

In Table 1 some of the systems studied during the last years are presented.

Table 1.
Anion adsorption studies by radiotracer technique on various oxides

Oxide	Anions	References
γ-Al$_2$O$_3$	HSO_4^-/SO_4^{2-}, Cl^-, TcO_4^-, $H_2PO_4^-$, CrO_4^-	93,174,95
AlOOH	HSO_4^-	175
hematite	HSO_4^-/SO_4^{2-}, TcO_4^-, $H_2PO_4^-$	92,176,94
FeOOH	HSO_4^-/SO_4^{2-}	175
Cr$_2$O$_3$	HSO_4^-/SO_4^{2-}	95
CuO	HSO_4^-/SO_4^{2-}, Cl^-	177
Bi$_2$O$_3$	HSO_4^-/SO_4^{2-}	97
TiO$_2$	HSO_4^-/SO_4^{2-}	96

1.6.1. Specific adsorption of anions induced by the protonation of the oxide surfaces

As it was emphasized in the previous sections, information concerning the specific interaction of radiolabeled anions with oxide surfaces, the adsorption of the labeled species, could be obtained in the presence of a great excess of a supporting electrolyte. Considering the low adsorbability of ClO_4^- ions, HClO$_4$ and NaClO$_4$ solutions were used as supporting electrolytes. (The problem of the distinction between specific and nonspecific adsorption will be discussed in the next section.) Most of the studies were focused on the effect of the protonation of the surface on the anion adsorption, determining the pH dependence of the adsorption of anions. In the case of sulfate adsorption very similar curves were obtained for various oxides.

A typical curve is shown in Fig. 4.

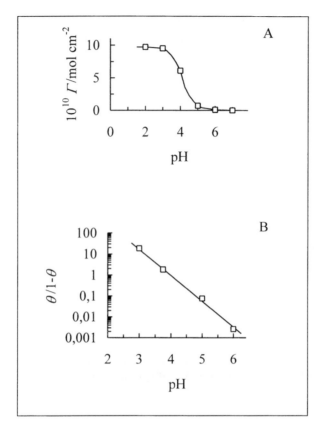

Fig. 4. (A) pH dependence of the adsorption of sulfate ions on hematite in 3×10^{-1} mol dm^{-3} NaClO$_4$ + 1×10^{-3} mol dm^{-3} labeled sulfate. (B) Plot of $-\lg\theta/(1-\theta)$ vs. pH from data obtained for the pH dependence of the adsorption

This type of behavior was interpreted in terms of the protonation of the oxide surface characterized by the equilibrium formulated by the equation

$$X + H^+ \rightleftarrows XH^+ \tag{23}$$

where X is a surface site. According to the assumptions the specific adsorption of anions takes place on the protonated surface sites. Considering Eq. 23 in the case of equilibrium:

$$k_a c_{H^+}\left(1 - \theta_{H^+}\right) - k_d\theta_{H^+} = 0 \tag{24}$$

where k_a and k_d are the rate constants of adsorption and desorption, respectively, and:

$$Kc_{H^+} = \frac{\theta_{H^+}}{\left(1 - \theta_{H^+}\right)} \tag{25}$$

where $K = \dfrac{k_a}{k_d}$ is the protonation equilibrium constant and θ_{H^+} is the coverage with respect to the protonated surface sites. For:

$$\frac{\theta_{H^+}}{1 - \theta_{H^+}} = 1 \tag{26}$$

$$\log K = \text{pH} \tag{27}$$

Thus, one can estimate the value of the protonation equilibrium constant the pH dependence ot the coverage at $\theta = 0.5$

$$-\lg\left(\frac{\theta_{H^+}}{1 - \theta_{H^+}}\right) = \lg K + \text{pH} \tag{28}$$

shown in Fig. 12/B.

According to the assumption outlined above the adsorption of sulfate species occurs only on the protonated surface sites $\left(\Gamma_{H^+}\right)$. In equilibrium at a given pH value,

$$k_s c_s \Gamma_{H^+}\left(1 - \theta_s\right) = k_s' \theta_s \tag{29}$$

At a fixed pH value, Γ_{H^+} is constant, denoting $k_s \Gamma_{H^+} / k_s'$ by D we obtain:

$$\frac{Dc_s}{1 + Dc_s} = \theta_s \tag{30}$$

This equation found to be in agreement with the experimental results.

Considering Eqs 28 and 30 the specific adsorption of sulfate ions can be given by the following equation:

$$\theta_s = \frac{\Gamma}{\Gamma_m} = \frac{ac_{H^+}}{1 + ac_{H^+}} \cdot \frac{bc_s}{1 + bc_s} \tag{31}$$

The simple model suggested for the case of the adsorption of labeled sulfate species cannot be unambiguously applied for other anions. For instance, in the case of phosphate anions the pH dependence of the adsorption of phosphate ions goes through a maximum as shown in Fig. 5. for the case of γ-Al$_2$O$_3$ and hematite (100). This phenomenon could be connected either with competitive adsorption of ClO$_4^-$ and phosphate species or the different adsorbability of H$_2$PO$_4^-$, HPO$_4^{2-}$ and PO$_4^{3-}$ ions. The former assumption concerning competitive processes was supported by results obtained from the study of the competitive adsorption of sulfate and phosphate ions.

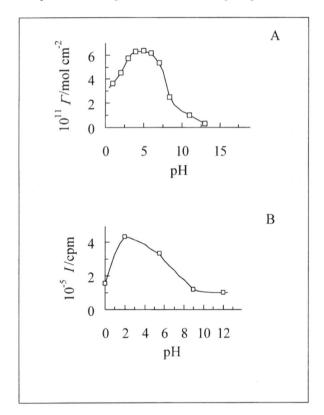

Fig. 5. (A) The pH dependence of the adsorption of labeled phosphate species (c = 3×10^{-4} mol dm^{-3} in 0.5 mol dm^{-3} NaClO$_4$) on γ-Al$_2$O$_3$. (B) The pH dependence of the adsorption of labeled phosphate species (c = 3×10^{-4} mol dm^{-3} in 0.5 mol dm^{-3} NaClO$_4$) on hematite.

1.6.2. *Radiotracer technique: a tool for the distinction between specific and non-specific adsorption*

On the basis of the comparative radiotracer study of the adsorption of sulfate and pertechnetate ions on γ-Al$_2$O$_3$ it was shown in Ref. 174 that the application of radiotracer technique allows us to distinguish specific and nonspecific adsorption without the knowledge and use of quantitative data as to the extent of the adsorption or the surface area of the adsorbent.

It follows from the considerations outlined in the previous section the charge determining species on the Al$_2$O$_3$ surface in acidic medium are the protons. Denoting by $\Gamma_{H^+}^b$ the surface excess concentration of protons bound to the surface, the charge balance characterizing the double layer as a whole should be

$$F\Gamma_{H^+}^b = F\sum z_i \Gamma_i \tag{32}$$

where Γ_i denotes the surface excess concentration of ionic components (cations and anions as well) believed to be present in the solution side of the double layer. (It is assumed that no charged species originating from bulk Al$_2$O$_3$ are involved.)

In the presence of two anions with equal charge, for instance, $|z_i| = 1$ for the case of non-specific adsorption according to the Gouy Chapman theory the following relationship can be given

$$\frac{c_1}{c_2} = \frac{\Gamma_1}{\Gamma_2} \tag{33}$$

where c_1 and c_2 are the concentrations of the anions in the bulk solution phase. The same refers for two cations.

For the sake of simplicity in Ref. 174 a system containing two cations and anions namely Na$^+$, H$^+$, ClO$_4^-$, TcO$_4^-$ was considered and the corresponding Γ values was denoted Γ_{Na^+}, $\Gamma_{H^+}^T$, $\Gamma_{ClO_4^-}$ and $\Gamma_{TcO_4^-}$. For the solution phase the following conditions were considered: $c_{TcO_4^-} \ll c_{ClO_4^-}$ thus $c_{Na^+} + c_{H^+} \approx c_{ClO_4^-}$.

On the basis of the double layer model the total surface excess of protons determinable by thermodynamic means $\Gamma_{H^+}^T$, can be conceived as composed of two components

$$\Gamma_{H^+}^T = \Gamma_{H^+}^b + \Gamma_{H^+}^s \tag{34}$$

where $\Gamma_{H^+}^s$ is the (negative) surface excess of protons in the solution side of the double layer. Taking into consideration Eqs. 32, 33 and 34 and neglecting the adsorption of Na^+ ions we obtain

$$\Gamma_{TcO_4^-} = \frac{\Gamma_{H^+}^b c_{TcO_4^-}}{c_{ClO_4^-}} \tag{35}$$

It was shown in Ref. 93 that in the case of the γ-Al_2O_3 a limiting value of the proton sorption $\left(\Gamma_{H^+}^L\right)$ is attained at pH values under 3. Thus at pH < 3 the following equation can be given

$$\Gamma_{TcO_4^-} = \Gamma_{H^+}^L \frac{c_{TcO_4^-}}{c_{ClO_4^-}} \tag{36}$$

However, even the application of this simplified relationship for the case of non-specific adsorption encounters significant difficulties owing to the very nature of the system. These problems are as follows:

i) In order to protonate the Al_2O_3 surface $HClO_4$ should be added to the solution phase, consequently ClO_4^- ions will appear in the solution phase.

ii) For the shifting of the pH to a preselected low value, for instance, to pH = 2.5 -3 (in order to obtain the limiting value in the protonation) $HClO_4$ $\left(ClO_4^- \text{ ions}\right)$ in a concentration of about 10^{-3} mol dm^{-3} should be added to the solution phase. Thus in the case of non-specific adsorption of labeled species $\left(TcO_4^-\right)$ present in concentration 10^{-4} mol dm^{-3} the extent of its electrostatic adsorption should be strongly influenced by the ClO_4^- ions.

It was shown in Ref. 174 that the errors caused by these factors can be minimized under appropriately chosen experimental conditions and a suggestion was made for the application of the radiotracer technique for the study of the double layer formed at protonated oxide surface using TcO_4^- as indicator species characterizing the behavior of the solution side of the double layer [174].

2. APPLICATION OF RADIOTRACERS IN CORROSION STUDIES

One of the first steps towards the application of radiotracer methods in fundamental corrosion studies is the investigation of the dissolution of metals and alloys in various media.

The fundamental questions connected with this topic were discussed in Section 1. presenting some illustrative examples and referring to the relevant literature. Important aspects of the corrosion of constructional materials of nuclear industry will be the one of main subjects of Chapter 10. In the present section only some of the new developments in the application of tracer method will be presented.

2.1. New methods for the study of corrosion and wear

For a long period neutron irradiation was the most important tool to produce radiotracers for study of wear and corrosion of various parts of a machine without the necessity of stopping and disassembling it.

In the last years, the most used methods have been the Thin Layer Activation (TLA) and Ultra Thin Layer Activation (UTLA) techniques [178, 179].

TLA is an ion beam technique. This method consists of an accelerated ion bombardment of the surface of interest of a machine part. Radioactive tracers are created by nuclear reactions in a well-defined volume of material. Loss of material owing to wear, corrosion or abrasion phenomena is characterized by monitoring the resulting changes in radioactivity.

There are two basic methods for measuring the material loss by TLA technique. One of them is based on remnant radioactivity measurements using a previously obtained calibration curve. The second is based on measuring the increasing radioactivity in the surrounding liquid phase.

The UTLA method is based upon the principle of recoil implantation by recoiling applied to radioactive heavy ions generated by a beam of light mass particles (p, d, ^{3}He, ^{4}He). These charged particles are easy to produce by an isocronous variable energy cyclotron. The most commonly used energies do not exceed 40 MeV.

A thin target (a few micrometers) of elementary composition **A** is bombarded by the primary beam and is activated following the nuclear reaction **A(a,b)B**. Some generated radioactive heavy ions **B** acquire sufficient kinetic energy (maximum energies of a few hundred keV to a few MeV) to recoil out of the target and be implanted in the material. Some of the of suitable nuclear reactions for the application of the UTLA method are as follows: ^{65}Cu(p, n)^{65}Zn, ^{56}Fe(p, n)^{56}Co, ^{48}Ti(p, n)^{48}V, ^{59}Co(p, pn)^{58}Co, ^{55}Mn(α, n)^{58}Co.

The UTLA method presents numerous advantages compared to direct activation methods [179]:

1) The activation being independent on the activated material composition and the deposition methods permitting to deposit a wide range of materials, the UTLA method may be applied to all kinds of materials. Moreover, the potentially damaging effects are far less than those induced by direct activation techniques using light or heavy ions.

2) The wide range of radioisotopes that can be generated and implanted allows carry out investigations corresponding to various experimental requirements (depth to activate, chemical nature of the radioelement in the case of selective corrosions, etc.).

3) The generated activities are very low (a few kBq). Radioprotection precautions are therefore considerably reduced.

4) The use of light particle beams (p, d, ^{3}He, α) is common to a vast majority of accelerators.

It is however a drawback that the implantation takes place in a vacuum chamber, the application field of this method is limited to reduced-dimensions samples.

An interesting application of the classical neutron activation method for the study of protective ability of paint coatings could be mentioned [180].

Investigations into the partial rates of corrosion and selective transfer of the metal substrate components and impurities into a given coating and medium are of extreme importance when developing corrosion proof steels and alloys as well as selecting constructive materials and coatings for equipment for the food and high purity material industries. Such data are crucial insofar as the corrosion mechanism and the effectiveness of the organic coating are concerned.

The experimental procedure involves the following steps.

1) The production of radionuclides by bombardment of the metal specimens with neutron flux in a reactor;

2) Application of coating on the radioactive specimens in accordance with common application techniques;

3) Corrosion-electrochemical tests of the coated and uncoated specimens in different media with regular sampling for the presence of labeled corrosion products;

4) Gamma-spectrometric analysis of corrosion products to obtain data on physicochemical characteristics of the processes.

2.2. Study of the adsorption of corrosion inhibitors
2.2.1. Inorganic inhibitors

Phosphate and pertechnetate:

A radiotracer study of the deposition of ^{32}P labeled phosphate and ^{99}Tc labeled pertechnetate (TcO$_4^-$) ions was reported in [181] on a 100 μm Al 1100 foil positioned at the bottom of the measuring cell. The behavior of two corrosion inhibiting oxy anions was compared as function of time and electrode potential.

In accordance with previous observations of the same group [182] it is assumed that the deposition of phosphate is initiated by the anodic dissolution of aluminum. The formation of technetium films was shown to proceed preferentially via cathodic processes and resulted in a greater extent of adsorption than that observed for phosphate.

It is assumed that the phosphate deposition is due to the result of adsorption, incorporation, and precipitation of insoluble phosphate compounds on the oxide film or corrosion products, which means that phosphate deposition is indirectly related to electrochemical reactions i.e., aluminum dissolution. The statement concerning phosphate deposition is in agreement with the results obtained with γ-Al$_2$O$_3$ [94] and is in correspondence with the reported pH dependence of phosphate adsorption on corroding Al and γ-Al$_2$O$_3$ [183].

On the other hand, the technetium deposition can be mainly ascribed to direct electrochemical reactions since pertechnetate adsorption is weak on aluminum oxide. The technetium deposit appears, in part to be a film of Tc(OH)$_4$ formed by the reduction of pertechnetate. It can be assumed that a stepwise reduction of Tc(VII) species occurs (presumably in one-electron steps) and the consecutive charge transfer processes are accompanied by (a) complexation reactions, (b) disproportionation, (c) precipitation or deposition processes and (d) transport phenomena.

A very rough approximation can be given by the following scheme (the oxidation state of the Tc species is marked in the simplest way as Tc(VII), Tc(VI) etc.):

disproportionation

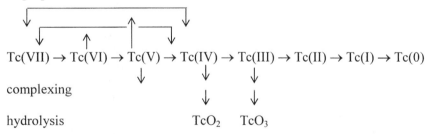

$$Tc(VII) \rightarrow Tc(VI) \rightarrow Tc(V) \rightarrow Tc(IV) \rightarrow Tc(III) \rightarrow Tc(II) \rightarrow Tc(I) \rightarrow Tc(0)$$

complexing

hydrolysis   TcO$_2$ TcO$_3$

For instance, the existence of two parallel pathways in the reduction of Tc(IV) species in H$_2$SO$_4$ supporting electrolyte unambiguously follows from this scheme. As the stability of the Tc(IV) complex formed is relatively high in acid media, no further cathodic reduction of this anion can be expected. However, the ratio of the amount of complex Tc(IV) species in the solution phase to that of deposited Tc depends on the ratio of the rate w_1 of complexation to the rate w_2 of further reduction of Tc(IV) ions formed at the electrode surface:

Tc(VII) (TcO_4^-)

$\downarrow$

Tc(IV) $\xrightarrow{w_1}$ complexation

$\downarrow w_2$

Tc deposition

A parallel process to the complexation could be the hydrolysis of the Tc(IV) species:

$$Tc^{4+} + 2H_2O \longleftrightarrow TcO_2 + 4H^+$$

The occurrence of this reaction explains the formation of $TcO_2 \cdot xH_2O$ deposit on the surface.

Adsorption of chromate:

As it is well known, chromate conversion coating is used as a pretreatment for aluminum alloys. Although the composition and structure of these coatings are very complex, it may be assumed that chromate ad- and absorption into Al_2O_3 formed on the surface and the adsorption processes on reduced chromate species, various Cr(III) oxides and hydroxides, could play an important role in the behavior of the coating. Thus in first approximation the study of chromate adsorption on Al_2O_3 and the investigation of anion adsorption on Cr_2O_3 may be suggested as models for adsorption studies.

The influence of solution pH on chromium(VI) deposition from 0.1 M NaCl + 0.1 mM Na_2CrO_4 solution onto Al 1100 alloy was studied under open circuit potential (OCP) condition using ^{51}Cr-labeled chromate species [184]. The data obtained from the in situ radiochemical analysis were complemented by X-ray photoelectron spectroscopy (XPS) and electrochemical studies. The results of these studies analyses indicated that both the kinetics and the extent of deposition are strongly pH dependent. Chromate deposition exhibited maximum surface coverage at pH 2. Chromate, even in 0.1 mM concentration, was found to effectively reduce the chloride content of the oxide film. In addition, the presence of chromate in the solution phase was found to alter the hydration and thickness of the passive film. XPS analyses revealed that chromate not only adsorbs but also electrochemically interacts with aluminum, creating a Cr(III,VI)-rich layer on the surface.

In order to gain a deeper insight into the role of aluminum and chromium oxides in the overall adsorption process an indirect radiotracer study of the adsorption of chromate ions on γ-Al_2O_3 and a direct study of the adsorption of sulfate ions on Cr_2O_3 using ^{35}S labeled H_2SO_4 were carried out [95].

The results obtained reflect the specific adsorption of sulfate ions, as ^{35}S labeled sulfuric acid was present in low concentrations ($c<10^{-3}$ mol dm^{-3}) in comparison to the large excess of perchlorate supporting electrolyte (0.25-1.0 mol dm^{-3}).

It has been found that the extent of adsorption is determined by the protonation of the surface sites, similar to the behavior of other oxides studied previously. A comparison of Cr_2O_3 and Al_2O_3 in this respect shows that the protonation of the former takes place at significantly lower pH values than that of the latter. In Fig. 6 the pH dependence of sulfate adsorption on the two oxides is compared on a relative scale. These curves reflect the protonation equilibrium of the oxides. It follows from the comparison of the two curves that the protonation on Cr_2O_3 takes place mainly in the pH range where Al_2O_3 attains its limiting value. Therefore, it could be expected that in the case of the mixed presence of the two oxides, the overall sorption behavior at pH values above 4 would be determined by the sorption behavior of Al_2O_3.

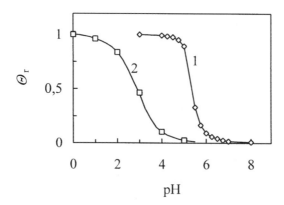

Fig. 6.   Comparison of the pH dependence of the adsorption of sulfate ions (relative coverage); the measured count-rate values were divided by the highest count rate obtained at lowest pH in the case of γ-Al_2O_3 (1) and Cr_2O_3 (2).

2.2.2. Organic inhibitors

Nowadays, the discharge of chemicals into rivers, lakes, or water reservoirs is controlled by strict environmental regulations. Owing to eutrophisation, environmental regulations have also been proposed to decrease the phosphorous content of natural water, which means that even the application of nontoxic inorganic phosphates should be reduced. One way to achieve this is the application of organic phosphono compounds instead of inorganic phosphorous materials. The reason for this is that C-P bonds are much more resistant to hydrolysis in comparison to P-O bonds [185]. A widely used model of organic phosphonic acid inhibitor is 1-hydroxy-ethane-1,1-diphosphonic acid (HEDP) [186,187]:

$$\begin{array}{ccc} HO & CH_3 & OH \\ \diagdown & | & \diagup \\ O = P - C - P = O \\ \diagup & | & \diagdown \\ HO & OH & OH \end{array}$$

The research on environmentally friendly inhibitor compounds is the center of the interest worldwide. One way is a possible decrease of the phosphorous content in anticorrosive molecules for water treatment while simultaneously retaining or increasing their inhibition efficiency. Such investigations involve the preparation of new molecules with lower phosphorous content. Phosphonate compounds containing the $-CH_2PO_3H_2$ fragment instead of, or together with, a carboxyl groups have a number of features that are due to their distinctive stereochemistry, their greater electro-negativity, and the greater potential of PO_4^{2-} than of COO^- for dentate formation. An important feature of phosphonates compared to aminocarboxyl chelates is their ability to form quite stable protonated MeHL complex [187].

A systematic study of the corrosion behavior of organic compounds was made in order to prove the importance of phosphonic and carboxyl groups in inhibitory processes and to optimize inhibitor concentration with minimum corrosion rate and maximum scale inhibiting activity [188]. As a result of these studies, a new class of environmentally friendly compounds have been developed based on amino acids [189]. The N,N-di(phosphonometyl)glycine (DPG) was found to be an excellent corrosion inhibitor [190]. Self-assembling molecules (SAMs) have been used recently as corrosion inhibitors [191].

The adsorption studies of ^{14}C labeled HEDP [192] on iron electrodes have shown that HEDP forms a loosely bound adsorption layer on iron oxide surface. The presence of zinc ([65]Zn) and calcium ([45]Ca) ions in aqueous 0.5 mol cm^{-3} NaClO$_4$ solutions leads to an increase in corrosion inhibition due to the formation of different weakly soluble complex compounds [193].

Coupled application of the *in-situ* radiotracer "foil" method and voltammetry gives preliminary information on the time, potential, concentration, and pH-dependence of HEDP adsorption on a polycrystalline gold electrode [194]. It has been stated that the adsorption of HEDP on polycrystalline gold is highly dependent on the electrode potential, as well as on the composition, concentration and the solution pH. The next step in the adsorption studies was the investigation of sorption phenomena taking place on the corrosion products i.e. on oxide surfaces.

The adsorption of DPG and HEDP was studied on hematite surface by indirect radiotracer technique using ^{35}S labeled H_2SO_4 as indicator species. As the adsorption of sulfate ions at hematite was studied in detail in [92] and a reversible adsorption was found ^{35}S labeled sulfate ions could be used as an ideal indicator species (see Fig. 6).

It was assumed that the adsorption of sulfate ions and the organic species can be described by a Langmuir isotherm and the following equations were given:

$$\theta_A = \frac{b_A c_A}{1 + b_A c_A + b_B c_B} \tag{37}$$

A and B denote the species studied and indicator species (sulfate), respectively

$$\theta_B = \frac{b_B c_B}{1 + b_A c_A + b_B c_B} \tag{38}$$

Transformation of Eq. 38 leads to the expression

$$\frac{1}{\theta_B} = 1 + \frac{1}{b_B c_B} + \frac{b_A c_A}{b_B c_B} \tag{39}$$

At a fixed concentration of the indicator species (B) and taking into account that $\frac{\Gamma_B}{\Gamma_B(max)} = \theta_B = \frac{I}{I(max)}$; θ_B should be proportional to the radiation intensity measured (I)

$$\theta_B = QI \tag{40}$$

where $Q = 1/I(max)$ is a proportionality factor characteristic for the system studied, we obtain:

$$\frac{1}{I} = D + F c_A \tag{41}$$

where $F = \dfrac{b_A Q}{b_B c_B}$; $D = Q\left(1 + \dfrac{1}{b_B c_B}\right)$.

From eqs 37 and 38:

$$\frac{\theta_A}{\theta_B} = \frac{b_A}{b_B} \cdot \frac{c_A}{c_B} \tag{42}$$

then

$$\theta_A = F' I c_A \tag{43}$$

By using eq. (43) for were the calculation of the θ values the adsorption isotherms of DPG and HEDP were obtained. This result could contribute to the better understanding of the mechanism of the inhibition in the presence of these species.

2.3. Comparative study of adsorption phenomena occurring on corroding metals and the corresponding oxides/ hydroxides

The role of various anions in the corrosion of metals is of great interest. Relatively few direct in situ information is available on the adsorption of anions in the course of corrosion processes.

Applying the version of thin layer technique elaborated for powdered adsorbents the problems caused by continuous transformation and dissolution of the metal can be avoided in the case of labeling with isotopes emitting soft β radiation and by using powdered (or less finely dispersed) metals as adsorbents.

In this case the continuous dissolution of the metal powder covered with adsorbed labeled species in acidic medium does not exert any influence on the radiation intensity measured if layer thickness remains to be higher than the self-absorption thickness, i.e. the detector "senses" the same thickness throughout the process in spite of the fact that part of the powder has been consumed. The same refers to the dissolving oxides.

A study of Zn, Al system was carried out in Ref. [88]. Fig. 7 shows how the extent of the specific adsorption of labeled sulfate ions on Zn and Al changes with the pH of the solution phase It may be seen from this figure that the extent of the adsorption goes through maximum. This behavior was interpreted by the assumption that anion adsorption takes place mainly on the protonated surface sites of the corresponding oxide/hydroxide system formed in the course of corrosion [88].

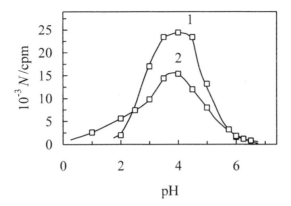

Fig. 7. pH dependence of the radiation intensity proportional to the adsorption of labeled
   sulfate ions on Zn powder (1) and on Al in the form of needles (2) in the presence of
   2×10^{-4} mol dm^{-3} labeled H_2SO_4 in 0.5 mol dm^{-3} $NaClO_4$. pH was changed by
   addition of NaOH and $HClO_4$.

Acording to Ref. 88 two competitive processes should be considered in acid
medium.
   1) Formation of surface oxides/hydroxides.
   2) Dissolution of the surface species according to elementary reactions
given by the following equations

a) $Me \rightarrow Me^{n+} + ne$
b) $2H^+ + 2e \rightarrow H_2$
c) $H_2O = H^+ + OH^-$
d) $Me^{n+} + nOH^- = Me(OH)_n$

The protonation of the hydroxide/oxide formed on the surface induce the
specific adsorption of sulfate ions, as demonstrated in Refs [91,92,93]. With
decreasing pH the surface hydroxide should dissolve on the metal and this leads
to the decrease in the number of adsorption sites for anions. Results of a separate
study of the adsorption of labeled sulfate ions on AlOOH [175] made plausible
this model. The pH dependence of the adsorption of anions onto AlOOH with
that of the same ions onto γ-Al_2O_3 was compared. It was demonstrated that the
character of the pH dependence of the adsorption is very similar in both cases at
pH values above pH = 2. It was found, however, that in contrast to the behavior
of γ-Al_2O_3, AlOOH dissolves at a significant rate at low pH values (pH < 2).
   The main conclusion drawn from the experimental results presented in
[88,175] demonstrate that the characteristic features of the overall sorption of

sulfate ions observed in the course of the corrosion of Al and Zn are similar to those of the corresponding oxides/hydroxides. The extent of sorption at a fixed pH value is determined by the amount of the corrosion products accumulated on the surface and owing to this situation there is no possibility of obtaining information concerning the adsorption on the "pure" metal sites, if any.

The very fact that the adsorption of sulfate ions on Al and Zn goes through maximum in the pH range studied can be interpreted in terms of factors determining the sorption capacity of the surface oxide layers formed. The specific adsorption of sulfate molecules is determined by the extent of protonation of the oxydes/hydroxides. The extent of protonation of AlOOH or Al_2O_3 tends to a limiting value (Γ_{max}) with decreasing pH and the most characteristic increase occurs in the pH range from 6 to 4. This pH range depends on the protonation equilibrium constant, K that can be determined from the pH dependence of the anion adsorption on protonated sites. At the pH value where $\Gamma = \Gamma_{max}/2$ or $I = I_{max}/2$: pH = lg K [92,93].

At very low pH values, the steady state coverage onto corrosion products tends to zero owing to their dissolution.

On the other hand, at higher pH values above pH=6 no significant protonation occurs thus the induced anion adsorption should be again very low.

As a continuation of the work with Zn and Al the adsorption of labeled sulfate and chloride ions on powdered Co and Fe was studied in 0.5 mol dm^{-3} NaClO$_4$ supporting electrolyte at various pH values in the course of the corrosion (dissolution) of the metals. The effect of the reduction of ClO_4^- ions (leading to the formation of Cl$^-$ ions) on the radiotracer measurements was also investigated. It was found that in both cases the adsorption of chloride and sulfate ions occurs in different pH ranges as shown in Fig. 8 in the case of Co. This phenomenon was explained by the assumption that the adsorption strength of the two species is very different on oxide covered and "pure" metal surfaces.

In strong acidic media where no significant coverage with respect to surface oxide species could be expected Cl$^-$ ions adsorb readily while sulfate adsorption occurs at pH values where the formation of surface oxide(hydroxide) layer is the predominant process. Thus the adsorption of Cl$^-$ ions is preferred on the metal surface sites while SO_4^--ions interact with the surface oxide sites.

This assumption is in agreement with the results obtained from radiotracer studies of anion adsorption on various oxides. The adsorbability of sulfate ions is significantly higher on γ-Al$_2$O$_3$ [93], hematite [92] and Cr$_2$O$_3$ [97] than that of Cl$^-$ ions. The specific adsorption of sulfate ions on these oxides can be observed in the presence of a great excess of Cl$^-$ ions.

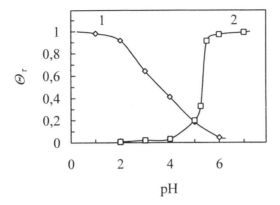

Fig. 8. Comparison of the pH dependence of SO_4^{2-} (2) and Cl^- (1) adsorption on Co (0.5 mol dm^{-3} NaClO$_4$ + 2×10^{-4} mol dm^{-3} labeled H$_2$SO$_4$ (2) or 1×10^{-4} mol dm^{-3} labeled HCl (1).

In this respect the effect of Cl^- ions in the pitting corrosion can be explained by the very plausible assumption that the Cl^- ions are not blocked by he surface oxide forming surface species. Thus they are able to penetrate inside the oxide film and interact with the "pure" metal sites accelerating the corrosion of the metal in form of pits.

ACKNOWLEDGEMENT

Financial support from the Hungarian Scientific Research Fund is acknowledged (OTKA Grants T 031703 and T 045888).

LIST OF ABBREVIATIONS

CV	Cyclic Voltammery
DPG	N,N-di(phosphonometyl)glycine
EQCM	Electrochemical Quartz Crystal Microbalance
HEDP	1-hydroxy-ethane-1,1-diphosphonic acid
HSA	Human Serum Albumin
IgG	Immunoglobulin
PANI	Polyaniline
PP	Polypyrrole
PPD	Poly(o-phenylenediamine)
PTCNQ	Ttracyanoquinodimethane polyester
PVF	Poly/vinyl ferrocene

RHE   Relative Hydrogen Electrode
SAMs   Self-Assembling Molecules
TLA   Thin Layer Activation
UHV   Ultra-High-Vacuum
UTLA   Ultra Thin Layer Activation
XPS   X-ray Photoelectron Spectroscopy

REFERENCES

[1]   A.J. Bard (ed.), Electroanalytical Chemistry, Vol. 3, Marcel Dekker, Inc., New York, 1969, pp. 135–197.
[2]   G. Horányi, Electrochim. Acta, 25 (1980) 45.
[3]   E. Yeager, J.O'M. Bockris, B.E. Conway and V.E. Kazarinov (eds.), Comprehensive Treatise on Electrochemistry, Vol. 9, Plenum Press, New York, 1984, pp. 393–443.
[4]   J.O'.M.Bockris, B.E. Conway, and R.E. White (eds.), Modern Aspects of Electrochemistry, Vol. 21, Plenum Press, New York, 1990, pp.65–119.
[5]   H. D. Abruna (ed.), Electrochemical Interfaces: Modern Techniques for In Situ Surface Characterization, VCH Publishers, New York, 1991, pp. 479–527.
[6]   J. Lipkowski and P.N. Ross (eds), Adsorption of Molecules at Metal Electrodes, Frontiers of Electrochemistry, VCH Publishers, New York, 1992, pp. 119–169.
[7]   G. Horányi, B. Electrochem., 5 (1989) 235.
[8]   Y.-E. Sung, A. Thomas, M. Gamboa-Aldeco, K. Franaszczuk and A. Wieckowski, J. Electroanal. Chem., 378 (1994) 131.
[9]   G. Horányi, Rev. Anal. Chem., 14 (1995) 1.
[10] J.J. Spivey (ed.) Catalysi, Specialist Periodical Report, Vol. 12. The Royal Society of Chemistry, Thomas Graham House, Cambridge 1996, pp. 254–301.
[11] A. Wieckowski (ed.), Interfacial Electrochemistry. Theory, Experiment, and Applications, Marcel Dekker, Inc, New York, 1999, pp. 477–491.
[12] A. Hubbard (ed.), Encyclopedia of Surface and Colloid Science, Marcer Dekker, New York, 2002, pp. 4423–4437.
[13] E. Matijević (ed.), Surface and Colloid Science, Vol. 16., Kluwer Academic, New York, 2001, pp. 341–393.
[14] G. Horányi and M. Wasberg, J. Electroanal. Chem., 404 (1996) 291.
[15] G. Horányi and M. Wasberg, J. Electroanal. Chem., 413 (1996) 161.
[16] G. Horányi, J. Electroanal. Chem., 417 (1996) 185.
[17] G. Horányi and A. Aramata, J. Electroanal. Chem., 434 (1997) 201.
[18] Gy. Horányi, ACH Models in Chem., 134 (1997) 33.
[19] G. Horányi, J. Solt, and F. Nagy, J. Electroanal. Chem., 31 (1971) 87.
[20] G. Horányi, J. Electroanal. Chem., 453 (1998) 113.
[21] Gy. Horányi, ACH Models in Chem., 135 (1998) 219.
[22] G. Horányi, J. Solid State Electrochem., 2 (1998) 237.
[23] G. Horányi, J. Solid State Electrochem., 4 (2000) 153.
[24] G. Horányi, J. Solid State Electrochem., 4 (2000) 177.
[25] A. Kolics and K.Varga, Electrochim. Acta, 40 (1995) 1835.
[26] Z. Németh, L. Erdei and A. Kolics, J. Radioanal. Nucl. Chem., Letters, 199 (1995) 265.
[27] Z. Németh, L. Erdei and A. Kolics, Corros. Sci., 37 (1995) 1163.

[28] K. Varga, P. Baradlai, W.O. Barnard, G. Myburg, P. Halmos and J.H. Potgieter, Electrochim. Acta, 42 (1997) 25.

[29] K. Varga, P. Baradlai and A. Vértes, Electrochim. Acta, 42 (1997) 1143.

[30] K. Varga, P. Baradlai, D. Hanzel, W. Meisel and A. Vértes, Electrochim. Acta, 42 (1997) 1157.

[31] G. Hirschberg, Z. Németh and K. Varga, J. Electroanal. Chem., 456 (1998) 171.

[32] G. Hirschberg, P. Baradlai, K. Varga, G. Myburg, J. Schunk, P. Tilky and P.Stoddart, J. Nucl. Mat., 265 (1999) 273.

[33] P. Waszczuk, P. Zelenay and J. Sobkowski, J. Electrochim. Acta, 40 (1995) 1717.

[34] M. Szklarczyk, N.N. Hoa and P. Zelenay, J. Electroanal. Chem., 405 (1996) 111.

[35] S. Smoliński, P. Zelenay and J. Sobkowski, J. Electroanal. Chem., 442 (1998) 41.

[36] P. Waszczuk, P. Zelenay and J. Sobkowski, Electrochim. Acta, 43 (1998) 1963.

[37] S. Smoliński and J. Sobkowski, J. Electroanal. Chem., 463 (1999) 1.

[38] P. Waszczuk, A. Wnuk and J. Sobkowski, Electrochim. Acta, 44 (1999) 1789.

[39] A. Wieckowski, J. Electroanal. Chem., 352 (1993) 313.

[40] W. Savich, S.-G. Sun, J. Lipkowski and A. Wieckowski, J. Electroanal. Chem., 388 (1995) 233.

[41] E. Herrero, J.M. Feliu, A. Wieckowski and J. Clavilier, Surface Science, 325 (1995) 131.

[42] A. Kolics, A.E. Tomas and A. Wieckowski, J. Chem. Soc., Faraday Trans., 92 (1996) 3828.

[43] A. Kolics, J.C. Polkinghorne and A. Wieckowski, Electrochim. Acta, 43 (1998) 2605.

[44] A. Kolics, J.C. Polkinghorne, A.E. Thomas and A. Wieckowski, Chem. Mater., 10 (1998) 812.

[45] A. Wieckowski, A. Kolics, J.C. Polkinghorne, E.D. Eliadis and R.C. Alkire, Corros. Sci. 54 (1998) 800.

[46] A. Wieckowski and A. Kolics, The 1999 Joint International Meeting, 196th Meeting of The Electrochemical Society, Hawaii, October 17-22, 1999; The Electrochemical Society; Honolulu, Hawaii, 1999, No 2126.

[47] A. Wieckowski and A. Kolics, J. Electroanal. Chem., 464 (1999) 118.

[48] J.B. Schlenoff, J.R. Dharia, H. Xu, L.Q. Wen and M. Li, Macromolecules, 28 (1995) 4290.

[49] J.B. Schlenoff, J. R. Dharia, H. Xu, L.Q. Wen and M. Li, Macromolecules, 28 (1995) 4290.

[50] J.B. Schlenoff and M. Li, Ber. Bunsenges. Phys. Chem. Chem. Phys., 100 (1996) 943.

[51] D. Poškus and G. Agafonovas, J. Electroanal. Chem., 393 (1995) 105.

[52] D. Poškus, G. Agafonovas and A. Žebrauskas, Chemija, (1996) 71.

[53] D. Poškus, G. Agafonovas and A. Žebrauskas, Chemija, (1996) 104.

[54] D. Poškus, G. Agafonovas and I. Jurgaitiene, J. Electroanal. Chem., 425 (1997) 17.

[55] D. Poškus, J. Electroanal. Chem., 442 (1998) 5.

[56] J.O'M. Bockris and K.T. Jeng, J. Electroanal. Chem. 330 (1992) 541.

[57] J. Jovancicevic, J.O'M. Bockris,J. L. Carbajal, P. Zelenay and T. Mizuno, J. Electrochem Soc., 133 (1986) 2219.

[58] P. Zelenay, M.A. Habib and J.O'M. Bockris, Langmuir 2 (1986) 393.

[59] J.O'M. Bockris,M. Gamboa-Aldeco, M. Szklarczyk, J. Electroanal. Chem., 339 (1992) 355.

[60] J.O'M. Bockris and Y. Kang, J. Solid State Electrochem., 1 (1997) 17.

[61] V.E. Kazarinov, V.N. Andreev, M.A. Spytsin and A.V. Shlepakov, Electrochim. Acta, 35 (1990) 899.
[62] V.E. Kazarinov, V.N. Andreev and M.A. Spitsyn, Electrochim. Acta, 41 (1996) 1757.
[63] V.N. Andreev, Russian J. Electrochem., 35 (1999) 735.
[64] V. N. Andreev, Russian J. Electrochem., 35 (1999) 740.
[65] J.-M. Herbelin, N. Barabouth and P. Marcus, J. Electrochem. Soc., 137 (1990) 3410
[66] S.E. Moulton, J.N. Barisci, A. Bath, R. Stella and G.G. Wallace, J. Coloid Interface Sci., 261 (2003) 312.
[67] M.V. ten Kortenaar, Z.I.Kolar, J.J.M. de Goeij and G. Frens, Langmuir, 18 (2002) 10279.
[68] G. Horányi, J. Electroanal. Chem., 370 (1994) 67.
[69] A. Kolics and G. Horányi, J. Electroanal. Chem., 374 (1994) 101.
[70] G.G. Láng, M. Ujvári and G. Horányi, J. Electroanal. Chem. 522 (2002) 179.
[71] G. Inzelt and G. Horányi, J. Electroanal. Chem., 491 (2000) 111.
[72] G. Horányi, J. Solt and F. Nagy, J. Electroanal. Chem., 31 (1971) 87.
[73] E.L. Krauskopf, K. Chan and A Wieckowski, J. Phys. Chem., 91 (1987) 2327.
[74] G. Horányi, E.M. Rizmayer and P. Joó, J. Electroanal. Chem., 149 (1983) 221.
[75] G. Horányi and E.M.Rizmayer, J. Electroanal. Chem. 176 (1984) 349.
[76] G. Horányi, E.M. Rizmayer and P. Joó, J. Electroanal. Chem., 154 (1983) 281.
[77] G. Horányi, E.M. Rizmayer and J. Kónya, J. Electroanal. Chem., 176 (1984) 339.
[78] G. Horányi and G. Vértes, Electrochim. Acta, 31 (1986) 1663
[79] G. Horányi and E.M. Rizmayer, J. Electroanal. Chem., 180 (1984) 97.
[80] G. Horányi and E.M. Rizmayer, J. Electroanal. Chem. 201 (1986) 187.
[81] G. Horányi, E.M. Rizmayer and P. Joó, J. Electroanal. Chem., 152 (1983) 211.
[82] G. Horányi and I. Bakos, J. Electroanal.Chem. 370 (1994) 213.
[83] G. Horányi and I. Bakos, J. Electroanal. Chem. 378 (1994) 143.
[84] K. Varga, E. Maleczki and G. Horányi, Electrochim. Acta, 33 (1988) 25.
[85] K. Varga, E. Maleczki, E. Házi and G. Horányi, Electrochim. Acta, 35 (1990) 817.
[86] K. Varga, E. Maleczki and G. Horányi, Electrochim. Acta, 33 (1988) 1167.
[87] K. Varga, E. Maleczki and G. Horányi, Electrochim. Acta, 33 (1988) 1775.
[88] G. Horányi and E. Kálmán, Corrosion Sci., 44 (2002) 899.
[89] G.G. Láng, M. Ujvári and G. Horányi, J. Electroanal. Chem., 522 (2002) 179.
[90] G.G. Láng, A. Vrabecz and G. Horányi, Electrochem. Comm. 5 (2003) 609.
[91] P. Joó and G. Horányi, J. Colloid Interface Sci., 223 (2000) 308.
[92] G. Horányi and P. Joó, J. Colloid Interface Sci., 227 (2000) 206.
[93] G. Horányi and P. Joó, J. Colloid Interface Sci., 231 (2000) 373.
[94] G. Horányi and P. Joó, J. Colloid Interface Sci. 247 (2002) 12.
[95] G. Horányi and L. Gáncs, J. Solid State Electrochem., 6 (2002) 485.
[96] G. Horányi, J. Colloid Int. Sci., 261 (2003) 580.
[97] G. Horányi, J. Solid State Electrochem. 7 (2003) 309.
[98] G. Inzelt and G. Horányi, J. Electroaal. Chem., 200 (1986) 405.
[99] G. Inzelt, G. Horányi, J.Q. Chambers and E.W. Day, J. Electroanal. Chem. 218 (1987) 297.
[100] G. Inzelt, G. Horányi and J.Q. Chambers, Electrochim. Acta, 32 (1987) 757.
[101] G. Inzelt and G. Horányi, J. Electroanal. Chem., 230 (1987) 257.
[102] G. Horányi and G. Inzelt, Electrochim. Acta, 33 (1988) 947.
[103] G. Horányi and G. Inzelt, J. Electroanal. Chem. 257 (1988) 311.
[104] G. Inzelt and G. Horányi, J. Electrochem. Soc. 136 (1989) 1747.

[105] G. Horányi and G. Inzelt, J. Electroanal. Chem. 264 (1989) 259.
[106] A.V. Slepakov, G. Horányi, G. Inzelt and V.N. Andreev, Elektrokhimiya, 25 (1989) 1280.
[107] K. Martinusz, G. Inzelt and G. Horányi, J. Electroanal. Chem., 395 (1995) 293.
[108] K. Martinusz, G. Inzelt and G. Horányi, J. Electroanal. Chem., 404 (1996) 143.
[109] A.J. Bard (ed.), Electroanalytical Chemistry, A Series of Advances, Vol. 18, Marcel Dekker, Inc., New York, 1994, pp. 89–241.

[110] M. Li and J.B. Schlenoff, Anal. Chem., 66 (1994) 824.
[111] R.M. Wang and J.B. Schlenoff, Macromolecules, 31 (1998) 494.
[112] R.S. Farinato and P.L. Dubin (eds.), Colloid-Polymer Interactions: From Fundamentals to Practice, John Wiley& Sons. Inc., 1999. pp. 225–250.
[113] S. Trasatti and R. Parsons, Pure and Appl. Chem., 58 (1986) .
[114] M. Gamboa-Aldeco and M. Szklarczyk, J. Electroanal. Chem., 281 (1990) 227.
[115] G. Horányi and A. Veres, J. Electroanal. Chem., 205 (1986) 259.
[116] G. Horányi and E.M. Rizmayer, J. Electroanal. Chem. 198 (1986) 379.
[117] M. Wasberg, J. Bácskai, G. Inzelt and G. Horányi, J. Electroanal. Chem., 418 (1996) 195.
[118] G. Horányi, React. Kinet. Catal. Let., 59 (1996) 211.
[119] L.M. Rice-Jackson, G. Horányi and A. Wieckowski, Electrochim. Acta, 36 (1991) 753.
[120] P. Zelenay, L.M. Rice-Jackson and A. Wieckowski, J. Electroanal. Chem., 283 (1990) 389.
[121] P. Zelenay, M. Gamboa-Aldeco, G. Horányi and A. Wieckowski, J. Electroanal. Chem., 357 (1993) 307.
[122] P. Zelenay, G. Horányi, C.K. Rhee and A. Wieckowski, J. Electroanal. Chem., 300 (1991) 499.
[123] A. Aramata, S. Taguchi, T. Fukuda, M. Nakamura and G. Horányi, Electrochim. Acta, 44 (1998) 999.
[124] G. Horányi and A. Aramata, J. Electroanal. Chem., 437 (1997) 259.
[125] K. Varga, P. Zelenay, G. Horányi and A. Wieckowski, J. Electroanal. Chem., 327 (1992) 291.
[126] G. Horányi and E.M. Rizmayer, J. Electroanal. Chem., 218 (1987) 337.
[127] G. Horányi and E.M. Rizmayer, J. Electroanal. Chem., 215 (1986) 369.
[128] G. Horányi and E.M. Rizmayer, J. Electroanal. Chem., 206 (1986) 297.
[129] M.I. Kulezneva, N.A. Balashova and V.E. Kazarinov, Elektrokhimiya,, 14 (1978) 128.
[130] G. Horányi, J. Electroanal. Chem., 36 (1972) 247.
[131] G. Horányi and E.M. Rizmayer, J. Electroanal. Chem., 169 (1984) 279.
[132] P. Zelenay and A. Wieckowski, J. Electrochem. Soc., 139 (1992) 2552.
[133] A. Wieckowski, P. Zelenay and K. Varga, J. Chim. Phys., 88 (1991) 1247.
[134] M.E. Gamboa-Aldeco, E. Herrero, P.S. Zelenay and A. Wieckowski, J. Electroanal. Chem., 348 (1993) 451.
[135] K. Varga, P. Zelenay and A. Wieckowski, J. Electroanal. Chem., 370 (1992) 453.
[136] J. Sobkowski, S. Smolinski, P. Zelenay, Coll. Surf. A., 134 (1998) 39.
[137] D. Poskus and G. Agafonovas, J. Electroanal. Chem., 493 (2000) 50.
[138] K. Varga, I. Szaloki, A. Somogyi, P. Baradlai, A. Aramata, T. Ohnishi and Y. Noya, J. Electroanal. Chem. 485 (2000) 121.
[139] I. Szalóki, K. Varga and R. Van Grieken, Spectrochim. Acta B 55 (2000) 1031.

[140]   G. Horányi, Electrochim. Acta 35 (1991) 1453)
[141]   A. Wieckowski and M. Szklarczyk, J. Electroanal. Chem., 142 (1982) 157.
[142]   G. Horányi and E.M. Rizmayer, J. Electroanal. Chem., 215 (1986) 369.
[143]   J.H. White, M.P. Soriaga and A.T. Hubbard, J. Phys. Chem., 89 (1985) 3227.
[144]   S.D. Rosasco, J.I. Stickney, G.N. Salaita, D.G. Frank, J.Y. Katekaru, B.C. Schardt,
        M.P. Soriaga, D.A. Stern and A.T. Hubbard, J. Electroanal. Chem., 188 (1985) 95.
[145]   J.L. Stickney, S.D. Rosasco, G.N. Salaita and A.T. Hubbard, Langmuir, 1 (1985) 66.
[146]   D.G. Frank, J.Y. Katekaru, S.D. Rosasco, G.N. Salaita, B.C. Schardt, M.P. Soriaga,
        D.A. Stern, J.L. Stickney and A.T. Hubbard, Langmuir, 1 (1985) 587.
[147]   B.C. Schardt, J.L. Stickney, D.A. Stern, D.G. Frank, J.Y. Katekaru, S.D. Rosasco,
        G.N. Salaita, M.P. Soriaga and A.T. Hubbard, Inorg. Chem., 24 (1985) 1419.
[148]   G. Horányi and E.M. Rizmayer, J. Electroanal. Chem., 248 (1988) 411.
[149]   G. Horányi and E.M. Rizmayer, Electrochim. Acta, 33 (1988) 1161.
[150]   G. Horányi, J. Electroanal. Chem., 51 (1974)163.
[151]   G. Horányi and E.M. Rizmayer, Electrochim. Acta, 30 (1985) 926.
[152]   D.A. Stern, H. Baltruschat, M. Martinez, J.L. Stickney, D. Song, S.K. Lewis, D.G.
        Frank and A.T. Hubbard, J. Electroanal. Chem., 217 (1987) 101.
[153]   P.N. Ross, J. Chim. Phys., 88 (1992) 1353.
[154]   A.R. Landgrebe, R.K. Sen and D.J. Wheeler (eds), Proceedings of the Workshop on
        Direct-Methanol-Air Fuel Cells, Vol, 92-14, The Electrochemical Society,
        Pennington, NJ. 1992. pp. 70-97.
[155]   M. Gamboa-Aldeco, P. Mrozek, C.K. Rhee, A. Wieckowski, P.A. Rikvold and Q.
        Wang, Surf. Sci. Letters, 297 (1993) L135.
[156]   M. Rubel, C.K. Rhee, A. Wieckowski and P.A. Rikvold, J. Electroanal. Chem., 315
        (1991) 301.
[157]   S. Smolinski, J. Sobkowski, Polish J. Chem., 75 (2001) 1493
[158]   A. Lukomska, S. Smolinski, J. Sobkowski, Electrochim. Acta, 46 (2001) 3111.
[159]   G. Horányi and E.M. Rizmayer, Electrochim. Acta, 32 (1987) 1057.
[160]   A. Wieckowski, Electrochim. Acta, 26 (1981) 1121.
[161]   G. Horányi, Electrochim. Acta, 35 (1990) 919.
[162] G. Horányi, J. Electroanal. Chem., 304 (1991) 211.
[163]   G. Horányi and S.B. Orlov, J. Electroanal. Chem., 309 (1991) 239.
[164]   G. Horányi, E.M. Rizmayer, E.P. Simon and J. Szammer, J. Electroanal. Chem., 323
        (1992) 329.
[165]   G. Horányi, E.M. Rizmayer, E.P. Simon and J. Szammer, J. Electroanal. Chem., 328
        (1992) 311.
[166]   S.E. Moulton, J.N. Barisci, A. Bath, R. Stella and G.G. Vallace, J. Colloid Interface
        Sci., 261 (2003) 312.
[167]   G. Horányi, Electrochim. Acta, 37 (1992) 2443.
[168]   G. Horányi, J. Electroanal. Chem., 344 (1993) 335.
[169]   P. Zelenay, P. Waszczuk, K. Dobrowolska and J. Sobkowski, Electrochim. Acta, 39
        (1994) 655.
[170]   G. Horányi and E.M. Rizmayer, Electrochim. Acta, 33 (1988) 113.
[171]   J. Lipkowski, L. Stolberg, S. Morin, D.E. Irish, P. Zelenay, M. Gamboa-Aldeco and A.
        Wieckowski, J. Electroanal. Chem., 355 (1993) 147.
[172]   G. Horányi, J. Electroanal. Chem., 284 (1990) 481.
[173]   J.B. Schlenoff, M. Li and H. Ly, J. Am. Chem. Soc., 117 (1995) 12528.
[174]   G. Horányi and P. Joó, J. Colloid Interface Sci., 243 (2001) 46.

[175] G. Horányi and E. Kálmán, J. Colloid Interface Sci., 269 (2003) 315.
[176] G. Horányi and P. Joó, Russ. J. Electrochem., 36 (2000) 1189.
[177] G. Horányi, J. Solid State Electrochem., 6 (2002) 463.
[178] O. Lacroix, T. Sauvage, G. Blondiaux, P. M. Racolta, L. Popa-Simil and B. Alexandreanu, Nucl. Inst. Meth. Phys. Res. A. 369 (1996) 427.

[179] O. Lacroix, T. Sauvage, G. Blondiaux and L. Guinard, Nucl. Inst. Meth. Phys. Res. B. 122 (1997) 262.
[180] V.G. Lambrev, N.N. Rodin and V.A. Koftyuk, Progr. Org. Coat., 30 (1997) 1.
[181] A.S. Besing, P. Waszczuk, A. Kolics and A. Wieckowski, Proc. Int. Symp., Corr. and Corr. Prof. San Francisco, 2001 pp. 393–401.
[182] A. Kolics, P. Waszczuk L. Gáncs, Z. Németh and A. Wieckowski, Electrochem. Solid-State Letters, 3 (2000) 369.
[183] Z. Németh, L. Gáncs, G. Gémes and A. Kolics, Corr. Sci., 40 (1998) 2023.
[184] L. Gáncs, A.S. Besing, R. Buják, A. Kolics, Z. Németh and A. Wieckowski, Electrochem. Solid-State Letters, 5 (2002) B16.
[185] E. Kálmán, European Federation of Corrosion Publ. No. 11. A Working Party Report on Corrosion Inhibitors, Institute of Materials, 1994, p. 12.
[186] E. Kálmán, B. Várhegyi, I. Bakó, I. Felhősi, F.H. Kármán and A. Shaban, J. Electrochem. Soc. 141 (1994) 3357.
[187] I. Felhősi, Z. Keresztes, F.H. Kármán, M. Mohai, I. Bertóti and E. Kálmán, J. Electrochem. Soc., 146 (1999) 961.
[188] J. Telegdi, E. Kálmán, F.H. Kármán, Corr. Sci., 33 (1992) 1099.
[189] M. Schütze (ed.), Materials Science and Technology, Vol. 19. Weinheim, Wiley-VCH 2000, pp. 471–537.
[190] J. Telegdi, F.H. Kármán, M. Shaglouf and E. Kálmán, 9SEIC, Ann. Univ. Ferrara NS Sez V Suppl N 11 (2000) 249.
[191] I. Felhősi, J. Telegdi, G. Pálinkás and E. Kálmán, Electrochim. Acta 47 (2002) 2335.
[192] L. Várallyai, J. Kónya, F.H. Kármán, E. Kálmán and J. Telegdi, Electrochim. Acta 36 (1991) 981.
[193] F.H. Kármán, E. Kálmán, L. Várallyai and J. Kónya, Z. Naturforsch., 46a (1991) 183.
[194] I. Felhősi, R. Ékes, P. Baradlai, G. Pálinkás, K. Varga and E. Kálmán, J. Electroanal. Chem., 480 (2000) 199.

Radiotracer Studies of Interfaces
G. Horányi (editor)

Chapter 5

Colloidal systems; solid/liquid interfaces

P. Joó[a] and K. László-Nagy[b]

[a]Department of Colloid and Environmental Chemistry, University of Debrecen, P.O. Box 31, H-4010 Debrecen, Hungary

[b]Department of Physical Chemistry, Budapest University of Technology and Economics,
H-1521 Budapest, Hungary

The application of radiotracer methods in colloid and surface science (and in nano-chemistry, as well) has been a subject of much interest in the last decades. Many examples have been reported since the initial work of Hevesy [1] "on the use of isotopes as tracers in the study of chemical processes" [2], and Paneth on the introduction of the radioactive indicator method [3].

Colloidal systems may be grouped into three general classifications [4]:
1. colloidal dispersions (these are thermodynamically unstable owing to their high surface free energy),
2. macromolecular colloids (these are thermodinamically stable true solutions of natural or synthetic macromolecular material),
3. association colloids (these are thermodinamically stable colloidal electrolytes).

Due to the fact that real interface exists only in the systems of colloidal dispersions, in this section we consider sorption phenomena only in such kind of systems. The particles in all colloidal dispersions have a large surface to volume ratio and for this reason interfacial chemistry/electrochemistry plays an important role in colloid science. At the interfaces between the solid dispersed phase and the liquid dispersion medium, for instance, characteristic surface properties such as adsorption (accumulation, uptake, binding), and electrical double layer effects are evident. In the past several years more and more investigations in the field of colloid and surface chemistry have focused on a variety of adsorbent- adsorbate aqueous systems.

1. ADSORPTION FROM AQUEOUS SOLUTION (P. Joó)

1.1. Labeled ion sorption at solid/liquid interfaces

In the majority of the experiments dealing with the adsorption of metal ions on various adsorbents, radioactive isotopes were used in tracer quantities. Adsorption from solution is important either theoretically or practically, including a special case of ion adsorption that of heterogeneous ion exchange. Adsorption processes are very important in chomatography, too, although it is not included in this section.

The Freundlich and Langmuir isotherm equations are frequently applied to describe adsorption from solution data. According to a series of papers of Imre [5,6] concerning the kinetics of adsorption and heterogeneous isotope-exchange processes using radioactive indicator method, at least two or three independent rate processes are assumed. In the initial stages of the uptake process the rate is decisively determined by the diffusion from the solution, then by structural properties of the solid phase and after a longer time by the isothermal recrystallization of the surface [7-10].

Muramatsu reviewed [11] the advantages of the radiotracer method to describe surface and colloidal phenomena and some of the earlier studies concerning dispersed systems.

Radiotracer studies of gold adsorption and desorption on heat-cleaned surfaces of polycrystalline molybdenum were carried out by von Goeler and Lüsher [12]. The desorption probability per unit time was found to obey a relation $w=w_o \exp (-E/kT)$ with $w_o=2 \cdot 10^{13}$ s^{-1} and $E=4.2\pm0.2$ eV. Adsorption characteristics of traces of Cr(III) and Cr(VI) on Pyrex, flint glass and polyethylene surfaces have been investigated by Shendrikar and West [13]. Adsorption was measured at various contact times by counting the γ-ray activity from ^{51}Cr radiotracer. Adsorption losses of Cr(VI) were observed less then 1% on all the three types of surfaces. Adsorption of Cs^+ and Sr^{2+} on synthetic hydrated manganese oxide (HMO) as a funtion of pH, salinity, metal ion and HMO concentrations has also been investigated by Singh and Tandon, using a radiotracer technique [14]. The adsorption of ions on HMO seems to be of a counter-ion type and the presence of K^+, Mg^{2+} and Ca^{2+} as well as sodium chloride at ocean level concentrations reduced the adsorption substantially. The surface and diffuse-layer charge at the TiO_2 - electrolyte interface were determined by Foissy, M' Pandou, Lamarche and Jaffrezic-Renault using a radiotracer technique, as well [15]. The experimental determination of the adsorption of Na^+ and Cl^- ions was shown to satisfy the Stern-Graham isotherm.

An overview of the status of solid-liquid interface science was presented by Bockris [16]. Organic and ionic adsorption isotherms and the radiotracer as well as other methods available for investigating the S-L interface were reviewed.

The need for applying in situ direct measurements to study the S-L interface was clearly required to extend our understanding in this field.

Study of Ag^+ uptake by kaolinite from an aqueous silver nitrate solution using the radiotracer technique, was also reported by Daniels and Rao [17]. The sorption proceeded in conformity with the Freundlich equation and occurred at the interlayer and intralayer positions as well as on the basal planes of the kaolinite microcrystals. The ion exchange isotherms of different diazonium cations have been determined on synthetic zeolites by Mohl, Fejes and Horváth, using ^{22}Na radiotracer [18]. A comparison between the amounts of ion exchanged and adsorbed cations has been made.

In two series of papers by Mishra and coworkers, the radiotracer technique has been used in ion adsorption studies as well as in radioactive waste management investigations of ion exchangers. Adsorption of labeled anions $(PO_4^{3-}, CrO_4^-, I^-)$ on metal powders (chromium, cobalt) and metal oxide powders (chromium(IV) oxide, titanium(IV) oxide, as well) in dilute aqueous solution, has been studied [19-23]. The kinetics of adsorption of anions follows the first order rate law and the process obeys the Freundlich isotherm. The rate of uptake of anions was fast initially, slowly decreased with time and finally approached a plateau. The adsorption behavior of labeled cations $(Ba^{2+}, Sr^{2+}, Hg^{2+}, Cd^{2+})$ on titanium(IV) oxide, sodium, potassium and lithium titanates, on chromium(IV) oxide as well as on hydrous ceric oxide, zirconium oxide has also been investigated [24-34]. The results show that the uptake of the cations studied follows first order rate law and equilibrium data fitted well for the Freundlich adsorption isotherm (for example see Fig.1.).

Hasany and Chaudhary also investigated the adsorption of strontium, cadmium and mercury on manganese dioxide from aqueous solutions [35-37]. The sorption of selenium(IV) at micromolar levels onto titanium oxide has been studied, as well [38]. Using linear regression analysis, sorption capacity and sorption free energy were evaluated for this system, too.

An interesting comparative study of the interactions of Tc(IV) and Tc(VII) with iron oxyhydroxides was carried out by Walton and coworkers [39]. Tc(IV) was rapidly removed by chemisorption using Fe(III) oxyhydroxides, but on the contrary, little sorption of Tc(VII) was observed.

Adsorption and desorption of hydrolysed hafnium, scandium and chromium species were measured using radioactive isotopes by Matijevic and coworkers [40-41]. The data were interpreted by combining aspects of the models of James and Healy and also of Anderson and Bockris.

The rates of the isotopic exchange between $BaSO_4$ and $PbSO_4$ have been investigated using $^{133}Ba^{2+}$ and $^{216}Pb^{2+}$ as radiotracers by Paige and coworkers [42], and appear to exhibit a fast, followed by a slow component. Equilibrium (exchange) and nonequilibrium (crystal growth) experiments yield different information about the interfacial kinetics. It is believed that the rate constants, that play the key role in crystal growth, are determined largely by the dehydration of cations.

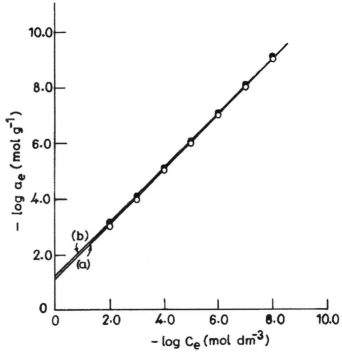

Fig. 1. Freundlich adsorption isotherm for Cd(II) on (a) potassium and (b) lithium titanate at 303 K [34]

The sorption data of Cs^+ and Ba^{2+} cations on magnesite were found to follow Freundlich type isotherms [43]. Laboratory experimental results of Catalette and coworkers [44], showed that magnetite has a high capacity of retention for Ba^{2+} and Eu^{3+}, while Cs^+ can only be sorbed if magnetite contains some impurities such as silica. A good fitting of experimental results has been obtained by using the surface complexation theory with the diffuse layer model (DLM).

Cesium adsorption on silica gel has been studied as a function of different particle sizes, cation concentrations and temperatures by using the radiotracer method [45]. The data were fitted both to Freundlich and Dubinin-Radushkevich isotherms (Fig. 2. and Fig. 3.).

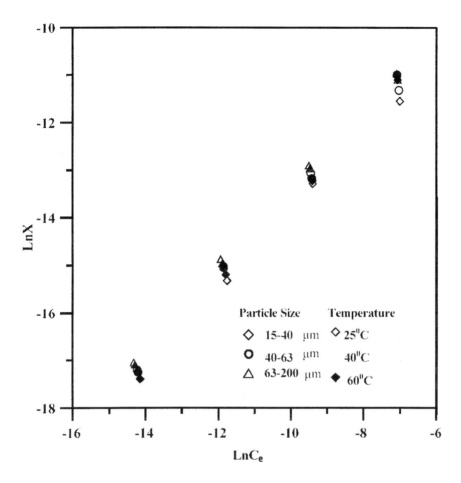

Fig. 2. Freundlich isotherm for cesium adsorption on silica gel at different temperatures [45].

The sorption and desorption of calcium were investigated on hydroxyapatite by Badillo-Almaraz and Ly [46]. Isotopic exchange of ^{45}Ca was utilized to obtain the solid/liquid distribution coefficients, K_d, of calcium, under conditions of solubility equilibrium. The experimental data were analyzed assuming a reversible component of the partition coefficient and a nonreversible sorption component that might result from the polycrystalline nature of the $Ca_{10}(PO_4)_6(OH)_2$ particle.

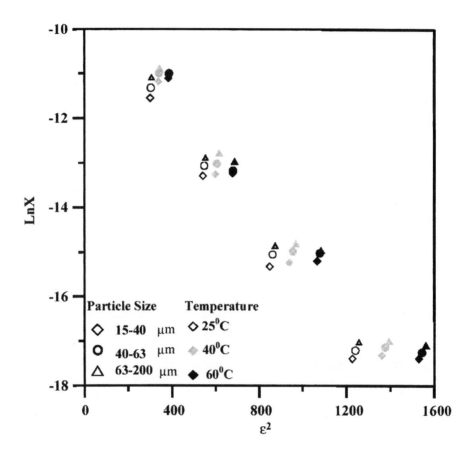

Fig. 3. D-R isotherms for cesium adsorption on silica gel at different temperatures [45].

1.2. Behavior of radioactive elements and radionuclides in the environment

Metal ions (either polluting or non-polluting) introduced into the environment can be bound on clay mineral or other components of soil by sorption, complexation and precipitation. Adsorption of Zn(II) on clay minerals (kaolinite, montmorillonite) and on bentonite has been investigated radiometrically, as well [47]. Adsorption isotherms fit the Freundlich type equation depending on pH value of each clay suspension. Sorption of Sr on bentonite was studied, too, using the batch technique. Thermodynamic parameters were determined and the sorption data were interpreted in terms of Freundlich, Langmuir and Dubinin-Radushkevich isotherms [48].

In a series of papers by Nagy, Kónya and coworkers, the adsorption and desorption of radioactive isotopes and heterogeneous isotope exchange processes on clay minerals and soils were studied [49-55]. The sorption of radioactive isotopes and complex formation can principally be evaluated on the basis of surface complexation models (SCM).

A special tailored aluminum nano-composite clay material was prepared for nuclear waste application, i. e. for radiostrontium removal [56]. Strontium uptake has also been studied by an aluminum-pillared montmorillonite in acidic aqueous solutions [57]. Desorption of Sr^{2+} was observed when the concentration of H_3O^+ in the solution raised. Sylvester and coworkers have investigated the adsorption of the uranyl ion (UO_2^{2+}) in contact with amorphous silica (SiO_2), γ-alumina (Al_2O_3), and montmorillonite surfaces in the pH range of 3.1-6.5 [58]. Results suggest that adsorption onto montmorillonite at low pH occurs via ion exchange, leaving the inner-sphere uranyl aquo-ion structure intact. At near-neutral pH, inner-sphere complexation with the surface predominates. Adsorption onto the silica and γ-alumina surfaces appears to occur via an inner-sphere, bidentate complexation with the surface.

A hydrochemical model succesfully describes the transport (and retention) of metallic radionuclides through upland catchments with acid, organic-rich soils, according to the work of Tipping [59]. The retention or transport depends on the strength of chemical interaction with the solid phases of the different soil horizons.

It is well known that the most stable form of Tc in the surface environment is TcO_4^-, which has got high mobility and availability to plants in the environment. Tagami and Uchida [60] carried out radiotracer experiments in order to investigate chemical transformation of Tc in soil during the change of soil water conditions.

Albrecht and collaborators have carried out a combined dye and radionuclide tracer study to investigate the link between preferential flow, leaching of surface applied substances and their distribution within structured and repacked soil monoliths [61]. The dye (Sulforhodamine B) served as a tracer for water movement within the soil and thus allowed linkage of the radiotracer (^{58}Co) with the flow pattern. Preferential flow in the stuctured monolith promoted the bypass and transport of both tracers. The homogenized monolith, on the other hand, showed mostly chomatographic infiltration.

^{65}Zn was used as a radiotracer to study sorption characteristics of zinc by untreated calcareous soils and by soils treated to remove $CaCO_3$ [62]. The sorption behavior of Zn before and after extraction of $CaCO_3$ from calcareous soils has provided supporting evidence with respect to the most significant sink for zinc in calcareous soil.

1.3. Effects of humic and fulvic subtances on the sorption

Contamination by toxic metals, radioactive elements and radionuclides is a serious problem in many areas of the world. Colloids have a high affinity for pollutants, and understanding the interaction of these contaminants with natural colloids, including humic and fulvic substances, is very important.

A humic ion binding model (HIBM) has been developed by Lofts and coworkers [63] to simulate the humic acid-solution partitioning of radiocesium and to model the role of humic acid in ^{134}Cs distribution in an upland peat soil. Among the artificial radionuclides released to the environment by nuclear activities, the transuranium elements are a major concern. Skipperud, Oughton and Salbu [64] performed dynamic tracer experiments where different plutonium-species were added to a soil-rainwater system to obtain information on distribution coefficient, K_D, values. The interaction of all Pu-species with soils seems to be rapid and follows a two-step reaction.

Hydrophilic and hydrophobic acids may complex inorganic contaminants and change their sorption behavior on geological materials, as well. Ticknor and coworkers [65] studied the effect of fulvic acid extracted from groundwater on the sorption of actinides and fission products on granite and selected materials. Their report indicated that, under aerobic conditions, the effect of dissolved organic material on sorption of radioisotopes depends on the radioisotope in question (i.e., ^{85}Sr, ^{137}Cs, ^{233}U, ^{238}Pu, ^{241}Am, ^{99}Tc and ^{125}I), and the concentration of the organic in solution.

A series of column experiments have been performed by Warwick and collaborators [66] to demonstrate the effect of humic substances on the transport of europium through porous media. Two approaches have been used to model the results: an approach based on an assumption of local chemical equilibrium, and one which accounts for kinetically hindered reactions. The interactions of a range of actinides (Th, U, Np, Pu, Am) with humic substances with average molecular weights determined by analytical ultracentrifugation around 49000, 15000 and 8000, were studied by gel permeation chromatography [67]. The three humic fractions vary greatly in their effectiveness and selectivity as ligands in the complexation for early actinides.

The influence of humic acids on the retention of Th(IV) as an analogue of tetravalent radionuclides (Zr, Hf, U, Np, Pu actinides) onto the surface of iron oxides (hematite and goethite) has also been investigated by Reiller and coworkers [68]. It was illustrated that the retention of thorium(IV) onto Fe oxides closely depends upon the ratio between humic and surface iron oxide sites.

The effects of an aquatic fulvic acid on the pH-dependent adsorption of mercury (^{203}Hg) and cadmium (^{109}Cd) to particulate goethite were studied in batch systems [69]. Competitive experiments were also performed using ^{65}Zn. In the absence of fulvic acid, the adsorption follows the expected behavior, however, in the presence of fulvic acid, additional adsorption sites are available by the organic molecule when it is associated to the goethite.

In a series of papers by Bors and collaborators [70-74], sorption of radioiodide ($^{125}I^-$), cesium ($^{134}Cs^+$), and strontium ($^{85}Sr^{2+}$) in organo-clays, soils, and hexadecylpyridinium-humate complexes were studied. In the case of (HDP_y^+) surfactant -modified humates, I^- ions exhibited increased adsorption, while the adsorption of Cs^+ was negligible.

To determine the Mg released by ammonium exchange on nondissolved humic acid in a seawater medium, a radiotracer technique, employing ^{27}Mg, has been used [75]. The $Mg \rightarrow NH_4$ exchange on humic acid comes to a steady-state value in < 18 min. The conditional equilibrium constant obtained for this reaction, $K_{cond}= 0.039 \pm 0.001\ M^{-1}$. The radiotracer technique can be extended to other geochemical solid phases in seawater, and it can be modified to study the behavior of the major cations by using ^{24}Na, ^{42}K, and ^{49}Ca.

The effects of ionic strength, pH, and humic substances on the sorption and desorption of Co(II) on alumina and silica were investigated by using the radiotracer ^{60}Co [76]. It was found that strong chemical bonds are formed between the bare and the coated alumina surface, as well, and Co(II) while a transition from the adsorption to the surface-induced precipitation of Co(II) on the bare alumina surface takes place.

The sorption processes in part determine the retardation of metal migration through the Geosphere. These sorption interactions may be complicated by the presence of organic matter, such as humic acids. The sorption of lanthanide europium onto colloidal goethite and boehmite and the influence of humic acid were examined by Fairhurst and Warwick [77]. Results showed that humic acid readily adsorbed to the minerals decreasing their zeta potential. In some cases, the humic acid was able to cause complete charge reversal on the minerals. Humic acid was shown to enhance europium adsorption at low pH, but reduce adsorption at intermediate and high pH.

The effects of ionic strength, fulvic acid and pH on sorption and desorption of Eu(III) and Yb(III) actinides on alumina were studied by using the batch technique and radiotracers $^{152+154}Eu$ and ^{169}Yb by Wang and coworkers [78]. It was found that the surface hydrolysis model can satisfactorily explain the results on the bare alumina, and that the competition among the soluble and sorbed fulvic acids and the complexations of free surface hydroxyl groups can also satisfactorily explain the observations on coated alumina. The adsorption of Am(III) on alumina, silica, and hematite, and the effects of humic substances, ionic strength, and pH were also investigated by Tao and coworkers [79]. It was found that a negative effect of fulvic acid on the adsorption on silica and a positive one existed on the adsorption on alumina. In contrast to the ^{241}Am sorption on alumina and silica, a tremendously high adsorbability on hematite was observed.

The uptake of transuranic radionuclides by humic and fulvic acids chemically immobilized on silica gel and their competitive release by complexing agents were also investigated by Bultman and coworkers [80].

1.4. Sorption of actinides onto colloidal dispersions

The surface complexation behavior of trivalent metal ions, Am(III) and Eu(III) on gamma-alumina has been studied by Rabung and collaborators [81]. Experiments were conducted at different pH (4-8) and ionic strength (0.001-0.1 M $NaClO_4$) values either in the presence or absence of CO_2. Using the two site surface complexation model, a good agreement was found between experimental data and modeling. Surface complexation is accompanied by a decreasing number of hydration water molecules in the first coordination sphere.

The surface sorption of Cm(III) onto aqueos suspensions of alumina at an ionic strength of 0.1 M $NaClO_4$ has been investigated by Stumpf and coworkers [82]. With increasing pH, two curium-alumina surface complexes were identified and characterized: $Al-O-Cm^{2+}(H_2O)_5$ and $Al-O-Cm^+(OH)(H_2O)_4$.

Hematite, montmorillonite and silica colloids were used for uptake experiments with redox-sensitive radionuclides $^{239}Pu(V)$ and $^{237}Np(V)$ [83]. The capacity of hematite to sorb Pu significantly exceeded that of montmorillonite and silica. Np uptake on all mineral phases was far less than Pu(V) sorption, suggesting that a potential Pu(V)-Pu(IV) reductive sorption process was involved. An analytical expression that evaluates the effects of the redox potential, E, and the pH was developed for studying and modelling the sorption of actinides onto colloidal Al_2O_3, FeOOH and SiO_2, by Alonso and Degueldre [84]. This includes surface complexation with one type of surface site and ion exchange processes. The model was applied to the sorption of americium, but it is also applicable to study the sorption of other redox-sensitive elements. The sorption of Am on chlorite and smectite/illite marl colloids was studied, as well [85]. In groundwater conditions (pH 8.6, $[Na^+]-[HCO_3^-]-2x10^{-2}$ M), americium sorbs onto marl groundwater colloids with an average size smaller than 500 nm.

A basic microscopic theory was developed by Abril [86,87] to describe the distribution, transfer and uptake kinetics of dissolved radionuclides by suspended particulate matter. Kinetic transfer coefficients are important parameters to understand the behavior of non-conservative radionuclides in aquatic environments. An experimental and modeling study of Pu uptake in natural aqueous suspensions of unfiltered waters from reservoirs and rivers has also been reported by Abril and collaborators [88]. The uptake curves were discussed with respect to numerical kinetics models. The data suggest that the main pathways for Pu uptake consist of two parallel and reversible reactions (fast physical adsorption reaction, i.e. Pu is fixed by electrostatic forces onto the external free surface of particles, and physical adsorption onto the inner surface of pores and free-edges), followed by a consecutive non-reversible reaction (a slower process) in which previously adsorbed Pu is transferred into the inner matrix of the particles, or forms other non-reversible unions.

The use of radiotracers for studies of metal sorption behavior in controlled laboratory experiments provides an ideal approach for investigating the role of individual parameters on processes, as well. Mc Cubbin and Leonard developed [89] a simplistic technique to produce a thorium radiotracer (^{234}Th) in order to determine the kinetics of sorption to marine particulate material. A number of laboratory experiments have been carried out [90] for the investigation of short-term oxidation and sorption by suspended particulate material, using ^{237}Np radiotracer, and artificial and natural seawater solutions. Results indicate that the uptakes of Np(IV) and Np(V) species by suspended sediment exhibit different kinetics. It has been shown that the major constituents of seawater can influence Np(IV) sorption and oxidation kinetics; a consistent redox equilibrium was attained within a relatively short period of time.

Sorption irreversibility and coagulation behavior of ^{234}Th with marine natural organic matter in the colloidal size range (1kDa-0.1μm) have extensively been investigated by Santschi, Honeyman and collaborators [91].

The uptake of ^{234}Th by colloidal organic matter is irreversible, and the Th-organic matter complex formed is strong enough. Coagulation experiments show that Th complexed with low molecular weight colloids is transferred to larger fractions. Results demonstrate that coagulation is the dominant step in the transport of ^{234}Th to the particulate phase and that colloidal matter is the dominant sorbent in marine surface water environments. If Th(IV) is desorbing very slowly or not at all, it can be due to the surface- active nature of the main colloidal carrier phase, which might sterically mask complexed metal ions and hinder the release of ^{234}Th. Metals are first scavenged by the smallest size fractions, and then transported via coagulation and aggregation into larger sinking particles. This kind of transport, the so called „colloidal pumping" (a Brownian – pumping one [92]) would then be able to transport ^{234}Th (and other metals like ^{59}Fe [93], ^{203}Hg [94], etc.) across the particle size spectrum during the coagulation of its surface- active carrier phase.

1.5. Tracing colloid-particle and colloid-colloid interactions (adagulation, heteroadagulation)

The term "adsorption" is not accurate fundamentally for the accumulation of lyophobic sols at a solid/liquid boundary layer, since it is a question of interaction between surfaces. For the uptake of lyophobic sols (colloidal dispersions with a size range of 1-500 nm [4]) at a solid/liquid phase boundary, von Smoluchowski [95] used the term adagulation (adhesive coagulation). If the material properties of the macroscopic adsorbent and the submicroscopic sol are not the same, then the correct term is heteroadagulation. The studies of von Buzágh [96] indicated that the heteroadagulation of lyophobic sols is affected by those factors, which determine the adhesion interaction between the sol particles and the solid surfaces. Mádi and coworkers investigated the heteroadagulation of gold sol particles, labeled with [198]Au, and coarse disperse barium sulfate by a radioactive tracer method [97,98]. The rate of heteroadagulation can be given by the following equation:

$n = n_\infty [1-\exp (-k (n_0/n_\infty) t]$, where n is the number of sol particles adhered at a given time t, n_∞ is the number of particles adhered in the equilibrium state, n_0 is the number of particles in the solution at t=0, and k is the rate constant. Heteroadagulation of chromium hydroxide sol, labeled with ^{51}Cr, and silicic acid, aluminum oxide and iron oxide has also been investigated [99-102]. The surface potential of the sol and adsorbent was changed through changing the pH. The effect of electrolyte on the interaction of surfaces was studied. The DLVO theory of sol stability, which is based on the calculation of the interaction between the particles [4], was found suitable for the explanation of the uptake of sol particles. Sorption of α-FeO(OH) colloidal particles was also studied on coarsely dispersed alumina and silica adsorbents using the radioisotopic tracer method by Párkányi-Berka and Joó [103]. The most important parameter determining the sorption of labeled sol particles is the surface property of the adsorbent, which is determined by the pH and by the hydrolysis products or complex forming agents, as well. The effect of indifferent electrolyte can be interpreted by the DLVO theory, and the linearised form of the Langmuir isotherm gives straight lines [104]. Sorption of chromium hydroxide sol on silica type adsorbents was extensively investigated by the means of a radioisotope tracer method by Csobán, Joó and coworkers [105-106]. This technique allowed for sensitive kinetic and equilibrium measurements in both directions: sorption and desorption, which are strongly dependent on solution conditions, such as pH, ionic strength or the presence of complexing species. Sorption and desorption of Cr(III) from freshly prepared and aged solutions were also studied, and they were followed by a radiotracer method. [51]Cr radioactive isotope was used [107]. Applying the same solution conditions,Cr(III) uptake and release are greater from the aged solution. Outer sphere complex formation can be deduced from the great ionic strength dependence at the onset of the sorption. Increasing ionic strength suppresses sorption onto silica and enhances sorption onto aluminum oxide [108].

A theoretical approach to the radiometric method for the characterization of particulate processes in stable colloidal suspensions was given by Subotic [109]. It was suitable for determining the change of the relative mean particle size during the ageing of systems. The sorption of gold colloids, present in solution along with dissolved gold (predominantly $Au(OH)H_2O^0$), onto pyrite and goethite has also been determined using radiotracer techniques by Schoonen and collaborators [110]. The removal of gold colloids is due either to the direct uptake of the Au colloids or to dissolution and subsequent adsorption. The difference in sorption efficiency between the pyrite surface, which is a much more efficient scavenger of gold than is goethite, and goethite surface reflects either: the lack of oxidizable groups on the goethite surface which could reduce dissolved Au(I) species to elemental gold, or the stability of the Au-S bond formed on the sulfide surface compared to the Au-O bond formed on the oxide surface. Growth of pyrite during gold sorption may explain the common occurrence of invisible gold in pyrite.

A radiotracer study of $^{139}Ce(III)$ and $^{54}Mn(II)$ uptake onto suspended particles in an estuary was carried out by Moffett [111]. The specific rates of oxidation for both elements, and the oxidation kinetics were studied, as well. The specific rates for Ce(III) and Mn(II) were very similar to each other.

The effect of size and geochemical properties on the binding of trace Cd and Cu metals to natural colloids and particles have been investigated by Lead and coworkers [112]. A simple one-site binding model provided a good description of the data. The sorption/desorption behavior of ^{85}Sr on Ca-montmorillonite and silica colloids, and the effect of the ionic strength of the solution on the sorption of Sr were studied by Lu and Mason [113]. The results suggest that clay and silica colloids facilitate the transport of Sr along a natural potential flowpath.

1.6. Colloid-facilitated tracer transport

Colloid-facilitated radionuclide transport in crystalline rock and groundwater is of general interest in contaminant hydrology. Their migration in porous rock in the presence of colloids, and effects of kinetic interactions were investigated by Li and Jen [114]. The adsorption processes for radionuclides with colloids and porous rock can be assumed as nonequilibrium and modeled by linear kinetic adsorption. Laboratory migration experiments were carried out with combinations of ^{85}Sr and ^{241}Am radionuclides and natural colloids within a fracture, in a large granite block, by Vilks and Baik [115]. Dissolved ^{85}Sr behaved as a moderately sorbing tracer, while dissolved ^{241}Am was completely adsorbed. The colloid and radionuclide retardation experiment was dedicated to the study of the in situ migration behavior of selected actinides and fission products (U, Th, Pu, Am, Np, Sr, Cs, I and Tc) in the absence and presence of bentonite colloids in a water-conducting feature (shear zone) (Möri and collaborators [116]). The field experiments were supported by an extended laboratory and modeling program. Experimental results are considered from the perspective of understanding the possible long-term behavior of a deep geological repository for radioactive waste.

Contaminant transport in subsurface environments is critical to the understanding of the risk associated with radionuclide migration. Site-specific information on the chemistry and natural colloid concentration of saturated zone groundwaters was combined with a surface complexation sorption model to evaluate the impact of colloids on calculated retardation factors [117].

Moridis and coworkers [118] reported 3-D site-scale studies of radioactive colloid transport under ambient conditions. The study of migration and retardation of colloids accounts for the complex processes in the zone investigated, and includes advection, diffusion, hydrodynamic dispersion, kinetic colloid filtration, colloid straining, and radioactive decay. The most important factors affecting colloid transport are the subsurface geology and site hydrology. The transport of colloids is strongly influenced by their size and by the parameters of the kinetic-filtration model used for the simulations.

The possible mechanisms of the generation and stability of bentonite colloids at the bentonite/granite interface of a deep geological radioactive waste repository have been investigated by Missana and coworkers [119]. It has been shown that solid particles and colloids can be detached from the bulk and mobilized by water flow. Gel formation and the intrinsic tactoid structure of the clay play an important role; and clay colloids can be formed even in quasi-static flow conditions. Structural chemistry of uranium associated (trapped) with Si, Al, Fe natural gels in a granitic uranium mine was also studied by Allard and collaborators [120].

The occurrence of a variety of storm runoff processes, including those associated with variable source areas have been demonstrated. When two radiotracers, ^{51}Cr and ^{82}Br, were used in a study of surface and subsurface storm runoff, irregular spatial flow patterns were observed [121]. A radiotracer study was undertaken to evaluate to what extent cadmium present in natural waters is associated to organic matter and/or to large inorganic molecules, using ^{109}Cd as radiotracer [122]. Iron colloids, and organic colloid matter associated transport of major and trace elements in small boreal rivers and their estuaries has been investigated by Pokrovsky and Schott [123]. Results suggest a preferential transport of trace elements as coprecipitates with iron-oxy(hydr)oxide colloids. Dissolved organic matter stabilizes these colloids and prevents their aggregation and coagulation. To simulate the behavior of radionuclides along a salinity gradient, in vitro sorption and desorption kinetics of Co, Mn, Cs, Fe, Ag, Zn and Cd were studied in fresh river water mixed with brackis estuarine waters [124].

1.7. Radiotracing in aquatic colloid chemistry

The generation and mobilization of colloids in aquatic systems (groundwater, surface water, seawater, estuarine water, etc.) and the sorptive behavior of the suspended particles and sediments are considered important issues in the transport of radioactive materials and the colloid-assisted transport of radioactive species sorbed on naturally occurring colloids, as well.

Stephenson and coworkers [125] carried out radiotracer dispersion tests in a fissured aquifer by using ^{51}Cr isotope. Aquifer characteristics in saturated zones, i.e. transmittivity and dispersivity, can be quite effectively determined in situ by means of artificial tracers, provided the behavior of such tracers is representative of that of the water in the aquifer. Radiotracer evaluation of groundwater dispersion in a multi-layered aquifer was also investigated using ^{131}I as tracer [126]. The dispersion coefficients for each separate layer and the whole aquifer were determined from a fit of the model to experimental data.

Water column and sediment data were combined with stable Pb and ^{210}Pb isotope data to trace the early diagenetic behavior and geochemical cycling of lead in lake environment [127]. The remobilisation of Pb appears to be caused by a pH-controlled desorption from the solid-phase sediment, which is consistent with a model describing surface complexation of Pb(II) on hydrous geothite surfaces. Lead is highly mobile in the sediment and may diffuse into the slightly alkaline lake water. A simple method for the determination of molybdenum and tungsten in surface water and seawater was reported, too [128]. Mo and W were concentrated on activated charcoal by adsorption.

Osmium concentrations and isotopic compositions (mean ^{187}Os/^{188}Os) were measured by a new method in seawater samples [129]. The data may be explained by the occurrence of two possible processes: the horizontal advection of a water mass of low Os concentration from the continental slope and/or the in-situ adsorption of Os onto sinking particles. Koschinsky and coworkers [130] investigated the role of different types of marine particles in the sorption of dissolved metal species in a disturbed deep-sea bottom water system using radiotracer and voltammetric measurement techniques. Compared to other deep-sea particles (like clay minerals, calcite, apatite), Mn and Fe oxides and oxyhydroxides were found to be far the most important phases in scavenging dissolved metals (Pb^{2+}, Cs^{+}, Mn^{2+}, Co^{2+}, etc.). The sorption experiments support a simple electrostatic model, and a comparison of the results of voltammetric and radiotracer techniques revealed that after a fast sorption process, isotopic exchange dominated reactions occur.

Herut and collaborators [131] examined the adsorption of phosphate, labeled with $^{32}PO_4^{3-}$, onto loess particles from surface water and deep seawater. The uptake and release of trace elements (^{109}Cd, ^{51}Cr, ^{60}Co, ^{59}Fe, ^{54}Mn, ^{65}Zn,) were also studied using estuarine waters and particles by Hatje and coworkers [132]. The kinetics of adsorption/desorption were investigated as a function of suspended particulate matter loading, pH, and salinity. Most of the metals showed relatively simple kinetics with an increase in uptake as a function of time and particle concentrations. The reversibility of sorption decreased in the order Co>Mn>Zn>Cd>Fe>Cr. The speciation and sorptive behavior of dissolved and particulate Ni in an organic-rich estuary has been studied by Turner and coworkers [133]. The uptake of nickel by estuarine particles was monitored in a series of laboratory simulations employing ^{63}Ni. Adsorption was dependent to some degree on salinity, pH, and particle type, but most importantly on the presence of dissolved organic matter.

Suspended particles are instrumental in controlling the reactivity, transport and biological impacts of substances in aquatic environments. The role of suspended colloidal particles in the biogeochemical cycling of trace constituents (trace metals, organo-metallic compounds and hydrophobic organic micropollutants) in estuaries by means of radiotracers has been reviewed by Turner and Millward [134]. Particle-water interactions encompass a variety of physical, biological, electrostatic and hydrophobic effects; the concentration of suspended particles and salinity.

Interactions between radioactively labeled colloids and natural particles were investigated, and evidence for a colloidal pumping mechanism was demonstrated by Santschi and coworkers [135]. A series of radiotracer experiments using natural colloidal organic matter from estuaries was carried out. Suspended particle uptake of colloidally bound trace metals occurred and the colloidal trace metals transferred into the particulate phase.

Trace metals in rivers can be present in different physicochemical forms, including low molecular mass ions, colloidal or particulate phases. The importance of suspended colloids and particles in the transport and fate of trace metals in rivers and their estuaries is well known. Time-dependent interactions between trace metals and freshwater sediments and their remobilization upon contact and mixed with seawater was investigated by Standring and coworkers [136]. River sediments (organic and inorganic) were labeled with $^{109}Cd^{2+}$, $^{65}Zn^{2+}$ and $^{54}Mn^{2+}$ radiotracers. Sorption of tracers occurred rapidly, followed by a slower approach to pseudoequilibrium. Sorption ratios from modeling data were greater for organic compared to inorganic sediments. High organic content in sediments increased initial sorption of tracers, but inhibited redistribution to more strongly bound fractions over time, resulting in greater remobilization of tracers when in contact with seawater.

Sequential leaching of surface sediment was carried out in order to assess potential mobility and the associations of heavy metals in the mineral phases by Koschinsky, et al. [137]. To distinguish the leaching behavior of aged and fresh sorbents, radiotracers were adsorbed on the solids prior to leaching. The sorption process was rapid, freshly sorbed metals were leached mostly in the same fractions as those in aged sorbents. Sediment trap samples were collected to examine the factors that control the flux of ^{234}Th, ^{228}Th, and ^{210}Pb. Total mass, organic carbon, biogenic opal, carbonate, and lithogenic fluxes were measured to compare with the radionuclide fluxes [138]. The flux of these particle-reactive radionuclides is not controlled by the total mass or the major sediment component fluxes.

The sorption behavior of Zn(II) on the mineral fractions of a marine sediment was also investigated using a sequential extraction-radiotracer technique [139], by Fujiyoshi and coworkers. The sorption isotherms of $^{65}Zn(II)$ on all the fractions fitted a Langmuir type equation.

1.8. Biocolloids

Sorption, attachment, and accumulation processes of biocolloids, such as cells and microorganisms, spheroidal or ellipsoidal bacteria, virus, microbe or protein particles, as well as adhesion phenomena of biomacromolecules (biopolymers) and other biomaterials have received considerable attention recently.

Colloidal phenomena have also been shown to be important in the association of dispersed protein or virus particles with metal oxides (magnetite, hematite) as well, which are also dispersed as colloid-sized particles [140]. Several experimental approaches reveal that the surfaces of biomaterials-proteins, nucleic acids and lipids, for instance- are covered with polar groups that attract water so as to create repulsive "hydration" forces when bodies are brought together. Polar groups then create repulsive or attractive physical forces between biocolloids [141].

^{14}C-Polyethyleneimine polymers have been used by Dixon and coworkers [142] to study the flocculation of silica, bacteria and algae. The degrees of flocculation, and the amounts of polymer adsorbed on the colloid surface and the corresponding equilibrium concentration of polymer in solution have been measured. Experimental results indicate that interaction of PEI polymers with anions in the algal and E. Coli nutrients, and with biopolymers formed in the growth of the latter are important in affecting flocculation.

Transport of bacteria over significant distances through aquifer sediments occurs primarily among bacteria with low affinity for sediment materials. Intra-population heterogeneity in biocolloid-collector affinity may be an important determinant of subsurface bacterial transport characteristics, with critical implications for pathogen transport and dispersal of bacteria for the remediation of hazardous waste. A new method, utilizing radiolabeled (^{3}H-leucine) cells and borosilicate glass beads was used to estimate bacterial biocolloid collision efficiency (α), defined as the fraction of colliding cells that adhere to the collector surface, by directly measuring the retention of cells in porous media [143].

The transport of colloids through porous media strongly depends on the kinetics of colloid deposition and release. Field studies suggest that mobile colloids can include clay minerals, Fe and Al-oxides or hydroxides, colloidal silica, organic matter and biocolloids [144].

Porous media column experiments were used by Jewett and collaborators to investigate Pseudomonas fluorescens strain P17 bacteria transport and retention using radiolabeled P17 [145]. Results were described using a steady-state transport model with bacterial attachment to soil as a pseudo-first-order process. The increased retention of bacteria at the gas-liquid and solid-liquid interfaces indicates the presence of the interface is an important factor in limiting pathogen migration, evaluating biocolloid-facilitated transport of pollutants.

Microorganisms can alter the stability, solubility, bioavailability and mobility of actinides in radioactive wastes as well as and in the natural environment. Under appropriate conditions, actinides can be solubilized or stabilized by direct (enzymatic) or indirect (nonenzymatic) microbial action. Biotransformation of various forms of uranium (ionic, inorganic and organic complexes) by aerobic and anaerobic microorganisms has been extensively studied, and the state of knowledge of the microbial transformation of actinides was also reviewed by Francis [146]. Bacteria are also playing an important role in the transport of radionuclides and other heavy metals in nature. Microbial activities can cause either dissolution and mobilization or immobilization of uranium in uranium mining waste piles. Aerobic and anaerobic strains of bacteria were used by Panak and coworkers [147] to quantify interactions with U(VI), to obtain more information on the effects of microbial activities on the mobilization/immobilization of radionuclides in geological environments. Fig.4. shows the bioaccumulation of uranium by two different strains of bacteria. Extraction studies with EDTA showed that only a small part of the accumulated uranium is adsorbed on the surface of the cell walls, whereas the main part is probably taken up by the cells.

Fundamental studies on the microbial transformations of actinides will be useful in developing novel biotechnological approaches to the treatment of waste streams and contaminated sites.

1.9. Miscellaneous

The adsorption behavior of iridium on an α-amino pyridine resin was studied by using the ^{192}Ir radiotracer technique [148]. A preconcentration method, which is also suitable for the analyses of trace Au and Pd in environmental and geological samples, was established. A radiotracer technique was used to follow desorption and exchange processes with vinyl 4 pyridine and quaternary ammonium membranes [149]. The cross-linking and hydration effects with NaCl and HCl solution were also investigated, and the importance of the deprotonation process in water and in NaCl solutions was pointed out, as well. A performance study of different cation-exchange membranes has been conducted by Boucher and collaborators [150]. Comparisons were made of sulfuric acid recovery rate, water transport, metal leakage and energy intake for membranes. Proton permselectivity and metal leakage, co-and counter-ion transport numbers were also investigated using a radiotracer method, too. It was documented that electrodialysis is suitable for the treatment of zinc hydrometallurgy.

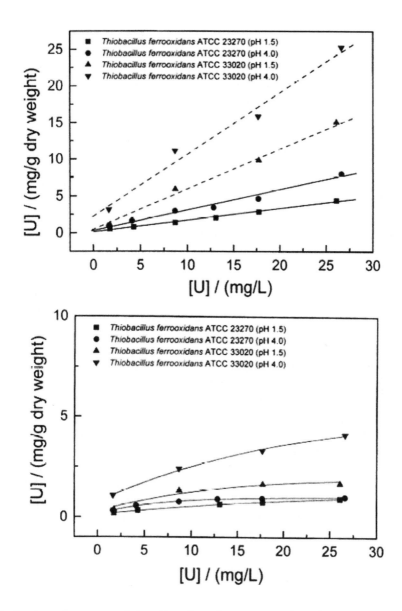

Fig.4. Total uranium accumulation (a) and uranium concentration adsorbed onto the surface (b) of Thiobacillus ferrooxidans ATCC 23270[T] and Thiobacillus ferrooxidans ATCC 33030 at pH 1.5 and pH 4.0 as a function of initial uranium concentration. The results are normalized to the dry weight of the bacteria. [147].

Radiotracer studies with ^{133}Ba, ^{45}Ca and ^{36}Cl were reported for PVC matrix ion-selective electrode membranes. The continued uptake of radioactive barium ions and the absence of evidence for the uptake of chloride ions confirm that membranes possess the permeation selectivity characteristics of ion-exchanger membranes. The extent of the incorporation of calcium ions was significantly less than that of barium ions [151].

Colloidal silica powders were chemically grafted with cyanosilane to study the interactions of Ag$^+$ ions with the grafted surfaces [152]. Electrophoretic mobilities and labeled ion adsorption (^{110m}Ag) were measured for bare and grafted silica. Adsorption of Ag$^+$ ions was appreciable on grafted silica powder due to the specific interactions of Ag$^+$ ions with the cyanogroups of the grafted silica.

Crystallization phenomena and floatability of sylvinite and kieseritic hard salt were studied, and a comparison was made between laboratory flotation results, adsorption measurements, radiotracer measurements, and measurements of adhesion forces as well as scanning electron microscopy investigations [153].

The uptake behavior of rice husk in the removal of Zn(II) ions has been studied as a function of contact time, adsorptive concentration, temperature, and pH, using the radiotracer technique employing a ^{65}Zn radiotracer [154]. A relatively slow uptake of metal ion was seen, which obeys the first-order rate law, and agrees well with the classical Freundlich adsorption isotherm. Evaluation of the thermodynamic data shows that the process involved is endothermic and apparently irreversible in nature, suggesting an ion exchange mechanism along with surface complexation. The adsorption behavior of rice husks for antimony ions from aqueous solutions has also been investigated [155], as a function of appropriate electrolyte, equilibration time, pH, amount of adsorbent, concentration of adsorbate, effect of divers ions, and temperature. The radiotracer technique was employed to determine the distribution of antimony (^{122}Sb), using a batch method. Studies show that the adsorption data follow the Freundlich isotherm, and the sorption process was found to be endothermic in the case of antimony uptake, as well.

A version of the in-situ radiotracer "foil" method has been used by Joó and coworkers to study the electroaggregation of silver radiosol labeled with ^{110m}Ag on a polycrystalline gold electrode and the surface electrochemistry of the resulting radiocolloid-modified electrode [156, 157]. It was found that the electroaggregation-disaggregation processes are time and electrode potential dependent.

LIST OF ABBREVIATION

DLM   Diffuse Layer Model
DLVO theory Derjaguin and Landau and Vervey and Overbeek theory

HDP	Hexadecylpyridinium
HIBM	Humic Ion Binding Model
HMO	Hydrated Manganese Oxide
PEI	Polyethyleneimine
SCM	Surface Complexation Model
S-L	Solid-Liquid

REFERENCES

[1] G. Hevesy, Radioactive Indicators, Interscience Publishers, New York, 1948.
[2] List of the Nobel Prize Laureates (1901-), Nobel Committees for Physics and Chemistry, The Royal Swedish Academy of Sciences, Stockholm, Sweden.
[3] F. Paneth, Radio-Elements as Indicators, Mc Graw Hill Book Co., New York, 1928.
[4] D. J. Shaw, Introduction to colloid and surface chemistry, Butterworths, London-Boston, 1976.
[5] L. Imre, Z. phys. Chem. A., 153 (1931) 261.
[6] L. Imre, Trans. Farad. Soc., 33 (1937) 571.
[7] L. Imre, Kolloid-Z., 131 (1953) 21.
[8] L. Imre, Kolloid-Z., 135 (1954) 161.
[9] L. Imre, Kolloid-Z., 154 (1957) 119.
[10] L. Imre, Kolloid-Z., 166 (1959) 122.
[11] M. Muramatsu, „Radioactive Tracers in Surface and Colloid Science" in E. Matijevic, Ed., Surface and Colloid Science, Vol. 6, Wiley, New York, 1973, p. 101.
[12] E. von Goeler and E. Lüscher, J. Phys. and Chem. of Solids, 24 (1963) 1217.
[13] A. D. Shendrikar and P. W. West, Anal. Chim. Acta, 72 (1974) 91.
[14] O.V. Singh and S. N. Tandon, Appl. Radiat. Isot., 28 (1977) 701.
[15] A. Foissy, A. M´ Pandou, J.M. Lamarche and N. Jaffrezic-Renault, Colloids and Surfaces, 5 (1982) 363.
[16] J. OM. Bockris, Materials Science and Engineering, 53 (1982) 47.
[17] E. A. Daniels and S. M. Rao, Appl. Radiat. Isot., 34 (1983) 981.
[18] M. Mohl, P. Fejes and G. Horváth, Appl. Radiat. Isot., 35 (1984) 408
[19] S. P. Mishra and T. B. Singh, Collect. Czech. Chem. Commun., 52 (1987) 960
[20] S. P. Mishra and T. B. Singh, Appl. Radiat. Isot., 37 (1986) 1121.
[21] S. P. Mishra and T. B. Singh, Appl. Radiat. Isot., 38 (1987) 541.
[22] S. P. Mishra, S. N. Singh and D. Tiwary, Appl. Radiat. Isot., 42 (1991) 1177.
[23] S. P. Mishra and N. Srinivasu, Appl. Radiat. Isot., 43 (1992) 789.
[24] S. P. Mishra and N. Srinivasu, J. Radioanal. Nucl. Chem. Art., 162 (1992) 299.
[25] S. P. Mishra and N. Srinivasu and D. Tiwary, Appl. Radiat. Isot., 43 (1992) 1253.
[26] S. P. Mishra and V. K. Singh, Radiochim. Acta, 68 (1995) 251.
[27] S. P. Mishra and N. Srinivasu, Radiochim. Acta, 61 (1993) 47.
[28] S. P. Mishra and D. Tiwary, J. Radioanal. Nucl. Chem. Art., 170 (1993) 133.
[29] S. P. Mishra and V. K. Singh, Appl. Radiat. Isot., 46 (1995) 75.
[30] S. P. Mishra and V. K. Singh, Appl. Radiat. Isot., 46 (1995) 847.
[31] S. P. Mishra, V. K. Singh and D. Tiwary, Appl. Radiat. Isot., 47 (1996) 15.
[32] S. P. Mishra, V. K. Singh and D. Tiwary, Radiochim. Acta, 73 (1996) 49.
[33] S. P. Mishra, V. K. Singh and D. Tiwary, Appl. Radiat. Isot., 48 (1997) 435.
[34] S. P. Mishra, V. K. Singh and D. Tiwary, Appl. Radiat. Isot., 49 (1998) 1467.
[35] S. M. Hasany and M. H. Chaudhary, Appl. Radiat. Isot., 32 (1981) 899.
[36] S. M. Hasany and M. H. Chaudhary, J. Radioanal. Nucl. Chem. Art., 89 (1985) 353.
[37] S. M. Hasany and M. H. Chaudhary, J. Radioanal. Nucl. Chem. Art., 100 (1986) 307.
[38] S. M. Hasany, A. M. Shamsi and M. A. Rauf, Appl. Radiat. Isot., 48 (1997) 595.
[39] F. B. Walton, J. Paquette, J. P. M. Ross and W. E. Lawrence, Nuclear and Chemical Waste Management, 6 (1986) 121.
[40] L. J. Stryker and E. Matijevic, Adv. Chem. Ser., 79 (1968) 44.
[41] B. Gray and E. Matijevic, Colloids Surf., 23 (1987) 313.

[42] C. R. Paige, W. A. Kornicker, O. E. Hileman, Jr. and W. J. Snodgrass, Geochim. et Cosmochim. Acta, 57 (1993) 4435.
[43] T. Shahwan S. Suzer and H. N. Erten, Appl. Radiat. Isot., 49 (1998) 915.
[44] H. Catalette, J. Dumonceau and P. Ollar, J. Contaminant Hydrology, 35 (1998) 151.,
[45] E. Bascetin, H. Haznedaroglu and A. Y. Erkol, Appl. Radiat. Isot., 59 (2003) 5.
[46] V. E. Badillo-Almaraz and J. Ly, J. Coll. Interface Sci., 258 (2003) 27.
[47] R. Fujiyoshi, A. S. Eugene and M. Katayama, Int. J. Radiat. Applic. and Instrument. A., 43 (1992) 1223.
[48] S. A. Khan, R-u-Rehman and M. A. Khan, Waste Management, 15 (1995) 641.
[49] N. M. Nagy, J. Kónya and Z. Urbin, Colloids Surf. A, 121 (1997) 117.
[50] J. Kónya and N. M. Nagy, Colloids Surf. A, 136 (1998) 297.
[51] J. Kónya and N. M. Nagy, R. Király and J. Gelencsér, Colloids Surf. A, 136 (1998) 309.
[52] N. M. Nagy, J. Kónya and T. Budai, Colloids Surf. A, 138 (1998) 81.
[53] N. M. Nagy and J. Kónya, Colloids Surf. A, 137 (1998) 231.
[54] N. M. Nagy, J. Kónya and I. Kónya, Colloids Surf. A, 137 (1998) 243.
[55] N. M. Nagy, J. Kónya and Gy. Wazelischen-Kun, Colloids Surf. A, 152 (1999) 245.
[56] D.T. Karamanis, X. A. Aslanoglou, P. A. Assimakopoulos and N. H. Gangas, Nucl. Inst. and Methods in Physics Researh, B, 181 (2001) 616.
[57] C. A. Papachristodoulou, P. A. Assimakopoulos, N. H. Gangas and D. T. Karamanis, Microporous and Mesoporous Materials, 39 (2000) 367.
[58] E. R. Sylwester, E. A. Hudson and P. G. Allen, Geochim. et Cosmochim. Acta, 64 (2000) 2431.
[59] E. Tipping, Environmental Pollution, 94 (1996) 105.
[60] K. Tagami and S. Uchida, Chemosphere, 38 (1999) 963.
[61] A. Albrecht, U. Schultze, P. B. Bugallo, H. Wydler, E. Frossard and H. Flühler, J. Environment. Radioactivity, 68 (2003) 47.
[62] W. Dong, Z. Guo, J. Du, L. Zheng and Z. Tao, Appl. Radiat. Isot., 54 (2001) 371.
[63] S. Lofts, E. W. Tipping, A. L. Sanchez and B. A. Dodd, J. Environment.Radioactivity, 61 (2002) 133.
[64] L. Skipperud, D. Oughton and B. Salbu, The Science of The Total Environment, 257 (2000) 81.
[65] K. V. Ticknor, P. Vilks and T.T. Vandergraaf, Appl. Geochem., 11 (1996) 555.
[66] P. W. Warwick, A. Hall, V. Pashley, N. D. Bryan and D. Griffin, J. Contam. Hydrol., 42 (2000) 19.
[67] Y. J. Zhang, N. D. Bryan, F. R. Livens and M.N. Jones, Environment. Pollution, 96 (1997) 361.
[68] P. Reiller, V. Moulin, F. Casanova and C. Dautel, Appl. Geochem., 17 (2002) 1551.
[69] M. Bäckström, M. Dario, S. Karlsson and B. Allard, The Science of The Total Environment, 304 (2003) 257.
[70] J. Bors, Radiochim. Acta, 51 (1990) 139.
[71] J. Bors and R. Martens, J. Environ. Radioact., 15 (1992) 35.
[72] J. Bors, St. Dultz and A. Gorny, Radiochim. Acta, 82 (1998) 269.
[73] J. Bors, S. Dultz and B. Riebe, Appl. Clay Sci., 16 (2000) 1.
[74] J. Bors, A. Patzkó and I. Dékány, Appl. Clay Sci., 19 (2001) 27.
[75] M. T. Von Beymann, C. A. Ungerer and E. Suess, Chem. Geology, 70 (1988) 349.
[76] X. K. Wang, W. M. Dong, G. Wang and Z. Y. Tao, Appl. Radiat. Isot., 56 (2002) 765.
[77] A. J. Fairhurst and P. Warwick, Colloids Surf. A, 145 (1998) 229.
[78] X. K. Wang, W. M. Dong, X. X. Dai, A. X. Wang, J. Z. Du and Z. Y. Tao. Appl. Radiat. Isot., 52 (2000) 165.

[79]   Z. Y. Tao, W. Li, F. Zhang, Y. Ding and Z. Yu, J. Coll. Interface Sci., 265 (2003) 221.
[80]   R. A. Bultman, F. Szabó, R. F. Clayton and C. R. Clayton, Waste Management, 174 (1998) 191.
[81]   T. Rabung, T. Stumpf, H. Geckeis, R. Klenze and J. L. Kim, Radiochim. Acta, 88 (2000) 711
[82]   T. Stumpf, T. Rabung, R. Klenze, H. Geckeis and J. L. Kim, J. Coll. Interface Sci., 238 (2001) 219.
[83]   W. Runde, S. D. Conradson, D. Wes Efurd, N. P. Lu, C. E. Van Pelt and C. D. Tait, Appl. Geochem., 17 (2002) 837.
[84]   U. Alonso and C. Degueldre, Colloids Surf. A. , 217 (2003) 55.
[85]   C. Degueldre, A. Bilewicz, W. Hummel and T. L. Loizeau, J. Environ. Radioact., 55 (2001) 241.
[86]   J. M. Abril, J. Environ. Radioact., 41 (1998) 307.
[87]   J. M. Abril, J. Environ. Radioact., 41 (1998) 325.
[88]   R. El Mrabet, J. M. Abril, G. Manjon and R. G. Tenorio, Water Research, 35 (2001) 4184.
[89]   D. Mc Cubbin and K. S. Leonard, The Science of The Total Environment, 173-174 (1995) 259.
[90]   D. Mc Cubbin and K. S. Leonard, Marine Chem., 56 (1997) 107.
[91]   M. S. Quigley, P. H. Santschi, L. Guo and B. D. Honeyman, Marine Chem., 76 (2001) 27.
[92]   B. D. Honeyman and P. H. Santschi, J. Mar. Res., 47 (1989) 951.
[93]   B. D. Honeyman and P. H. Santschi, Environ. Sci. Technol., 25 (1991) 1739.
[94]   M. C. Stordal, P. H. Santschi and G. A. Gill, Environ. Sci. Technol., 30 (1996) 3335.
[95]   M. Smoluchowski, Z. Physik. Chem., 92 (1918) 129.
[96]   A. Buzagh, Kolloid-Z., 125 (1952) 14.
[97]   I. Mádi, T. Varró and A. Bolyós, Electroanal. Chem. and Interfacial Electrochem., 60 (1975) 109.
[98]   I. Mádi, Colloid Polym. Sci., 252 (1974) 337.
[99]   M. Párkányi-Berka and I. Mádi, Magy. Kém. Folyóirat, 92 (1986) 424.
[100] M. Párkányi-Berka and I. Mádi, Magy. Kém. Folyóirat, 92 (1986) 437.
[101] M. Párkányi-Berka and I. Mádi, Acta Chim. Hungarica, 125 (1988) 695.
[102] M. Párkányi-Berka and I. Mádi, Acta Chim. Hungarica, 125 (1998) 705.
[103] M. Párkányi-Berka and P. Joó, Colloids Surf., 49 (1990) 165.
[104] M. Párkányi-Berka, K. Csóbán and P. Joó, Kémiai Közlemények, 76 (1993) 83.
[105] K. Csóbán, M. Párkányi-Berka and P. Joó, Magy. Kém. Folyóirat, 102 (1996) 89.
[106] K. Csóbán and P. Joó. J. Radioanal. Nucl. Chem., 219 (1997) 19.
[107] K. Csóbán, M. Párkányi-Berka, P. Joó and Ph. Behra, Colloids Surf. A., 141 (1998) 347.
[108] K. Csóbán and P. Joó, Colloids Surf. A., 151 (1999) 97.
[109] B. Subotic, Powder Technol., 24 (1979) 35.
[110] M. A. A. Schoonen, N. S. Fisher and M. Wente, Geochim. et Cosmochim. Acta, 56 (1992) 1801.
[111] J. M. Moffett, Geochim. et Cosmochim. Acta, 58 (1994) 695.
[112] J. R. Lead, J. Hamilton-Taylor, W. Davison and M. Harper, Geochim. et Cosmochim. Acta, 63 (1999) 1661.
[113] N. P. Lu and C. F. V. Mason, Appl. Geochem., 16 (2001) 1653.
[114] S. H. Li and C. P. Jen, Waste Management, 21 (2001) 569.
[115] P. Vilks and M. H. Baik, J. Contaminant Hydrology, 47 (2001) 197.

[116] A. Möri, W. R. Alexander, H. Geckeis, W. Hauser, T. Schäfer, J. Eikenberg, Th. Fierz, C. Degueldre and T. Missana, Colloids Surf. A., 217 (2003) 33.
[117] J. S. Contardi, D. R. Turner and T. M. Ahn, J. Contaminant Hydrology, 47 (2001) 323.
[118] G. J. Moridis, Q. Hu, Y. S. Wu and G. S. Bodvarsson, J. Contaminant Hydrology, 60 (2003) 251.
[119] T. Missana, U. Alonso and M. J. Turrero, J. Contaminant Hydrology, 61 (2003) 17.
[120] T. Allard, P. Ildefonse, C. Beaucaire and G. Calas, Chem. Geology, 158 (1999) 81.
[121] D. H. Pilgrim and D. H. Huff., J. Hydrology, 38 (1978) 299.
[122] L. Maggi, V. Caramella-Crespi, N. Genova and S. Meloni, Appl. Radiat. Isot., 33 (1982) 217.
[123] O. S. Pokrovsky and J. Schott, Chem. Geology, 190 (2002) 141.
[124] P. Ciffroy, J.-M. Garnier and Lakhdar Benyahya, Marine Pollution Bulletin, 46 (2003) 626.
[125] D. Stephenson, W. A. J. Paling and A. S. M. De Jesus, J. Hydrology, 110 (1989) 153.
[126] B. R. Meyer, C. A. R. Bain, A. S. M. De Jesus and D. Stephenson, J. Hydrology, 50 (1981) 259.
[127] A. Widerlund, P. Roos, L. Gunneriusson, J. Ingri and H. Holmström, Chem. Geology, 189 (2002) 183.
[128] H. A. v. d. Sloot, G. D. Wals and H. A. Das, Anal. Chim. Acta, 90 (1977) 193.
[129] O. B. Woodhouse, G. Ravizza, K. Kenison Falkner, P. J. Statham and B. Peucker-Ehrenbrink, Earth and Planetary Sci. Letters, 173 (1999) 223.
[130] A. Koschinsky, A. Winkler and U. Fritsche, Appl. Geochem., 18 (2003) 693.
[131] B. Herut, T. Zohary, R. D. Robarts and N. Kress, Marine Chem., 64 (1999) 253.
[132] V. Hatje, T. E. Payne, D. M. Hill, G. Mc Orist, G. F. Birch and R. Szymczak, Environ. Internat., 29 (2003) 619.
[133] A. Turner, M. Nimmo and K. A. Thuresson, Marine Chem., 63 (1998) 105.
[134] A. Turner and G. E. Millward, Estuarine, Coastal and Shelf Sci., 55 (2002) 857.
[135] L. S. Wen, P. H. Santschi and D. Tang, Geochim. Cosmochim. Acta, 61 (1997) 2867.
[136] W. J. F. Standring, D. H. Oughton and B. Salbu, Environ. Internat., 28 (2002) 185.
[137] A. Koschinsky, U. Fritsche and A. Winkler, Deep Sea Research Part II., 48 (2001) 3683.
[138] J. M. Smoak, W. S. Moore, R. C. Thunell and T. J. Show, Marine Chem., 65 (1999) 177.
[139] R. Fujiyoshi, T. Gomei and M. Katayama, Appl. Radiat. Isot., 47 (1996) 165.
[140] V.L. Vilker, J. Lyklema, W. Norde and B. J. M. Verduin, Water Research, 28 (1994) 2425.
[141] V. A. Parsegian, Adv. Coll. Interface Sci., 16 (1982) 49.
[142] J. K. Dixon, R.C. Tilton and J. Murphy, Water Research, 8 (1974) 659.
[143] M. J. Gross, O. Albinger, D. G. Jewett, B. E. Logan, R. C. Bales and R. G. Arnold, Water Research, 29 (1995) 1151.
[144] R. Kretzschmar and H. Sticher, Phys. and Chem. of the Earth, 23 (1998) 133.
[145] D. G. Jewett, B. E. Logan, R. G. Arnold and R. C. Bales, J. Contam. Hydrology, 36 (1999) 73.
[146] A. J. Francis, J. Alloys and Compounds, 271-273 (1998) 78.
[147] P. Panak, B. C. Hard, K. Pietzsch, S. Kutschke, K. Röske, S. Selenska-Pobell, G. Bernhard and H. Nitsche, J. Alloys and Compounds, 271-273 (1998) 262.
[148] X. Dai, Z. Chai, X. Mao, J. Wang, S. Dong and K. Li, Anal. Chim. Acta, 403 (2000) 243.
[149] F. Sshue, J. Sledz, J. Molenat and J. Denamganou, Europ. Polym. J., 31 (1995) 1067.

[150] M. Boucher, N. Turzotte, V. Guillemette, G. Lantagne, A. Chapotot, G. Pourcelly, R. Sandeaux and C. Gavach, Hydrometallurgy, 45 (1997) 137.
[151] B. Doyle, G. J. Moody and J. D. R. Thomas, Talanta, 29 (1982) 257.
[152] A. De and N. Jaffrezic-Renault, Colloids Surf., 27 (1987) 159.
[153] H. Stechemesser, K. Volke and Th. Jung, Colloids Surf. A., 88 (1994) 91.
[154] S. P. Mishra, D. Tiwari and R. S. Dubey Appl. Radiat. Isot., 48 (1997) 877.
[155] N. Khalid, S. Ahmad, A. Toheed and J. Ahmed, Appl. Radiat. Isot., 52 (2000) 31.
[156] P. Joó, V. Holló, K. Varga and G. Hirschberg, Magy. Kém. Folyóirat, 106 (2000) 148.
[157] P. Joó and K. Varga, Colloids Surf. A., 193 (2001) 161.

2. ADSORPTION OF NEUTRAL SPECIES AND SURFACTANTS FROM AQUEOUS AND ORGANIC MEDIA (K. László-Nagy)

In this chapter, we focus on neutral small molecules, molecules of biological relevance, polymers and, finally surfactants, independently of their ionic or non ionic character. Neutral species represent a different group, as the interactions are mainly governed by van der Waals or dispersion forces. In the majority of cases monolayer coverage occurs. As already highlighted, the principal advantage of tracer techniques is their great sensitivity. A further advantage is the possibility of investigating competitive adsorption/desorption processes.

Although the isotopic effect can be or is neglected in most cases where the tracer technique is applied, the adsorption distribution coefficient of isotopically labeled large molecules can vary appreciably with the nature of the label and its position in the molecule. This feature can be used for the separation of compounds that differ only in the presence or position of an isotopic label by adsorption chromatography. A study of the effects of isotopic labeling on sample adsorption energy may provide fundamental insight into the nature of the adsorption process [1, 2]. As with any labeling procedure, both the statistical occurrence of the label and any specific chemical affinity must be considered [3].

2.1. Small molecules

Self-diffusion of water and simple alcohols in heteroionic forms of various zeolites was thoroughly investigated by radiotracer techniques as well. Tritiated water HTO was employed to follow water mobility in heteroionic zeolites of various porosities. The radiochemical technique used enables estimates to be made of the extent to which water molecules can be fixed in the zeolite framework. Natrolite, stilbite and analcime were compared. HTO mobility measurements showed that the migration of water cannot be restricted by ion exchange in the framework of natrolite with narrow pores [4]. This unsuitability for HTO containment stems from the fibrous nature of the natrolite structure, which expands by relaxation of the chains of Al-Si tetrahedra that are the building blocks of the framework. This contrasts with the lack of similar flexibility in the analcime framework, whose 3D nature enables restrictions to water movement to be greatly increased by the presence of small numbers of large cations. When open-structured zeolites, e.g., stilbite, are employed, the HTO can be immobilized by the introduction of small amounts of Co, Ni or Zn cations [5]. ^{14}C-n-decane was used in both the liquid and gas phase to follow the diffusion properties of the heteroionic forms of Linde Type A (LTA) zeolite [6]. Dyer *et al.* reported the use of ^{14}C-methyl, -ethyl and –n-butyl alcohol to follow molecular diffusion through Mg, Ca, Sr, Zn, Co, Ni, Li, K and H exchanged LTA crystallites. The diffusion rate of labeled methanol was very rapid in the case of Sr and Co exchanged LTA, but values of the self-diffusion coefficient D through NaNiA were reported in the range 10^{-11}-10^{-13} m^2/s. A two-stage diffusion process was observed in the case of ^{14}C-ethanol and n-butanol. From the energy barrier of butanol migration it was concluded that the mobility of unbound alcohol molecules contributed to the initial diffusion process and that the second process was associated with alcohol molecules in some form of solvatation shield around the cations present in the larger zeolite cages. According to the self-diffusion coefficients calculated from the early part of the diffusion (Table 1) close to ambient temperature, the total cation

Table 1
Self-diffusion coefficients D (10^{-19} m^2/s) for ^{14}C-ethanol and ^{14}C-butanol at 298 K in heteroionic zeolites (Ref. 7)

Zeolite	^{14}C-ethanol	^{14}C-butanol
MgNaA	12.8	4.91
CaNaA	24.0	0.91
SrCaA	8.4	1.33
ZnNaA	7.9	0.58
LiNaA	6.1	3.74
KNaA	2.6	2.18
HNaA	16.9	0.94
NaA	13.1	0.96

volume in the unit cell of LTA zeolite is a major factor in controlling the rates of progress of organic molecules through the zeolite cavities [7].

The surface properties of sparingly soluble oxides in contact with aqueous electrolyte solutions are of great importance from several aspects, including environmental problems, mineral processing technologies, catalysts, pigments, adsorbents, etc. Silica and metal oxides have several common features. For an accurate description of the oxide/electrolyte solution system in mixed solvents, information about the adsorption of the components of the solvent is required. The adsorption of n-alcohols is particularly interesting, especially from liquid mixtures. A serious problem related to this kind of study is the residual water on the surface of oxides that can be removed only by very drastic means. Fig. 1 shows a typical plot of methanol adsorption on silica measured by a radiotracer method as a function of pH and ionic strength [8]. The adsorbed organic co-solvent reduces the number of water molecules associated with the surface and thus the number of =SiOH groups responsible for the formation of the surface charge. The adsorption of methanol was found to be practically independent of the ionic strength.

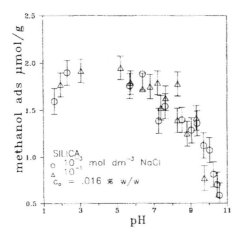

Fig. 1. Methanol adsorption on silica as a function of pH and ionic strength (Ref. 8)

The adsorption of acetone and allyl alcohol from solutions on Raney nickel was studied by three methods involving ^{14}C-labeled molecules, acetone and allyl alcohol [9]. In the case of acetone the catalyst was removed from the solution and the radiocarbon recovered by a chemical process. The adsorption was irreversible and as low as 0.1 mg per gram of nickel. In the case of allyl alcohol the adsorbed amounts were about 100 times larger, but the adsorption was readily followed by isomerization and hydrogenation. The resulting molecules have the same carbon chain, and radioactivity measurements made it possible to determine their total number in solution and at the catalyst surface either from individual samples or by direct counting of the effluent from the adsorption cell. The advantages of the different procedures applied were compared by the authors.

The effect of six different humic substances on the bioconcentration of ^{14}C-labeled pyrene by bacterivorous thread worms (nematodes) was investigated and sorption coefficients were calculated from reductions in bioconcentration. These data showed good agreement with sorption coefficients measured with a fluorescence quenching technique [10] indicating that the biologically available fraction of radiopyrene was comparable to the amount determined as being freely available by the fluorescence method.

The concentration of four pesticides 2,4-dichlorophenoxyacetic acid (2,4-D), atrazine, *O,O*-diethyl *S*-ethylthiomethyl phosphorodithioate (phorate) and *S*-*tert*-butylthiomethyl *O,O*-diethyl phosphorodithioate (terbufos) in soil solution during sorption experiments was measured using UV spectrophotometry, gas liquid chromatography, high performance liquid chromatography and radiotracer technique. The use of the latter three techniques gave a good estimate of the concentration of pesticides in soil solution, and can be used as methods for measurement of these pesticides in sorption experiments [11]. The pesticides remaining in soil could be quantitatively analyzed after extracting with a toluene- or dioxane-based scintillation solution.

The adsorption of the pesticides on clay minerals from their aqueous solution was investigated in the presence of surfactants [12]. ^{14}C-labeled pesticides 1,1'-dimethyl-4,4'-bipyridinium dichloride (paraquat) and biphenyl and tritiated dodecyltrimethylammonium bromide (DTAB) were used in the experiments. Adsorption measurements were completed by calorimetry and X-ray diffraction. It was found that the preferentially adsorbing di-cationic paraquat partially replaced the preadsorbed mono-cationic DTAB on the bentonite surface, while the adsorption of the hydrophobic biphenyl was enhanced in the presence of preadsorbed DTAB due to the hydrophobization of the bentonite surface. Adsorption of ^{14}C-labeled biphenyl in the presence of both tritiated DTAB and tritiated dodecyl octaethylene glycol ether was also investigated on four different layer silicates, kaolinite, illite, montmorillonite and bentonite [13]. The uptake of the biphenyl was limited by the specific surface area of the clay minerals and it clearly increased when the surface was weakly hydrophobized by surfactants, i.e., the surfactant concentration did not exceed the critical micelle concentration (cmc). Above the cmc solubilization by the surfactant and adsorption by the clay mineral compete for the pollutant molecules, reducing the adsorption of the biphenyl.

[14]C-labeled methyl parathion was used in a series of experimental investigations of pesticide retention on various fabrics including 100 % cotton and 65 % polyester/35 % cotton fabrics [14-17]. The effect of chemical finishing of the fibers as well as laundering was studied as well [15, 16]. Kinetic transport from contaminated fabric through a model skin was also measured in order to study the pesticide protective behavior of the fabrics. Adsorption and retention of the pesticide within the fabric structure was found to be an important mechanism for reducing dermal exposure [17].

Radiolabeled hair dyes *p*-phenylenediamine and bis-(5-amino-1-hydroxyphenyl)-methane were applied *in vitro* to analyse percutaneous penetration and dermal absorption of hair dyes on excised pig skin. *p*-[U-^{14}C]Phenylenediamine dihydrochloride and [rings-U-^{14}C]radiolabeled bis-(5-amino-1-hydroxyphenyl)-methane dihydrochloride were assessed under simulated use conditions and were analyzed in representative formulations including the specific conditions for oxidation hair dyes. To be able to differentiate between topically adsorbed and systemically available amounts, the bioavailability of the hair dyes is defined as the amount penetrated and/or remaining in the exposed skin after removing the stratum corneum. Less than

1% of the assessed topically applied dyes remained in the exposed skin after removing the stratum corneum in the presence of hydrogen peroxide, typically added to oxidation hair dyes prior to applications (bioavailable amount). 85–89% of the exposed hair dyes were found in washing solutions and are excluded as non-bioavailable. The adsorbed test substances removed with the horny layer of the exposed skin contained an additional 3% in both cases and should also be taken as non-bioavailable [18].

2.2. Biologically relevant molecules

Radiotracers, having practically the same physico-chemical and biological properties as the non labeled molecules, allow non-invasive means of detection and measurement of the fate of biologically relevant compounds inside more or less complex experimental systems or living organisms.

Haghshenas and coworkers studied the exchange between dissolved fats and fat crystals in dispersions by quantifying the rate of the exchange between the fats in solution and in crystals, using tripalmitin (PPP) in the β polymorphic state as model fat [19]. The dispersions contained PPP crystals and dissolved radiolabeled PPP (^{14}C-PPP, used as the probe) in medium chain triglyceride oil.

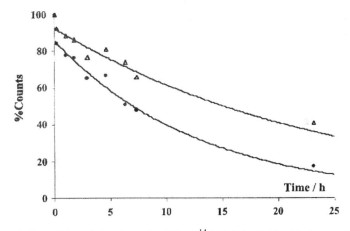

Fig. 2. The variation of the relative intensity (%) of ^{14}C-PPP in the liquid phase at 20 mg PPP in 5 cm^3 oil saturated by ^{14}C-PPP (Δ) and at 40 mg PPP in 5 cm^3 oil saturated by ^{14}C-PPP ($\bullet$). Solid lines exponential fit, suggesting first-order kinetics. (Ref. 19)

The exchange process between the dissolved ^{14}C-PPP and unlabeled PPP crystals was followed using a radio-detector. The effect of the crystal concentration on the exchange rate was investigated (Fig. 2). They found that the rate of the exchange is dependent upon both the dissolution of the surface of the crystals and the diffusion and crystallisation of the dissolved compounds. On average, the surface of the crystals renewed within about 5 h. A monolayer of PPP contains about 4–6 mg/m^2. The exchange between dissolved and solid fats is relatively fast if the crystals are small, but slow when crystals were large, which is important when the stability of semi-solid fat products is discussed.

The distribution between the aqueous and solid phase of 17β-estradiol in the ng-range in waste-water was investigated by spiking experiments with a radio-labeled hormone [2,4,6,7-^{3}H]-17β-estradiol. The distribution was measured by liquid scintillation counting. The major part of 17β-estradiol remained in the aqueous phase and did not adsorb to the sewage sludge. This behavior could not be predicted from the physico-chemical properties, such as the octanol/water partition [20], as they do not reflect the situation at very low concentrations. Adsorption isotherms have to be used to describe the equilibrium within the real conditions in the 0.5-50 ng/l range.

Baquey and co-workers [21] reported *in vitro* protein adsorption studies, applying simultaneous radioactivity measurement and imaging. The method they described allowed them to study competitive adsorption by using two or more radiotracers in the same experiment. Scintigraphic and autoradiographic imaging at a macroscopic and microscopic level, respectively, were applied to reveal homogeneity defects of either morphological or chemical origin on the material surfaces. They used radiolabeled monoclonal antibodies directed towards the proteins to characterize *in situ* the conformational state of the adsorbed protein and its functionality. Their *ex vivo* method made possible a non-invasive follow-up of phenomena occurring at a flowing blood – material interface. Kinetics of the deposition of several biological species could be easily assessed and embolization of platelet aggregates could be detected.

Protein adsorption on metal surfaces was studied by ellipsometry combined with the use of radiolabeled protein [22]. For these experiments the radioactively labeled β-lactoglobulin was prepared by reductive methylation using ^{14}C-formaldehyde [23]. As only the ε-amino groups of lysyl residues and the α-NH$_2$ terminus are methylated in the presence of sodium cyanoborohydride [24], three of the 15 lysine residues of β-lactoglobulin were labeled. Comparisons between electrophoresis, isoelectric focusing and heat precipitation of the labeled and unlabeled β-lactoglobulin showed that labeling had only a minor effect on the properties of the protein [25]. From the investigation of sequential and competitive adsorption of κ-casein and labeled β-lactoglobulin on hydrophyilic and hydrophobic surfaces it was concluded that κ-casein adsorbs after the saturation value of adsorption of β-lactoglobulin was reached. β-lactoglobulin, however, did not adsorb after the plateau value of κ-casein was established. When the two proteins were added simultaneously, the surface energy of the substrate influenced both the total adsorbed amount and the composition of the adsorbed layer [26].

The adsorption of human lysozyme to bare mica and mica hydrophobized with a $C_{20}NH_2/C_{20}OH$ Langmuir-Blodgett film was quantified by means of ESCA, radiotracer technique and total internal reflection fluorescence (TIRF). The human placenta lysozyme was labeled with carrier-free [125]I. Although the isotherm shapes were similar (Fig. 3), the absolute adsorption values obtained

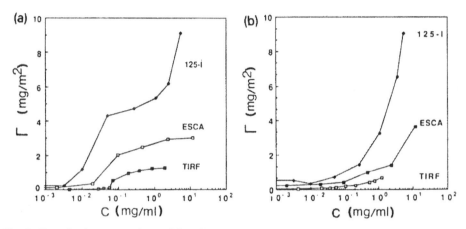

Fig. 3. Quantitative comparison of the adsorption isotherms obtained by ESCA, TIRF and radiotracer experiments. Adsorption on (a) hydrophilic and (b) hydrophobic mica. (Ref. 27)

from TIRF and radioactive measurements were considerably different from those obtained from ESCA. The lower apparent adsorption values obtained from TIRF using fluorescein-labeled lysozyme can be rationalized by a change in fluorescence quantum yield on adsorption or by preferential adsorption of unlabeled lysozyme. The apparent higher adsorption obtained from ^{125}I labeling measurements can be explained by an apparent sample area that is larger than the geometric area due to adsorption of labeled lysozyme in the interstices between mica layers. Relative binding and relative cleavage rate was larger on the hydrophobized mica than on bare mica, similar at low and high surface concentrations [27].

2.3. Polymers

Sensitive techniques are usually required to measure equilibrium polymer concentrations corresponding to the steeply rising part of the adsorption isotherm at very low equilibrium concentrations. Radiotracer techniques with their very high inherent sensitivity are therefore an excellent candidate for this purpose [3, 28].

A monograph on the behavior of polymers at interfaces has recently been released by Fleer *et al.* [29]. In this book the authors recall a rather early study on the adsorption of polymers with a bimodal distribution in chain length. In dilute solutions of polydisperse polymer, the long chains are found preferentially on the surface because they lose less translational entropy (per unit mass) in the solution, while they gain about the same (total) adsorption energy. Howard and Woods reported an experimental illustration. They studied the adsorption of a binary mixture of polystyrene samples of two different molar masses from cyclohexane to silica [30]. The total adsorbed amount was measured by the depletion method and the amount of the individual fractions by radiotracer techniques. The small molecules were found to adsorb only at low polymer concentrations, where the surface is unsaturated, but they are displaced when the amount of the long molecules is high enough to cover the surface. .

Schrader *et al.* [31] applied radiotracer techniques for the measurement of remaining poly(3-aminopropylsiloxane) coating over a glass surface as a function of exposure time to water. They discovered that only hot water could remove very quickly a first layer of coating.

Desorption and exchange kinetics of polymers were investigated by several methods [32]. Pefferkorn *et al.* measured the exchange between radiolabeled and non-labeled samples of polyacrylamide [33]. The exchange was found to be rather slow and proportional to the solution concentration of the polymeric displacer. This latter indicated that the rate of exchange was limited by the attachment of a chain to the surface. The same group also observed a slow exchange for polystyrene (M=360K) adsorbed on silica from carbon tetrachloride solution [34].

2.4. Surfactants

Surfactants are generally long-chain molecules containing both hydrophilic and lipophilic moieties, i.e., amphiphilic structures. They behave in a distinctive manner in water as they significantly modify interfacial properties already at very low concentration by adsorption. Provided that a suitably labeled sample is available or can be synthesized, solution radiotracer techniques can be used to obtain solution concentrations of surfactants. The method has the advantage of being highly sensitive and specific. It is particularly useful either for very low concentrations or for a single component of a mixed system [35-37].

Pearson *et al.* [35] investigated the adsorption of cetyl trimethylammonium bromide (CTAB) by H-kaolinite from aqueous and from mixed surfactant solutions. The uptake of anionic and non-ionic surfactants was measured using a radiochemical method based on the use of 1^{-14}C-CTAB (Fig. 4). Experiments were performed under conditions of varying mixture composition, pH and temperature. From the combination of these results with electrophoretic mobility and surface area data, they concluded that CTAB alone adsorbs initially by ion-exchange on to negatively charged sites on the H-kaolinite crystal faces. Van der Waals attractive forces are responsible for the build-up of subsequent layers with a changeover to adsorption, yielding a step-wise isotherm. In the presence of anionic surfactant sodium dodecyl sulfate (SDS) or sodium dodecyl benzene sulfonate, adsorption is reduced. The highly charged anionic micelles compete with the H-kaolinite for the available long

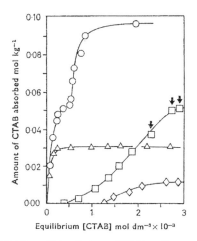

Fig. 4. The adsorption of CTAB on H kaolinite at 298 K. o from water;   from 0.01 M sodium dodecyl sulfate (SDS); ◊ from 0.1 M SDS; Δ from 0.2 % w/v Triton X-100 (TX-100). Arrows indicate turbid mixtures (Ref. 35)

cetyl trimethylammonium cation. In the presence of the non-ionic TX-100 and cetomacrogol 1000, cation adsorption on the crystal faces occurs preferentially, producing a monolayer: the corresponding isotherm in Fig. 4 is of Langmuir type. The competitive effect of the non-ionic micelles becomes marked only when the negative surface charge has been neutralized.

Adsorption of tritiated dodecyltrimethylammonium bromide (DTAB) and tritiated dodecyl octaethylene glycol ether was investigated on four different layer silicates, kaolinite, illite, montmorillonite and bentonite [13]. Unexpectedly, no great differences were found between the uptake of the two surfactants. However, the adsorption mechanisms up to a monolayer formation are fundamentally different (ion exchange and physisorption).

Dixit and Biswas studied the quantitative characterization of adsorption and flotation as a function of pH in case of the zircon – aqueous sodium oleate system in order to improve mineral flotation efficiency [36]. The oleate was labeled with ^{14}C. Both the adsorption (Fig. 5) and the flotation show a bell-shaped pH dependence, due to the different role of the hydroxyl ions in the acidic

$$(HOl + OH^- \underset{\text{I}}{\overset{\text{I}}{\rightleftharpoons}} Ol^- + H_2O \text{ and } Ol^- \underset{\text{II}}{\overset{\text{II}}{\rightleftharpoons}} Ol - M)$$

and alkaline

$$(M - OH + Ol^- \rightleftharpoons M - Ol + OH^-)$$

pH range (Ol^- and M represent the oleate anion and the mineral surface, respectively).

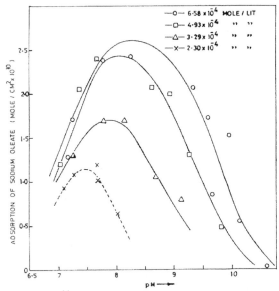

Fig. 5 Adsorption of sodium ^{14}C-oleate on zircon at various pH values. Experimental points (symbols) were fitted by assuming Langmuir-type adsorption (lines) (Ref. 116)

The pH variation of adsorption of the collector surfactant on the mineral surface was described quantitatively by equations developed on the basis of various adsorption isotherm equations. It was, however, found difficult to predict the bell-shaped pH variation of flotation recovery from the adsorption curve, although the two quantities are correlated.

A direct "slurry" method of measuring solution adsorption on a solid was developed by Nunn *et al.* [38] in order to overcome the difficulties of the conventional immersion method [39] for the measurement of adsorption at high concentration or low surface area. This method can also be useful for multicomponent mixtures where it might be desirable to minimize the composition changes that result from substantial adsorption. The increased amount of the component in question has to be measured in the separated solids, rather than the depletion of the equilibrium liquid phase. The method was tested for the adsorption of ^{14}C-labeled sodium laureate and n-hexanol from aqueous and n-decane solution, respectively. Radiolabeling makes it possible to detect the adsorbed material in the presence of the substrate. Fig. 6 illustrates that the slurry and conventional methods give identical values for the surface excess. By suitably arranging experimental conditions, one can obtain the adsorption of a component with an arbitrarily small change in solution composition.

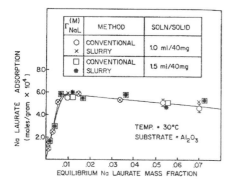 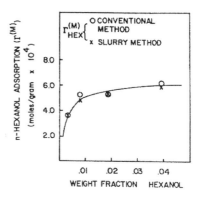

Fig. 6. Comparison of adsorption on alumina by conventional and slurry method at 30 °C. Left: Na laureate from aqueous solution; right: n-hexanol from n-decane solutions. (Ref. 38)

The adsorption of nonionic radioactive dimethyl dodecyl phosphine oxide on hydrophobic and hydrophilic polymer powder and fiber surfaces and carbon black was studied in the temperature range 1.5–40 °C. The isotherms measured were all of Langmuir type and no significant change in adsorption was observed as a function of temperature [40]. On polyethylene powder and nylon the maximum adsorption was reached near the cmc. The maximum adsorption on polyethylene powder and polyester fiber was similar to the value for carbon black, while the maximum adsorption capacity on microporous polystyrene beads was reached well above the cmc. It was only the half of the value found for carbon black. The adsorption capacity expected from the nitrogen adsorption surface area was also higher. Probably not all the surface was accessible to the surfactant. From the extremely high adsorption capacity and extremely long equilibration time in the case of nylon fibers it was concluded that the surfactant was soluble in the nylon.

The adsorption of doubly labeled radioactive ethoxylated dodecanol with an average of 9.5 ethoxy groups was studied by Gordon *et al.* on the surface of nylon and polyester fibers [41]. The alkyl chains were labeled with ^{3}H, and the ethoxy chains contained [14]C. The isotherm was of Langmuir type on polyester.

From the analysis of both the liquid phase and the adsorbent it was concluded that the surfactant components with the shorter ethoxy chains were preferentially adsorbed. Adsorption on nylon gave similar results as in [40], indicating that this surfactant is also soluble in nylon.

2.5. Isotopic exchange

Since the pioneering work of Paneth and Vorwerk [42], heterogeneous isotopic exchange processes have been widely used in the investigation of the solid surface, including either its structure or composition. This technique offers a rare opportunity to study adsorption phenomena from pure liquids on solid adsorbents. In these processes, generally the systems studied are in equilibrium except for the isotope ratio, which makes the exchange reaction heteroentropic and this causes some variation in the Gibbs free energy,

$$\Delta G = -T\Delta S,$$

if the isotope effect can be ignored. The experimental technique is different from the tracer technique. First the solid/liquid adsorption system is equilibrated and then the bulk liquid phase is changed to another one of the same chemical composition but of different isotope ratio. The time dependence of this isotope ratio is then investigated. A simple way to determine the kinetics of the isotope ratio in the liquid phase is the application of the radiotracer technique either by using fractional sampling or a flow system. Both single and multi-component liquid phases can be used. Care must be taken to replace the same volume of liquid at the same concentration as was removed.

Nagy *et al.* applied this technique on porous adsorbents [43-45] using ^{14}C labeled ethanol or benzene and ^{14}C-benzene – n-heptane mixtures as liquid phases. Two types of exchange curves were obtained. Linear exchange curves were obtained on macroporous systems, i.e., on adsorbents with less than 1-2 mm diffusion (or pore) length, they obtained linear exchange. A complex exchange curve was obtained on porous adsorbents of large specific surface area, where the total exchange is a superposition of processes with different rates. From the decomposition of the exchange curves they concluded that in the case of porous carbons the exchange rate is determined by adsorption – mainly chemisorption – interactions and by the diffusion in pores. The latter was illustrated by exchange curves obtained on activated carbon of the same quality but different grain size.

László *et al.* investigated isotopic exchange processes at 298 K with ^{14}C-labeled toluene on carbonized and activated carbons prepared from various raw materials [46]. Samples with identical average grain size were used for their experiment. The apparent half-life values $t_{1/2}$ deduced from the "fast" and "slow" section of the complex exchange curves obtained are compiled in Table

2. Comparison of the ratios of the carbonized and activated samples of the same raw material shows that activation increases the microporosity and thus the length of the diffusion path. The lower half-life ratio of the lignocellulosic carbon reflects the more open pore structure compared to the polymer-based carbons, as was concluded also from gas adsorption and small angle X-ray scattering (SAXS) measurements [46, 47].

Table 2
Comparison of isotopic molecular exchange and low temperature nitrogen adsorption data on carbonized and activated carbons from different precursors

Precursor	Form	$t_{1/2,fast}$ [min]	$t_{1/2,slow}$ [min]	$t_{1/2,slow} / t_{1/2,fast}$	BET surface area [m^2/g]	Total pore volume [cm^3/g]
Polyacrylo-	carbonized	7.5	62.3	8.3	7	0.004
nitrile	activated	3.8	90.1	23.6	544	0.278
Poly-ethylene-	carbonized	15.5	76.5	4.9	242	0.139
terephthalate	activated	5.0	124.9	24.8	1190	0.624
Cellulose	carbonized	11.6	91.9	7.9	415	0.212
	activated	5.7	64.8	11.3	1139	0.655

Among the various physical experimental methods developed for estimating the mean surface hydroxyl group concentration of porous silicas, isotopic exchange is a suitable technique for the quantitative determination of the surface hydroxyl groups. Tritium-labeled water is preferably employed instead of heavy water in isotopic molecular exchange reaction studies on porous silica because the radioactive tracer provides a simple means of detection. When the silica, outgassed at 473 K under vacuum, is exposed to tritium-labeled water the following exchange reaction takes place at the surface:

$$\equiv Si\text{-}OH + HTO \rightleftharpoons \equiv Si\text{-}OT + H_2O$$

A complete exchange of surface hydroxyl groups into OT groups was achieved when an excess of HTO was employed and the equilibration took place at high temperatures. Nevertheless, correction for the isotopic effect has to be taken into account [48].

Schlupen *et al.* carried out systematic studies [49] in order to investigate the conditions and the extent of irreversible sorption and sorption hysteresis of a hydrophobic compound on a mineral surface. Adsorption and desorption isotherms of pyrene on zeolite were established from various solvents, such as surfactant solution, 2-propanol, 2-propanol/water or a microemulsion. Different adsorption and desorption isotherms were found for pyrene physisorbed on the mineral surface. The dynamics of the equilibrated system were studied by radio-tracer experiments using pyrene[4,5,9,10-^{14}C] and isotopic exchange technique. After the preceding desorption process a portion of pyrene-containing solution was replaced by an equal amount of radiolabeled pyrene solution of the same concentration. No exchange could be detected on the time scale of the desorption experiments. The results were explained in terms of metastable equilibrium.

Isotopic exchange also happens in natural colloidal systems, e.g. in groundwater. During the circulation of water isotopic exchange with rocks should be mentioned. Such exchange takes place only for oxygen because rocks do not contain sufficient hydrogen to modify the isotopic composition of water appreciably. The kinetics is very slow, therefore high temperatures (e.g. 300 °C) are necessary for substantial modification [50].

LIST OF ABBREVIATIONS

cmc   critical micelle concentration
CTAB   Cetyl Trimethylammonium Bromide
DTAB   dodecyltrimethylammonium Bromide
ESCA   Electron Spectroscopy for Chemical Analysis
LTA   Linde Type A
PPP   Tripalmitin
SAXS   Small Angle X-ray Scattering
SDS   Sodium Dodecyl Sulfate
TIRF   Total Internal Reflection Fluorescence

REFERENCES

[1] L. R. Snyder, Principles of Adsorption Chromatography. The Separation of Nonionic Organic Compounds, Marcel Dekker, New York, 1968.

[2] P. D. Klein, Adv. Chromatogr., 3 (1966) 3.

[3] E. Parsonage, M. Tirell, H. Watanabe and R. G. Nuzzo, Macromolecules, 24 (1992) 2007.

[4] A. Dyer and H. Faghihian, Micropor. Mesopor. Mater., 21 (1998) 27.

[5] A. Dyer and H. Faghihian, Micropor. Mesopor. Mater., 21 (1998) 39.

[6] M. Bülow, P. Struve and L. V. C. Rees, Zeolites 5 (1985) 113.

[7] A. Dyer and S. Amin, Micropor. Mesopor. Mater., 46 (2001) 163.

[8] M. Kosmulski, J. Colloid Interface Sci. 156 (1993) 305.

[9] P. Fouilloux, M. Reppelin and P. Bussiere, Appl. Radiation Isotopes, 23 (1972) 567.

[10] M. Haitzer, H. G. Lohmannsroben, C. E. W. Steinberg and U. Zimmermann, J. Environ. Monitor., 2 (2000) 145.

[11] S. Baskaran and N. S. Bolan, Commun. Soil Sci. Plant, 29 (1998) 369.

[12] T. Rheinländer, E. Klumpp, M. Rossbach and M. J. Schwuger, Progr. Coll. Polym. Sci., 98 (1992) 190.

[13] T. Rheinländer, E. Klumpp and M. J. Schwuger, J. Disp. Sci. Techn., 19 (1998) 379.

[14] S. K. Obendorf, R. S. Kasunick, V. Ravichandran, J. Borsa and C. W. Coffman, Arch. Environ. Contam. Toxicol. 21 (1991) 10.

[15] I. Rácz, J. Borsa and S. K. Obendorf, Textile Res. J. 68 (1998) 69.

[16] E. Csiszár, J. Borsa, I. Rácz and S. K. Obendorf, Arch. Environ. Contam. Toxicol. 35 (1998) 129.

[17] S. K. Obendorf, E. Csiszár, D. Manefuangfoo and J. Borsa, Arch. Environ. Contam. Toxicol. 45 (2003) 283.

[18] W. Steiling, J. Kreutz and H. Hofer, Toxicol. in Vitro, 15 (2001) 565.

[19] N. Haghshenas, P. Smith and B. Bergenståhl, Coll. Surf. B, 21 (2001) 239.

[20] M. Fürhacker, A. Breithofer and A. Jungbauer, Chemosphere, 39 (1999), 1903.

[21] Y. F. Missirlis and W. Lemm (eds.), Modern Aspects of Protein Adsorption on Biomaterials, Kluwer, The Netherlands, 1991.

[22] T. Nylander, Proteins at the metal/water interface. Adsorption and solution behaviour, PhD Thesis, Lund, 1987.

[23] D. Dottavio-Martin and J. M. Ravel, Anal. Biochem., 87 (1978) 562.

[24] N. Jentoft and D. G. Dearborn, J. Biol. Chem., 254 (1979) 4359.

[25] B. O. Rowley, D. B. Lund and T. Richardson, J. Dairy Sci., 62 (1979) 533.

[26] T. Arnebrant and T. Nylander, J. Coll. Interf. Sci., 111 (1986) 529.

[27] C. G. Gölander, V. Hlady, K. Caldwell and J. D. Andrade, Colloids Surf., 50 (1990) 113.

[28] T. Sato, T. Tanaka and T. Yoshida, J. Polymer Sci., B 5 (1967) 947.

[29] G. J. Fleer, M. A. Cohen-Stuart, J. M. H. M. Scheutjens, T. Cosgrove and B. Vincent, Polymers at Interfaces, Chapman and Hall, London, 1993.

[30] G. J. Howard and S. J. Woods, J. Polym. Sci., A2 10 (1972) 1023.

[31] M. E. Schrader and A. Block, J. Polym. Sci., Part C (1971) 281.

[32] J. C. Dijt, Kinetics of polymer adsorption, desorption and exchange, PhD Thesis, Wageningen, 1993.

[33] E. Pefferkorn, A. Carroy and R. Varoqui, J. Polym. Sci., Polym. Phys. Ed. 23 (1985) 1997.

[34] E. Pefferkorn, A, Haouam and R. Varoqui, Macromolecules 22 (1989) 2677.

[35] J. T. Pearson and G. Wade, J. Pharm. Pharmacol., 24 Suppl. (1972) 132.
[36] S. G. Dixit and A. K. Biswas, AIChE Sym. Ser. 71 (1975) 88.
[37] R. D. Kulkarni and P. Somasundaran, Coll. Surf. 1 (1980) 387.
[38] C. Nunn, R. S. Schechter and W. H. Wade, J. Coll. Interf. Sci., 80 (1981) 598.
[39] G. Schay, Pure Appl. Chem., 48 (1976) 393.
[40] R. C. Mast and L. Benjamin, J. Coll. Interf. Sci., 31 (1969) 31.
[41] B. E. Gordon and W. T. Shebs, 5th Int. Congress on Surface Activity, Barcelona 3 (1968) 155.
[42] F. Paneth and W. Vorwerk, Z. Phys. Chem., 101 (1922) 445.
[43] L. G. Nagy, G. Török, H. Mester, T. Zanati and L. Puskás, Kem. Kozl., 49 (1978) 309.
[44] L. G. Nagy, Period. Polytech. Chem. Eng., 20 (1976) 25.
[45] L. G. Nagy, G. Fóti and G. Schay, J. Coll. Interf. Sci. 75 (1980) 338.
[46] K. László, A. Bóta and L. G. Nagy, Carbon, 38 (2000) 1965.
[47] A. Bóta, K. László, L. G. Nagy and T. Copitzky, Langmuir, 13 (1997) 6502.
[48] K. K. Unger: Porous Silica. Its properties and use as support in column liquid chromatography, Elsevier, Amsterdam, 1979.
[49] J.Schlüpen, F.-H. Haegel, J. Kuhlmann, H. Geisler and M. J. Schwuger, Coll. Surf. A 156 (1999) 335.
[50] R. Bowen, Groundwater. Elsevier, London, 1986 2nd Edition

Radiotracer Studies of Interfaces
G. Horányi (editor)
© 2004 Elsevier Ltd. All rights reserved.

Chapter 6

Study of liquid/fluid interfaces

G. Horányi

Institute of Materials and Environmental Chemistry, Chemical Research Center, Hungarian Academy of Sciences, PO Box 17, H-1525, Budapest, Hungary

Liquid/gas and liquid/liquid interfaces belong to this group of systems.

An outstanding problem in the area of interfacial phenomena is the determination of molecular ordering and structure at liquid/gas and liquid/liquid interfaces. These interfaces play an important role in many chemical and biological systems in addition to being interesting model systems for the study of the statistical physics of interfaces and membranes. The industrial utilization of materials with liquid/fluid interfaces, either in processing or in the finished product, is vast.

Biopolymers, especially proteins and hydrocolloids, are being used in foods to stabilize dispersed systems, such as food foams and oil-in-water emulsions, and also to control rheological and textural properties of foods. Because proteins are amphiphilic and exhibit a high tendency to adsorb to air/water and oil/water interfaces, they primarily function as emulsifiers in foods.

It is known that some hydrocolloids, such as gum arabic, guar gum, gum tragacath, and methylcellulose, adsorb to air/water and oil/water interfaces and exhibit surface/interfacial activity as evidenced from a reduction in the interfacial tension.

Direct determination of the amount of species adsorbed per unit area of liquid/gas or liquid/liquid interface, is not generally undertaken because of the difficulty of isolating the interfacial region from the bulk phase. Instead, the amount of material adsorbed per unit area of interface is calculated indirectly from surface or interfacial tension measurements. A plot of the surface (or interfacial) tension as a function of (equilibrium) surfactant concentration in one of the liquid phases, rather than an adsorption isotherm, is generally used to describe adsorption at these interfaces. From such a plot the excess amount of the surfactant adsorbed per unit area of interface can be calculated by the Gibbs adsorption equation.

Assuming that activities may be given by concentrations (dilute solutions), the surface excess may be obtained from the Gibbs Eq. (1)

$$\Gamma = -\frac{1}{nRT}\frac{d\gamma}{d\ln(c)} \tag{1}$$

where $n = 1$ for non-ionic surfactants, neutral molecules or ionic surfactants in the presence of excess electrolyte, and $n = 2$ for 1:1 ionic surfactants, assuming electrical neutrality of the interface. Here Γ is the equilibrium surface excess, R the gas constant, T the Kelvin temperature and c the bulk surfactant concentration. The adsorption isotherm, Γ vs. c, can therefore be obtained by measuring the surface tension γ at various bulk surfactant concentrations.

The first direct tests of the surface densities predicted by the Gibbs adsorption isotherm were made with McBain's microtome method and with foam concentration methods.

The first applications of radiotracer technique were of indirect nature measuring the activity of samples collected from the surface layer of solutions. Two of these methods will be mentioned here.

The application of the so-called platinum ring method was suggested by Hutchinson [1] for the study of the adsorption of [22]Na labeled sodium dodecyl sulfate. Radioactive sodium, [22]Na, was exchanged with some of the normal sodium in a pure sample of the detergent from which the aqueous solutions were prepared.

Platinum wire of about 0.5 mm diameter was bent into the shape of a circle provided with a stirrup for convenience in handling. The ring was immersed, after gentle flaming, into a solution of the labeled species, and then carefully raised through the surface, so that a film of the solution was formed on the ring. Once free of the surface of the solution the ring plus film was rapidly transferred to a weighing bottle and the weight of the solution in the film determined. The contents of the weighing bottle were rinsed out and the quantity of radiosodium in the rinsings determined in the conventional way.

By comparing the count of the film material with an equal weight of the solution *drawn from the interior of the bulk solution* the excess sodium present in the surface of the film was readily determined.

Assuming that the number of sodium ions adsorbed on the surface of the film is exactly the same as the number of dodecyl sulfate ions, as seems reasonable, the quantity of sodium dodecyl sulfate adsorbed per unit area was calculated.

Another attempt to apply radiotracer technique was reported by Shinoda [2] who elaborated the "bubble" or "foam" method, using ^{35}S labeled sodium pdodecylbenzene sulfonate. The main characteristics of the method are as follows:

Aqueous solutions of surfactant mixture were prepared by dissolving known amounts of radioactive and radioinactive surfactants accurately weighed on a microbalance. A solution of known concentration and composition was introduced into an apparatus where, bubbles were generated by the action of a circulating pump. The size and the number of bubbles generated per minute were controlled by changing the position of a vibrator, and the frequency of the intermittence of the current. Air was passed through a soda lime absorbent at the beginning of the experiment to eliminate errors, which might be caused by carbon dioxide in the air.

Bubbles attained adsorption equilibrium while they moved from one end to the other end of a nearly horizontal glass tube. The foams were well drained while they gradually moved upwards through a glass tube. This glass tube was inclined to facilitate drainage of the foams. The well-drained foams were collapsed while they expanded or contracted at the end of the tube and collected in the small collector.

Fixed volumes of original solution and of collapsed foams were taken with micro pipet. The samples were dried and their radioactivity was determined.

A really direct in situ method is offered by the surface count radiotracer technique. The species whose adsorption is to be measured should be labeled by an isotope emitting soft β-radiation (^{3}H, ^{14}C, ^{35}S). A survey of the relevant literature before 1973 can be found in the review by Muramatsu [3]

The first attempts to use labeled compounds for the in situ measurement of adsorption at liquid/gas interface date back to the end of forties and beginning of the fifties of the last century [4-7]. The principle of the method is very simple. A radiation detector is placed above the solution (typically approx. 0.5 mm) and because of strong absorption of β-radiation by the solution the signal reaching the detector is dominated by the signal from the surface region. In order to determine the extent of the adsorption three measurements are required: (i) the measurement of the radiation coming from the sample (I_t); (ii) determination of the radiation intensity of a calibration specimen to scale the signal; and (iii) determination of the background to take account of the radiation from the bulk solution (I_b).

The background sample must be one in which the radioactive material is *not* surface active, i.e. it is uniformly distributed through the solution. If the energy of the radiation is not too high, then the I_b radiation intensity is

$$I_b = \frac{\alpha A c s}{\mu} \qquad (2)$$

where α is a geometric factor, A is the exposed area of solution, c is the concentration of the emitter in the solution, s is the specific activity of the solute, and μ is the absorption coefficient. Because of its surface activity the surfactant itself cannot be the solute used in this background measurement unless the solvent is changed. However, a change of the solvent may alter μ. For the sample solution we can write the total activity, I_t, as the sum of the activity from the surface excess, I_e, and from the bulk, I_b.

$$I_t = I_e + I_b \tag{3}$$

where

$$I_e = \alpha' \Gamma \tag{4}$$

where Γ is the surface excess and α' is a proportionality factor to be determined by calibration ($\alpha' = \alpha As$). A number of means have been used to calibrate the experiment, the most typical of which is to measure the activity of a spread monolayer labeled similarly to the surfactant. Sometimes, by adding large quantities of electrolyte to the sub-phase, an insoluble layer of the actual surfactant under study may be used for calibration.

It follows from Eqs (2), (3) and (4) that the I_e/I_b ratio depends on the ratio of Γ and c/μ. μ strongly depends on the energy of the β radiation as shown in Fig. 1. (E_{max} values of ^3H, ^{14}C and ^{35}S are 0.018, 0.155 and 0.167 MeV respectively.)

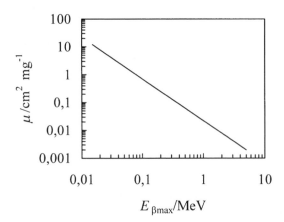

Fig. 1. Relationship between massabsorption coefficient (μ) and the maximal energy of β-radiation ($E_{\beta max}$).

In Table 1 data characteristic for the radiation emitted by ^{3}H and ^{14}C are compared.

Table 1
Comparison of characteristic data of the radiation emitted by ^{3}H and ^{14}C.

Tracer	Half-life year	Max. energy keV	Max. energy range mg cm^{-2}	μ_m cm^2 mg^{-1}
^{3}H	12.26	17.95	0.5	11.5
^{14}C	5568	155	28	0.26

Thus at a given solute concentration the background radiation is the lowest in the case of tritiated compounds while for ^{14}C (or ^{35}S) labeling there is an increase of one order of magnitude in the background radiation. Thus in the case of ^{14}C labeled solute at a concentration of 10^{-3} mol dm^{-3} c/μ is about 3×10^{-9} mol cm^{-2} (aqueous solution: $\rho \approx 1$ g cm^{-3}) i.e. almost an order of magnitude higher value than that expected for the maximum in the case of monolayer adsorption of an organic species.

In order to avoid the synthesis of surface active agents in radioactive form an indirect method was suggested by Judson et al [7] more than fifty years ago. Their attempt is an interesting example of the early application of the "indirect radiotracer method".

Their approach was based on the very idea that in the case of ionic agents the number of counter-ions adsorbed should be equivalent to the number of adsorbed surface-active ions. Thus the measurement of the adsorption of the counter-ions would offer a relatively simple method of estimating of the adsorption of different surface-active agents without requiring the preparation of their labeled form.

The adsorption of ^{35}S labeled sulfate ions was studied by the "surface count" method in the presence of various cationic, anionic and non-ionic, surface-active agents:

di-*n*-octylsodium sulfosuccinate,

γ-stearamidopropyl-2-hydroxyethyldimethylammonium chloride and sulfate

di-2-ethylhexylsodium sulfosuccinate

cetylpyridinium chloride

Interesting results were obtained concerning the competitive adsorption of different species involving the competition between sulfate and chloride ions. (As to the details see ref [7].)

The application of the in situ "surface count" method opened an important way to study the kinetics of adsorption in the interface. This aspect was discussed in detail in one of the first papers based on the application of the method [6]. Fig. 2 shows, for instance, the count-rate vs. time curve obtained in the course of the adsorption ^{35}S labeled di-n-octyl sodium sulfosuccinate into the aqueous solution/air interface from a solution containing 9×10^{-7} mol dm^{-3} labeled species.

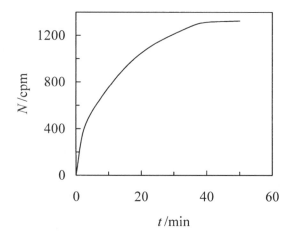

Fig. 2. Count-rate vs. time curve obtained in the course of the adsorption of ^{35}S labeled di-n-octyl sulfosuccinate (c = 9×10^{-7} mol dm^{-3}).

In Ref. 6 on the basis of the assumption that diffusion is the rate-controlling step, the diffusion coefficient of the labeled compound was calculated.

The radiotracer techniques presented and discussed above were applied for the study of the liquid/air (water/air) interfaces. Although the extension of these methods to liquid/liquid interfaces encounters significant difficulties there are some sporadic examples in the literature dealing with this question. Graham and Phillips in the series of their papers [8-10] dealing with the adsorption of ^{14}C labeled proteins report an interesting version of the application of the surface count method to the sorption of proteins into oil/water interface [11].

The adsorption of ^{14}C-labeled species at the oil/water interface was monitored by a scintillation-counting technique. The oil phase was toluene containing 1×10^{-1} wt% of a non-surface-active scintillator [BBOT, 2,5-di(5-butyl-2-benzoxazolyl) thiophene]; the oil-soluble BBOT was excited by protein molecules at and close to the interfacial excess region. A photomultiplier tube was positioned close to the oil surface to count the scintillations, and after suitable calibration, the concentration of protein adsorbed at the oil/water interface was determined.

Another interesting new methodology for studying labeled protein adsorption at oil/water interfaces was elaborated recently by Sengupta and

Damodaran [12] extending their technique used for the investigation of sorption phenomena of labeled proteins at air/water interface [13].

The measuring system was set up such that the interfacial tension as well as the signal originating from the labeled species could be monitored simultaneously during the entire course of the experiments.

A 1000 Å thick triglyceride oil film (triolein and trilinolein) was created on the surface of the water phase.

A stock solution of the oil in hexane-chloroform solvent was spread slowly on water surface. The thickness of the triolein film thus formed after evaporation of the solvent was estimated to be about 1000 Å. The film was allowed to equilibrate and the solvent to evaporate for another 1-2 hours.

The protein concentration at the oil/water interface was monitored by measuring interfacial radioactivity using a rectangular gas proportional counter at a distance of about 4-5 mm from the water surface.

An interesting version of the radiotracer technique was elaborated recently by Schlenoff and his coworkers for the investigation of polymer adsorption and generally, for the radiochemical study of polymers [14-17]. The water/polymer interface to some extent can be considered as a special case of the liquid/liquid interface (water/amorph solid state).

The first important step in these studies was the combination of ion exchanger and plastic scintillator. The surface of polystyrene-based plastic scintillator was sulfonated to yield an ultrathin cation-exchanging layer of well-defined geometry. Radioisotopes absorbed into this layer cause efficient scintillation, permitting in situ measurement of ion content, selectivity coefficients, and exchange kinetics. Initial experiments were focused on absorption measurements of $^{45}Ca^{2+}$ as a probe ion. The study of the adsorption of other species can be carried out via the displacement of adsorbed labeled Ca^{2+} species (indirect technique) or labeling the species to be studied and measuring the sorption of these species directly.

In the next step the kinetics of the adsorption of radiolabeled poly(styrene sulfonate) on the scintillating polymer was studied.

Finally the measurement of the adsorption of a radiolabeled random hydrophylic/hydrophobic copolymer at liquid/liquid interface was carried out. Adsorption of a water-soluble ^{35}S labeled random copolymer of styrenesulfonate and tert-butylstyrene (20 mol %) to the toluene/water interface was followed in situ using the organic phase as a scintillator.

Closing the survey on the experimental methods two important fields of the application of the radiotracer technique will be presented in the following sections:

a) investigation of adsorption of sodium dodecyl sulfate and similar alkyl sulfates

b) study of the adsorption of proteins.

1. INVESTIGATION OF THE ADSORPTION OF SODIUM DODECYL SULFATE AND SIMILAR COMPOUNDS

One of the first attempts in 1949 to determine the adsorption of sodium dodecyl sulfate was connected with the application of the Pt-ring method mentioned in the previous section [1]. The results obtained, however, did not attain those furnished by the microtome method. This problem was presumably connected with the very technique and the application of ^{22}Na isotope. Two years later Anianson used the "surface count" method carried out with ^{35}S labeled dodecyl sodium sulfate and hexadecyl sodium sulfate [4].

The adsorption of the mixture of the two components was studied in detail. In addition, the coadsorption of ^{45}Ca labeled calcium ions with sodium dodecyl sulfate was investigated. (To the problem of Ca^{2+} coadsorption we return later.) The selective adsorptivity at air/solution interface was studied by Shinoda [2] in the following systems:

sodium p-dodecylbenzene sulfonate^{35}S – sodium dodecyl sulfate,
potassium p-dodecylbenzene sulfonate^{35}S – potassium dodecanoate,
potassium p-dodecylbenzene sulfonate^{35}S –potassium tetradecanoate,
potassium hexadecanoate1-^{14}C – potassium dodecanoate and
potassium hexadecanoate1-^{14}C – potassium tetradecanoate.

Pioneering work in the systematic study of the adsorption of alkyl compounds was carried out by Tajima and coworkers in the seventies. They studied the adsorption of several tritium labeled compounds at the gas/solution interface. As a first step they elaborated a reliable version of the "surface count" method for the accurate mesurement of the adsorption of tritiated sodium dodecyl sulfate [18]. This technique offers a real in situ mesurement of the adsorption.

The validity of the Gibbs adsorption equation was checked and confirmed. It was demonstrated that in the absence of added inorganic salt the adsorption can be described by the Gibbs adsorption isotherm applied for the aqueous binary strong electrolyte solution when the activity in place of concentration is used. Adsorption isotherm and adsorption kinetics were determined. An equation of state for the ionic adsorbed film was proposed.

This work was followed by the study of the effect of excess salt on the adsorption of SDS [19]. The adsorption isotherm of SDS was determined in the presence of a great excess of NaCl (i.e. an salt containing common cation with SDS). The validity of the Gibbs equation for this case was also confirmed. As a continuation of this work the effect of NaCl concentration on the adsorption behavior of SDS was investigated. The validity of the Gibbs adsorption isotherm was found on the basis of the assumption that the adsorption of Cl⁻ ions does not take place.

The effect of the pH on the adsorption of an ampholytic surfactant, N-dodecyl-β-alanine at nitrogen/solution interface was studied in detail by the same group arriving to the same conclusion that there is a good coincidence of data obtained by calculation and direct radiotracer measurements.

Although the experimental results reported by Tajima and coworkers are thirty years old they serve as a basis for the experimental validation of theoretical considerations presented recently [20-23]. A new theoretical model based on the concept that the air/water interfacial layer of adsorbed ionic surfactants constitutes a nonautonomous phase was presented by Berlot et al. [20]. The analysis of the consequences of the model was carried out in the light of the data by Tajima concerning the effect of sodium chlorid on the adsorption of SDS.

The comparison of the applicability of van der Waals and Frumkin isotherms for SDS at various salt concentrations was discussed by Kolev et al [22] fitting the data reported by Tajima's group to the data calculated on the basis of Frumkin and van der Waals models.

Also in the case of models reported in refs [21] and [23] the authors rely on the data published in [18,19, 24, 25].

Some of the results obtained for the adsorption maximum of SDS by various authors using different labeling are shown in Table 2.

Table 2
Extent of adsorption of sodium dodecyl sulfate obtained by using radiotracer technique

Label	Extent of adsorption at saturation, 10^{10} Γ mol cm^{-2}	Reference
^{3}H	3.19	18
^{3}H	5.0	26
^{3}H	5.89	27
^{3}H	5.4	28
^{3}H	3.1	29
^{35}S	3.7±1	30
^{35}S	3.35	31
^{35}S	7.5	32
^{35}S	4.8±1	33

It should be noted that the problem of the rate of hydrolysis of ^{35}S labeled SDS according to the reaction

$$C_{12}H_{25}OSO_3^- + H_2O \rightarrow C_{12}H_{25}OH + HSO_4^-$$

was studied by Muramatsu [34] determining the rate of the formation of $^{35}SO_4^{2-}$. A pH dependent slow reaction was found in the pH range from 1 to 4 at $1-4\times10^{-3}$ mol dm^{-3} SDS concentrations.

The authors arrived at the conclusion that discrepancies connected with time dependence of surface tension of SDS solution cannot be ascribed to the hydrolysis at least under the conditions of their experiments.

The adsorption of ^{45}Ca labeled calcium ions at the air/solution interface of an aqueous solution of sodium dodecyl sulfate containing a small amount of calcium ions was studied by Shinoda [35]. A preferential adsorption of Ca^{2+} ions over Na^+ ions was found. The selectivity (S) defined by the equation

$$\frac{X_{ads}}{1 - X_{ads}} \Big/ \frac{X_{bulk}}{1 - X_{bulk}}$$ attains 200, where X_{ads} and X_{bulk} are the fraction of Ca^{2+}

species referred to the total amount of the cationic species in the adsorption layer and bulk liquid phase, respectively. (1 - X) gives the fraction of the Na^+ ions as only Ca^{2+} and Na^+ ions are present in the system studied.

The adsorption of calcium at the air-solution interface from aqueous solutions of a non-ionic agent, octylglucoside, was also determined. It was found that there is some adsorption of calcium ions at the monolayer of the non-ionic agent, but much less, as compared with that of the ionized layer. This situation is well visualized by the data presented in [35]. In a solution of dodecyl sulfate (1×10^{-2} mol dm^{-3}) containing 1×10^{-6} mol dm^{-3} of calcium chloride the ratio of calcium ions to surface active ions in the adsorbed layer was about 0.1, whereas in solution of octyl glucoside (2.5×10^{-2} mol dm^{-3}) containing 1×10^{-2} mol dm^{-3} of calcium chloride the ratio of calcium to surface active molecules at the surface was again about 0.1. The ability of the adsorbed layer of these two solutions to adsorb calcium differs by a factor of about 10^4. The ratio of calcium ions to dodecyl sulfate ions at the surface was about 100 times greater than the corresponding ratio in the bulk of the solution, whereas the ratio of calcium ions to octyl glucoside at the surface was about half of that in the solution.

It should be noted that Aniansson as early as in 1951 clearly demonstrated the adsorption competition between Ca^{2+} and Na^+ ions [4]. He studied the behavior of a system containing sodium dodecyl sulfate (SDS) and $CaCl_2$ labeled with ^{45}Ca. Changing the concentration of former species at a fixed very low $CaCl_2$ concentration (10^{-6} mol dm^{-3}) and measuring the adsorption of Ca^{2+} ions a remarkable competition between Na^+ and Ca^{2+} counter-ions was found. In Fig. 3. the $\Gamma_{Ca^{2+}}$ vs. SDS concentration curve, created on the basis of data reported in [4] is shown. It may be seen from this curve that Na^+ ions displace Ca^{2+} ions only at high $c_{Na^+}/c_{Ca^{2+}}$ concentration ratios.

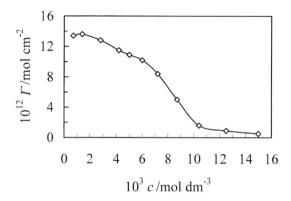

Fig. 3. Effect of SDS concentration on the adsorption of Ca^{2+} ions ($c_{Ca^{2+}} = 10^{-6}$ mol dm^{-3}).

This result calls the attention to the problem that impurities involving, for instance, Ca^{2+} counter-ions in very low concentration could modify the behavior of SDS, and falsify certain experimental data.

In [36] the co-adsorption of tritium labeled N-dodecyl-β-alanine (DBA) and various ^{35}S labeled alkylsulfates (SAS as sodium decyl-, dodecyl and tetradecyl sulfates) was studied at the surface of their equimolar solutions.

The results obtained confirm the assumption that equimolar complex of DBA and SAS is present in the bulk solution phase and in the surface layer.

2. STUDY OF THE ADSORPTION OF PROTEINS

In the case of proteins and other macromolecules both the equilibrium and kinetic measurements are equally important. The importance of the latter follows from the very fact that the adsorption of protein and other macromolecular species is often strongly history dependent owing to the slow relaxation of non-equilibrium structures. The rate at which molecules adsorb is exquisitely sensitive to, and thus a sensitive measure of the interfacial structure; adsorption kinetics may therefore serve both to identify and to quantify history dependence.

The adsorption of ^{14}C labeled proteins was studied during the last decade by a number of authors. The radiolabeling in most cases was carried out by reductive methylation of the amino groups using cyanoborohydride and ^{14}C-formaldehyde. In [8-10] [1-^{14}C]acetylated proteins were used. Table 3, without attempting completness, gives a survey on proteins involved in radiotracer sorption studies carried out at fluid/fluid interfaces.

Table 3.
Proteins involved in radiotracer studies at fluid/fluid interfaces

Name	References
α(s)-casein	37
α-lactalbumin	12, 38, 39, 40, 41
β-casein	8, 9, 10, 12, 13, 37, 42, 43, 44, 45
β-lactoglobulin	9, 38, 39, 41, 46
bovine serum albumin	8, 9, 10, 47
human immunoglobulin G	48, 49
lipases	50
lysozyme	8, 9, 10, 12, 13, 42, 45, 51, 52

It may be seen from this table that β-casein and lysozyme are the most frequently studied proteins.

In a study of the competitive adsorption of β-casein and gum arabic glycoprotein (GAGP) at the air/water interface the reductive methylation procedure was followed for labeling [43]. From the specific radioactivity values of the labeled compounds the stoichiometry of the labeling could be calculated.

It was found that the specific radioactivity values correspond to incorporation of 1.9 and 2.7 moles of $^{14}CH_3$ per mole of β-casein and GAGP, respectively. This means that, in each case one amino group has been modified to $-N(^{14}CH_3)_2$ derivative. It was assumed and experimentally confirmed that the reductive alkylation of amino groups in proteins using ^{14}C-labeled formaldehyde does not alter protein structure and the net charge of the protein is also not altered because the pK of $-N(^{14}CH_3)_2$ is similar to that of $-NH_2$. Therefore, ^{14}C-labeling does not exert influence on the adsorption behavior of β-casein and GAGP.

Competitive adsorption of β-casein and GAGP from solution mixtures was studied as follows: to monitor adsorption of the β-casein component, stock solutions of $[^{14}C]$-β-casein and unlabeled GAGP were mixed with the buffer to the required final bulk concentration ratio. The solution was poured into a Teflon through ($19 \times 5.2 \times 1.27$ cm^3) and the surface of the solution was swept using a capillary tube attached to an aspirator. Adsorption was then allowed to proceed from the bulk phase to the air/water interface. Although adsorption of both β-casein and GGP occurred simultaneously, the measured surface radioactivity corresponded only to the concentration of radiolabeled β-casein in the interface. To determine the surface concentration of GAGP at the interface, the experiment was repeated by mixing stock solutions of $[^{14}C]$-GAGP and unlabeled β-casein with the buffer. In the competitive adsorption studies reported, the bulk concentration of GAGP was fixed at 39 μg ml^{-1} and the concentration of β-casein was varied from 0.2 to 6.0 μg ml^{-1}. In most experiments, neither the surface concentration nor the surface pressure changed significantly after 10 h of adsorption. In order to be consistent, equilibrium adsorption values were determined from surface radioactivity values after 16-24 h of adsorption.

Adsorption of proteins and polymers at an interface is generally assumed to follow a Langmuir-type adsorption model when the bulk protein/polymer concentration is below the critical concentration at which formation of a saturated monolayer occurs. The equilibrium interfacial concentration, Γ_{eq}, under these conditions is given by the relationship

$$\Gamma_{eq} = \frac{KC}{1 + KC(1/\Gamma_{sat})} \tag{5}$$

where K is the equilibrium constant and C is the bulk concentration of proteins. The adsorption isotherms of β-casein and GAGP were analyzed according to Eq. (1). It is often argued that because polymers undergo structural rearrangements at interfaces, they cannot follow a Langmuir type adsorption model. However, it was found that adsorption of these polymers to the air/water interface reasonably follow the Langmuir model. The equilibrium binding constants were 1.25 and 0.174 cm, respectively, for β-casein and GAGP, indicating that the affinity of GAGP to the air/water interface was about an order of magnitude smaller than that of β-casein.

According to the ideal Langmuir model for competitive adsorption of two non-interacting polymers at an interface, the equilibrium composition of the mixed polymer film at the interface is given by

$$\frac{\Gamma_1}{\Gamma_{tot}} = \frac{K_1 C_1}{K_1 C_1 + K_2 C_2} \tag{6}$$

and

$$\frac{\Gamma_2}{\Gamma_{tot}} = \frac{K_2 C_2}{K_1 C_1 + K_2 C_2} \tag{7}$$

where Γ_1 and Γ_2 are interfacial concentrations, C_1 and C_2 are bulk concentrations, and K_1 and K_2 are equilibrium binding constants of polymer 1 and polymer 2, respectively. With the knowledge of K_1 and K_2 from the adsorption isotherms of polymers 1 and 2 in single-component systems, the composition of the mixed polymer film can be predicted for any combination of C_1 and C_2. However, if the polymers negatively interact with each other and exhibit thermodynamic incompatibility of mixing in the interfacial film, then the binding constants K_1 an K_2 in the binary system would not be the same as those in single component systems. Consequently, experimental Γ_1/Γ_{tot} and Γ_2/Γ_{tot} values will not be the same as those predicted by Eqs. (6) and (7). The extent of this deviation is a direct measure of the degree of thermodynamic incompatibility of mixing between the polymers.

It was found that the experimental results obtained from competitive adsorption studies departed very significantly from the predicted ones on the basis of Eqs (6) and (7). The radiotracer study of the adsorption of proteins at water/air interfaces was extended to water/lipid(oil) interfaces.

The adsorption behavior of proteins at water/oil interfaces should follow the same general physical picture, however, specific differences have to be expected caused by the protein structure and chemistry. At the water/air interface the protein molecules try to expose the hydrophobic parts upon the air phase, which leads to an unfolding of the molecules. This process can proceed, as long and as far as there is time and space at the interface available. Thus, the process of unfolding is restricted by a competitive adsorption of other molecules, proteins or surfactants. Once a molecule is spread over a free interfacial area (unfolded) it is unclear to what extent these molecules can refold, and what will be the conformation of the molecules after refolding.

For water/oil interfaces the situation is different, as the adsorbing protein molecules can penetrate into the hydrophobic oil phase with the hydrophobic parts of the molecule. This means that this type of unfolding can proceed even at a comparatively strong competition at the interface due to adsorption of other molecules, as the unfolding does not happen at the interface but within the oil bulk phase. Again the question of refolding due to increased competition or compression of the interfacial layer is difficult to answer.

A methodology has been developed for studying the adsorption behavior of proteins at oil/water interfaces by radiotracer technique monitoring adsorption of ^{14}C-labeled proteins at the oil/water interface [12]. A 1000 Angstrom thick tryglyceride oil film was formed on the water surface, β-casein was used to generate a standard curve for relating interfacial radioactivity (μCi m^{-2}) to cpm (count rate) at the oil/water interface. Adsorption isotherm of β-casein was determined in the bulk protein concentration range 1.5×10^{-5}-3.8×10^{-3} % by weight of solution. The saturated monolayer coverage was found to be about 7.3 mg m^{-2}. This value was quite different from other values reported in the literature. Adsorption studies with another protein, lysozyme, at the oil/water interface also revealed a high surface concentration of 3.0 mg m^{-2}. The most significant difference between the adsorption of β-casein at the oil/water and air/water interfaces was the lack of an induction period for the development of interfacial pressure in the former. This difference may be attributed to the attractive dispersion interaction between protein molecules and the oil phase.

Adsorption dynamics and interfacial properties of α-lactalbumin in native and molten globule state conformation at air/water interface was studied by Cornec et al using radiotracer method [38 - 41].

A computer-controlled Langmuir minitrough with a Wilhelmy plate was employed for adsorption dynamics and spreading experiments.

A solution of ^{14}C labeled protein in phosphate buffers was poured into the Langmuir minitrough. The radiation detector was immediately positioned and both the surface pressure (π) and counts per minute (cpm) of β-radiation were monitored. Data were automatically taken and stored through a computer interface. Adsorption experiments were carried out for 20 h. In order to convert cpm to the surface concentration the detector was calibrated with radioactive samples of known surface and bulk concentrations.

The surface hydrophobicity of molten globule conformation of α-lactalbumin (prepared by lowering pH to 2) was found to increase by 15-fold compared to the protein in is native form (pH 7) with increased flexibility as evidenced by a much smaller hysteresis area of surface pressure isotherms. This higher surface hydrophobicity resulted in (i) higher surface activity of α-lactalbumin in its molten globule conformation compared to the native form inspite of a much higher electrical energy barrier due to much higher net charge and (ii) diffusion controlled adsorption for much longer times and surface pressures than the native form. The diffusion coefficient of α-lactalbumin in molten globule state was lower than that for native form (2.22×10^{-10} m^2 s^{-1} vs. 4.65×10^{-10} m^2 s^{-1}) and increased for the former at higher ionic strength. The area per molecule during the dynamics of adsorption was found to be the same at a fixed surface pressure for both the conformations and lower than the corresponding value for fully denatured protein thus indicating that α-lactalbumin does not fully unfold at the interface in both the conformations. Even though α-lactalbumin adsorbed much more in its molten globule conformation, its interfacial rheological parameters were low compared to the native form. Ionic strength did not influence the relationship between the interfacial elasticity and surface concentration for both the conformations.

Adsorptive behavior of a globular protein with a monoglyceride monolayer spread on the aqueous surface was studied by the same author with the aim to study lipid-protein interaction [46].

The dynamics of surface pressure and of surface concentration Γ of a ^{14}C radiolabeled β-lactoglobulin, a globular protein, adsorbed onto a spread monolayer of 1-monopalmitoyl-rac glycerol (named commonly as a monoglyceride), at the air/water interface were measured. The adsorption of ^{14}C labeled β-lactoglobulin was enhanced at short times when C-16 monoglyceride of 73.4 and 24.7 μg was spread on the aqueous surface. However, the amounts of protein adsorbed at the steady state (after 10 h) were lowered with the values being 0.8 and 1.2 mg m^{-2}, respectively. Spreading of such amount of monoglyceride that forms a dense packed monolayer onto air/water interface led to complete displacement of the protein from the aqueous surface after 2.5 hours, possibly because of the surface pressure and exclusion effects.

The adsorption of lysozyme at the air/buffer interface has been studied by Xu and Damodaran by a radiotracer method [42, 51]. A major experimental feature of these studies was that the amount of protein adsorbed is a significant fraction of the protein in the bulk at the beginning of the adsorption process. On the basis of these data several kinetic models of adsorption were derived by R Douillard [53], which take into account the bulk protein concentration and a possible attraction of the protein in solution by the protein molecules forming the interfacial adsorption layer. The data were analyzed assuming either an exponential regime (the rate of adsorption is controlled by the distance from equilibrium) or a diffusive regime (the rate of adsorption is controlled by the diffusion of protein molecules in the solution). The full model (depletion of protein from the bulk and attraction by the previously adsorbed molecules) leads to an expression for the rate of adsorption that gives a good fit to the experimental data. It was concluded that the adsorption of lysozyme at the air/buffer interface is very probably a 'cooperative' process.

A kinetic study on the adsorption of compact, water-soluble proteins onto aqueous surfaces was reported by Cho and Cornec [39]. Two compact sized globular proteins, β-lactoglobulin and α-lactalbumin were kinetically characterized at the aqueous solution surface with the measurement of surface pressure (π) and surface concentration (Γ) *via* a radiotracer method. The adsorption kinetics was of diffusion control at early times, the rates of increase of π and Γ being lower at longer times due to growing energy barrier. At low concentrations, an apparent time lag was observed in the evolution of π for β-lactoglobulin but not for α-lactalbumin which was shown to be due to the non-linear nature of the π-Γ relationship for the former. The area per molecule of an adsorbed β-lactoglobulin during adsorption was smaller than that for spread monolayer since β-lactoglobulin was not fully unfolded during the adsorption. For α-lactalbumin, however, no such difference in the molecular areas for adsorbed and spread monolayer was observed indicating thereby that α-lactalbumin unfolded much more rapidly (has looser tertiary structure) than β-lactoglobulin. Surface excess concentrations of α- lactalbumin was found to evolve in two steps possibly due to the change in the orientation of the adsorbed protein from a side-on to an end-on orientation.

Although the electrochemistry at the interface between two immiscible electrolyte solution (ITIES) gained significant interest during the last two decades [54], to our knowledge, no application of the radiotracer technique for the investigation of the interfacial layer were reported in the literature. Presumably this lack of approach is connected with the enormous technical problem to create a liquid/liquid interface and simultaneously to control electrochemical parameters and measuring the radiation coming from the radioactive species present in the interface.

Nowadays neutron reflection methods are widely used for the study of the air/water interface. This method in certain respects is similar to the radiotracer method. Similar to the latter, in the case of neutron reflection technique also the "labeling" of the surfactant is required by isotopes (deuterium). As in the case of the radiotracer technique the reliability of the method strongly depends on the reliability of the separation of background signal from the signal originating from the species present in the interface. A calibration of the signals is required. For neutron reflection the ratio of background to the signal corresponding to the sorption could be significantly lower than in the case of the radiotracer technique. Nevertheless, there are many problems to be solved where the application of radiotracer technique should be preferred to neutron reflection.

For the labeling in the case of radiotracer technique it is sufficient to substitute just one atom in the molecule studied by its radioactive isotope, while at least six deuterium atoms are necessary to obtain satisfactory signal in the case of neutron reflection.

It is a significant drawback of the neutron reflection technique that species without replaceable hydrogen atoms cannot be studied. The situation is the same when the number of replaceable atoms is not enough. Radiotracer technique is the only possibility, for instance, for the investigation of the adsorption of Ca^{2+} ions.

3. MISCELLANEOUS

3.1. Reactions in insoluble monolayers

An interesting aspect of the behavior of insoluble monolayers is their reactivity. Surface pressure and radioactivity of insoluble monolayer of methyl-^{14}C palmitate were determined simultaneously by Muramatsu and Ohno [55] in order to monitor the hydrolysis of the ester. A continuous significant decrease in the apparent molecular area of the ester was found. This phenomenon was ascribed to the evaporation and hydrolysis of the ester.

A similar study was devoted to the investigation of the oxidation of oleic acid monolayer on aqueous permanganate solution [56]. ^{14}C labeled oleic acid was spread on aqueous permanganate solution and the time dependence of surface radiation and surface pressure were determined simultaneously.

On the basis of the analysis of experimental data a mechanistic picture was suggested. According to this mechanism the main processes involve the relative quick formation of *cis*-epoxyoctadecanoic acid (*cis*-epoxyacid) in the monolayer followed by the rate determining step, the desorption of *cis*-epoxy acid into the underlying subphase. In parallel with the latter step there is an interfacial reaction leading to the formation azelaic and nonanoic acids. It was emphasized in Ref. 56 that for interpretation of the overall process the autooxidation and desorption of oleic acid, occurring even in the absence of permanganate, should be taken into consideration. Thus the overall process was characterized by the following scheme:

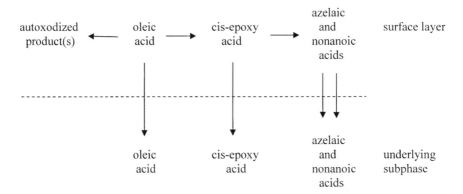

3.2. Adsorption of polymers at liquid/fluid interfaces

A detailed discussion of the problems related to this field is available in a review article by Schlenoff [17]. In the present chapter only one interesting point will be presented mainly on the basis of results reported by Schlenoff and coworkers [16].

The adsorption of ^{35}S labeled water-soluble random copolymer of styrene sulfonate and *tert*-butylstyrene (20 mol %) to toluene/water interface was followed by scintillation technique. The organic phase served as scintillator (see previous sections).

Adsorption isotherms were determined for water/toluene and 1 mol dm^{-3} NaCl aqueous solution/toluene interfaces. In both cases at copolymer concentrations above 10^{-5} g/ml the interfacial excess approaches to a limiting value. (1.4 mg m^{-2} for 1 mol dm^{-3} NaCl and 0.24 mg m^{-2} for the salt free solution.)

According to the assumptions of the authors of Ref. 16 the conformation of the adsorbed polymer involves all the *tert*-butyl groups inserting into toluene, while the styrenesulfonates form short loops in the water. The adsorption is rapid it takes only few minutes, for instance, at 7.2×10^{-5} g/ml concentration.

The assumption that adsorption occurs under equilibrium was proved by the study of the exchange (self-exchange) of labeled species adsorbed in the interface with a 10-fold excess of unlabeled copolymer. The exchange of a significant part of the adsorbed polymer is rapid (takes some minutes) but the remaining part may be exchanged over the course of several hours. It is assumed that the slow part reflects desorption of longest chains.

A comparison of the adsorption of copolymer at the toluene/water and polystyrene/water interfaces shows that the adsorption isotherms are similar in these cases. The exchange (self-exchange) with nonlabeled species is slower at the solid/liquid interface than in the case of liquid/liquid interface.

ACKNOWLEDGEMENT

Financial support from the Hungarian Scientific Research Fund is acknowledged (OTKA Grants T 031703 and T 045888).

LIST OF ABBREVIATIONS

DBA	N-dodecyl-β-alanine
GAGP	gum arabic glycoprotein
SAS	sodium alkylsulfates
SDS	sodium dodecyl sulfate

REFERENCES

[1]   E. Hutchinson, J. Colloid Sci., 4 (1949) 599.
[2]   K. Shinoda and K. Mashio, J. Phys. Chem., 64 (1960) 54.
[3]   E. Matijevic (ed). Surface and Colloid Science Vol. 6., John Wiley & Sons. Inc., New York, 1973. pp. 101–184.
[4]   G. Aniansson, J. Phys. Chem., 55 (1951) 1286.
[5]   J.K. Dixon, A.J. Weith, A.A. Argyle, D.J. Salley, Nature, 163 (1949) 845.
[6]   D.J. Salley, A.J. Weith Jr., A.A. Argyle and J.K. Dixon, Proc. Roy. Soc. (London), A203 (1950) 42.
[7]   C.M. Judson, A.A. Lerew, J.K. Dixon and D.J. Salley, J. Phys. Chem., 57 (1953) 916.
[8]   D.E. Graham and M.C. Phillips, J. Colloid Interface Sci., 70 (1979) 403.
[9]   D E. Graham and M.C. Phillips, J. Colloid Interface Sci., 70 (1979) 415.
[10] D.E. Graham and M.C. Phillips, J. Colloid Interface Sci., 70 (1979) 427.
[11] D.E. Graham, L. Chatergoon and M.C. Phillips, J. Phys. E.: Sci. Inst. 8 (1975) 696.
[12] T. Sengupta and S. Damodaran, J. Colloid Interface Sci., 206 (1998) 407.
[13] S. Xu, and S. Damodaran, J. Colloid Interface Sci., 157 (1993) 485.

[14] M. Li and J.B. Schlenoff, Anal. Chem., 66 (1994) 824.
[15] J.B. Schlenoff and M. Li, Ber. Bunsenges. Phys. Chem., 100 (1996) 943.
[16] R.M. Wang and J.B. Schlenoff, Macromolecules, 31 (1998) 494.
[17] R.S. Farinato and P.L. Dubin (eds.), Colloid-Polymer Interactions: From Fundamentals to Practice, John Wiley& Sons. Inc., 1999. pp. 225–250.
[18] K. Tajima, M. Muramatsu and T. Sasaki, Bull. Chem. Soc. Japan, 43 (1970) 1991.
[19] K. Tajima, Bull. Chem. Soc. Japan, 43 (1970) 3063.
[20] I. Berlot, P. Letellier, J.C. Moutet and D. Segal, J. Phys. Chem. A., 103 (1999) 1463.
[21] S. Karakashev, R. Tsekov and E. Manev, Langmuir, 17 (2000) 5403.
[22] V.L. Kolev, K.D. Danov, P.A. Kralchevsky, G. Broze and A. Mehreteab, Langmuir, 18 (2002) 9106.
[23] Z. Adamczyk, G. Para and P. Warszynski, Bull. Polish Acad. Sci. Chem., 47 (1999) 175.
[24] K. Tajima, Bull. Chem. Soc. Japan, 44 (1971) 1767.
[25] K. Tajima, M. Iwahashi and T. Sasaki, Bull. Chem. Soc. Japan, 44 (1971) 3251.
[26] G. Nilsson, J. Phys. Chem., 61 (1957) 1135.
[27] M. Muramatsu, K. Tajima and T. Sasaki, Bull. Chem. Soc. Japan, 41 (1968) 1279
[28] G.G. Jayson and G. Thompson, J. Colloid Interface Sci., 111 (1986) 65.
[29] A.W. Cross and G.G. Jayson, J. Colloid Interface Sci. 162 (1994) 45.
[30] R. Matuura, H. Kimizuka, S. Miyamoto and R. Shimozawa, Bull. Chem. Soc. Japan, 31 (1958) 532.
[31] J.M. Corkill, J.F. Goodman, D.R. Haisman and S.P. Harold, Trans. Faraday Soc., 57 (1961) 821
[32] R. Ruyssen and J. Maebe, Mededeel. Koninkl. Vlaam-Acad. Wetenschap. Belg. Kl. Wetenschap., 15 (1953) 5.
[33] E. Hutchinson, J. Colloid Sci., 4 (1949) 599.
[34] M. Muramatsu and M. Inoue, J. Colloid Interface Sci., 55 (1976) 80.
[35] K. Shinoda and K. Ito, J. Phys. Chem., 65 (1961) 1499.
[36] A. Nakamura and M. Muramatsu, J. Colloid Interface Sci., 62 (1977) 165.
[37] S. Damodaran and T. Sengupta, J. Agr. Food Chem., 51 (2003) 1658.
[38] M. Cornec and G. Narsimhan, J. Agr. Food Chem., 46 (1998) 2490.
[39] D. Cho and M. A. Cornec, Bull. Korean Chem. Soc., 20 (1999) 999.
[40] M. Cornec, D.A. Kim and G. Narsimhan, Food Hydrocolloids, 15 (2001) 303.
[41] M. Cornec, D. Cho and G. Narsimhan, J. Colloid Interface Sci., 214 (1999) 129.
[42] S. Xu and S. Damodaran, Langmuir, 10 (1994) 472.
[43] S. Damodaran and L. Razumovsky, Food Hydrocolloids, 17 (2003) 355.
[44] J.R. Hunter, P.K. Kilpatrick and R.G. Carbonell, J. Colloid Interface Sci., 142 (1991) 429.
[45] J.R. Hunter, R.G. Carbonell and P.K. Kilpatrick, J. Colloid Interface Sci., 143 (1991) 37.
[46] D. Cho and M. Cornec, Korean J. Chem. Eng. 16 (1999) 371.
[47] D. Cho, G. Narsimhan and E. I. Franses, J. Colloid Interface Sci., 191 (1997) 312.
[48] D. Cho, G. Narsimhan and E. I. Franses, J. Colloid Interface Sci., 191 (1997) 312.
[49] A. Baszkin, M.M. Boissonnade, A. Kamyshny, S. Magdassi, J. Colloid Interface Sci., 244 (2001) 18.
[50] C.S. Rao and S. Damodaran, Langmuir, 18 (2002) 6294.
[51] S. Xu and S. Damodaran, Langmuir, 8 (1992) 2021.

[52] R. Hunter, P.K. Kilpatrick and R.G. Carbonell, J. Colloid Interface Sci., 137 (1990) 462.
[53] R. Douillard, Thin Solid Films, 292 (1997) 169.
[54] F. Reymond, D. Fermín, H. J. Lee and H. H. Girault, Electrochim. Acta, 45 (2000) 2647.
[55] M. Muramatsu and T. Ohno, J. Colloid Interface Sci. 35 (1971) 469.
[56] M. Iwahashi, K. Toyoki, T. Watanabe and M. Muramatsu, J. Colloid Interface Sci., 79 (1981) 21.

Radiotracer Studies of Interfaces
G. Horányi (editor)
© 2004 Elsevier Ltd. All rights reserved.

Chapter 7

Study of phenomena occurring at solid/solid interfaces Investigation of interface diffusion and segregation by the radiotracer technique

G. Erdélyi, D.L. Beke and I.A. Szabó

Department of Solid State Physics, University of Debrecen, H-4010 Debrecen, PO Box 2. Hungary

1. INTRODUCTION

The last more than eighty years have indeed been the golden age of the radiotracer technique; the method has played a very important, determining role in diffusion research. In this chapter we try to give a review of this period focusing on studies of grain boundary (GB) and interface diffusion and segregation. First, we will give a historical retrospection of the most important milestones of the radiotracer technique in studies of phenomena occurring at solid/solid interfaces (Sec. 2). Then we outline the basic phenomenological equations for grain boundary and interface diffusion. The classification of the diffusion kinetic regimes will be presented together with illustrations of the most important technical realizations of the more and more sophisticated measurements (Sec. 3). In Sec. 4. our present knowledge on the mechanisms of diffusion and the grain-boundary structures will be summarized, while in Sec. 5. the possible ways of determination of the segregation parameters will be reviewed. Sec. 5. contains results on interface diffusion experiments and Sec. 6. exposes some open questions and future challenges especially in the light of the huge demand produced by the enormously growing field of nanotechnology.

2. HISTORICAL RETROSPECTION

As von Hevesy states in his Nobel lecture: "The conception of the diffusion of a substance into itself, self-diffusion, was introduced by Maxwell. No further use was made of this concept until fifty years later, when the method of radioactive (isotopic) indicators was introduced."

Austin, in 1896, performed the first quantitative determination of diffusion rates in solids [1]. His contemporaries assumed that the rate of mixing, i.e.

diffusion, is nearly the same for any element in an alloy. The results for gold have indicated that diffusion in solid metals is a comparatively rapid process even at low temperatures.

Radioactive tracers made new techniques possible. One could measure true self-diffusion, and also the sensitivity was much better than any other method available. The first tracer diffusion experiment was performed in 1920, in Budapest by Gróh and von Hevesy [2]. They measured the self-diffusion coefficient, D, of lead in molten lead, using a lead isotope, ThB, member of the Thorium decay series. They found D=2.2 cm^2/day ($2.5 \cdot 10^{-9}$ m^2/s). Gróh and von Hevesy determined the self-diffusion of solid lead in 1921 [3]. They had to select another active lead isotope, the RaD with longer decay-time, because the diffusion was much slower. Annealing at 280°C, which is only 46°C below the melting point, for more than a year, they determined $D < 10^{-4}$ cm^2/day (10^{-13} m^2/s), which is at least 300 times smaller than the results of gold in solid lead at 251°C [1]. They concluded that there is a substantial difference between the diffusion of two different metals in each other and self-diffusion in solid metals. The long annealing times needed for solid state diffusion could only be reduced with more microscopic sectioning techniques.

The role of grain boundary diffusion in tracer experiments was first noticed in 1925 by G. von Hevesy and A. Obrutsheva in Copenhagen [4]. Historically, this was the opening of sophisticated investigations of solid-solid interfaces (grain-boundaries or interfaces) by radiotracer techniques, which is the subject of the present contribution. They performed three self-diffusion experiments with the tracer ThB in lead. They compared the diffusivity in single crystal lead with polycrystalline foils. They wrote: "The results found indicate that even the slow rate of diffusion observed just below the melting point is not due to an exchange of place in crystals of appreciable size, but in the "amorphous" material, which is found between the crystals and must necessarily show a less regular structure than the material composing the individual crystals, and thus will be more capable of allowing an exchange in the position of neighboring atoms." Thus, the role of grain boundaries was recognized, but the mechanism was still quite obscure. Von Hevesy had always been very reserved in proposing a model, and he stated only that it is necessary for this transport that the lattice is made looser.

In 1920 he discussed the possibility of relating the electrolytic conductivity of salts with diffusion, according to Einstein's relation [5]. Comparing the conductivity of a transparent NaNO$_3$ crystal with that of a solidified NaNO$_3$ melt, as well as that of a calcite crystal with that of marble, von Hevesy found that the polycrystalline material conducted considerably better than the single crystal by about a factor of 50 [6]. The deviation from the ideal structure of the crystal at the grain boundaries favored the transport of matter. He was ready to withdraw his model, proposed in the same year, in the light of the new experimental results: "As to the mechanism of conductivity it can be concluded from the above results that it is not based on the shift of columns of ions through the crystal; such a shift would in a single crystal take place easier than in polycrystalline material. If on the contrary, one assumes that the transport of matter in the crystal is connected with the presence of loosened sites, then one can expect in advance that deviations of the ideal lattice and also the irregular ordering of ions at the boundary of the single crystallites in the polycrystalline material is favorable to the transport of matter."

The discovery of artificial radioactivity greatly enlarged the possibilities for the determination of self-diffusion rates. Different experimental methods were also developed, which extended the applicability of the tracer technique further to the study of boundaries and interfaces. However, very little progress was made until the early 1950s when improved sample preparation techniques and a wider availability of radioactive tracers led to some significant discoveries concerning the influence of boundary structure on boundary diffusion.

In 1920, demonstrating that Cu diffusion is enhanced at Ni grain boundaries, Barnes [7] also observed that twin boundaries do not act as fast diffusion pathways. These experiments were carried out by annealing bimetallic strips and revealing the interdiffusion zone by etching. Barnes also showed that the boundaries moved during diffusion, which was an early observation of diffusion-induced grain-boundary migration (DIGM). Also in 1950 Beck et al. [8] deduced from studies of grain growth in Al that grain-boundary diffusion was orientation dependent. In 1951, Achter and Smoluchowski [9], using a similar technique to that of Barnes, studied the depth to which Ag penetrated boundaries in [100] columnar Cu. They detected no enhanced diffusion in low-angle boundaries but found an increasing diffusivity in high-angle boundaries, showing a maximum at a misorientation of about 45°.

The first modern type of experiment, using radiotracers and bicrystals of controlled misorientation, were performed by Couling and Smoluchowski and reported in 1954 [10]. They have measured the penetration depth of Ag radiotracers into [100] tilted Cu bicrystals using autoradiography. They discovered that, for intermediate misorientations (~20°), diffusion was faster parallel to the <100> tilt axis than when perpendicular to it. These observations, they believed, supported the boundary structure concepts of Burgers (dislocation arrays) at intermediate misorientations and Motts (islands of coherence) at high misorientations.

The first investigation of self-diffusion as a function of misorientation in bicrystals was performed by Turnbull and Hoffman in Ag [11]. They demonstrated that there is enhanced diffusion even for low-angle grain boundaries. The dislocation pipe model accounted for the new observations and Turnbull and Hoffman could determine the Arrhenius parameters for the pipe diffusion coefficient.

The development of thin film technologies from 1960 led to the widening of the range of materials studied. Grain-boundary diffusion could be studied only at relatively low temperatures (less than half of the melting temperature), where the extent of lattice diffusion is very limited. As a result, much attention was given to developing the so-called microsectioning techniques, which could determine depth profiles on a micron or sub-micron scale. Although the electrochemical method was the first useful microsectioning technique for certain materials, as Ta, Nb and W, it was restricted in applicability. In 1973 Gupta [12] developed a microsectioning method based on ion erosion in an r.f. plasma and applied it to the depth profiling of radiotracers in Au. This method was subsequently applied to diffusion in oxides by Atkinson and Taylor [13]. The related technique of erosion using an ion beam was developed by Mehrer and co-workers [14] and applied to self-diffusion studies of Ni. The ion erosion techniques potentially suffer from problems of ion-induced mixing (knock-on) and changes in surface topography. Nevertheless, they were shown to be capable of depth resolution of better than 5 nm and can maintain acceptable surface topography to depths of several microns in most cases. The classical method of autoradiography was also developing. In 1976, Ishida et al. [15] showed how resolution and sensitivity could be improved for the radioisotopes that emit Auger electrons. During this period the increasing use of thin films in the microelectronic industry gave added impetus to the studies of grain-boundary diffusion, because it was the main source of degradation of devices.

The classical permeation measurements of von Hevesy were also revitalized in 1976 by Tompkins and Pinnel [16]. They have studied the permeation of Cu through Au thin films by using Auger electron spectroscopy (AES). These studies revealed that permeation experiments are sensitive to boundary conditions at the surface, but the technique does have the advantage that it can be performed *in situ* during the diffusion process.

The diffusion mechanism along grain boundaries was clarified further by the measurements of the effect of hydrostatic pressure on tracer grain boundary diffusion. From such experiments, first Martin et al. in 1967 [17] and later on more precisely Erdélyi et al. in 1987 [18] deduced an activation volume for Ag GB self-diffusion as well as for Zn GB heterodiffusion in Al. The values were close to that for lattice diffusion, which was known to occur by a vacancy mechanism. Further support for the vacancy mechanism of grain-boundary diffusion in Ag was provided by the isotope-effect measurements of Robinson and Peterson [19].

Direct observations of grain boundaries by field-ion and transmission electron microscopy revealed that the boundary 'width' was, at most, a few lattice spacing in extent. Thus, according to the experiments, grain boundaries can be visualized as narrow slabs, in which diffusion takes place by point defects, whose energies of formation and migration are lower in the boundary region than in the lattice.

Increasing attention has also been paid to the effect of segregation on impurity diffusion in recent years. As it is known, segregation produces enrichment of one or several components (impurities) at grain boundaries, phase boundaries and terminal surfaces. As we will see, the impact of segregation on grain boundary diffusion is considerable and it depends on the orientation of the grain boundary.

Early experiments on 'nanocrystalline materials' were described by Horvath et al. [20]. They produced Cu with a grain size of only 8 nm by consolidating Cu powder at high pressure and observed grain-boundary diffusion in this material at 80 and 120°C. The diffusivity was much greater than expected for normal grain boundaries at these temperatures, being more alike than expected for surface diffusion. Recent investigations have shown that, in order to receive reliable results, it is very important to use materials with stable and well-defined microstructures. These types of nanomaterials have become available only recently. The new results and techniques on nanomaterials will be summarized in the last part of the chapter.

3. BASIC EQUATIONS, KINETIC REGIMES FOR GRAIN-BOUNDARY AND INTERFACE DIFFUSION

3.1 Introduction

Early diffusion experiments, carried out by radiotracer technique, clearly show that the volume self-diffusion coefficients, measured in polycrystalline materials at temperatures $T \leq (0.6 - 0.7)T_m$, where T_m is the melting point of the matrix, are generally higher than the coefficients obtained at the same temperature measured in single crystals of the same material [21]. The lower the temperature, the more pronounced the effect becomes. It is generally accepted that the diffusion coefficients obtained by using polycrystalline samples are apparent or effective diffusion coefficients, including some contribution of diffusion along grain boundaries as well.

In order to estimate the volume fraction of grain boundaries in an annealed pure metal, two input parameters are necessary: the grain boundary thickness (δ) and the grain size (d). With the values of δ=0.5 nm, and d=1mm, one can get 10^{-6} for the grain boundary fraction. However, this almost negligible fraction of atoms in boundaries, for example in silver, doubles the silver diffusion coefficient at 650 °C [21,22].

The diffusion coefficient of an atom (D) characterizing the random movement of atoms in solids, assuming vacancy mechanism, is proportional to the product of the jump frequency and the vacancy concentration [22]. In the above example, this product for atoms diffusing in boundaries should be at least 10^6 times higher than for the rest of the atoms in the bulk. Similarly, the diffusivity is also enhanced by other lattice defects such as dislocations, phase boundaries and free surfaces. The enhanced diffusion is generally termed as *short-circuit diffusion*.

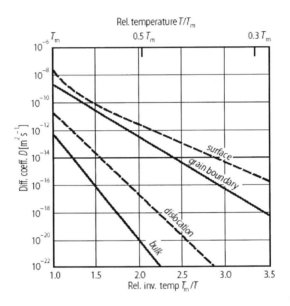

Fig. 1. The diffusivity *vs.* reduced reciprocal temperature for volume diffusion and for different types of short-circuit diffusion [23].

In the literature, the regions of high diffusivity (grain boundaries, dislocations, phase boundaries) are termed as *high diffusivity paths*. The diffusion coefficients obey an Arrhenius-type temperature dependence, but for short-circuit diffusion the activation energies are definitely smaller than the activation energy characterizing the volume (bulk) diffusion (see Fig.1). Experimental results tell us [24, 25] that the ratio of the GB and volume diffusion activation energies (E_{GB}, E) for metals can be given as:

$$\frac{E_{GB}}{E} \approx 0.4 - 0.6. \tag{1}$$

The majority of phase transformations, reactions and structural changes are diffusion-controlled phenomena. Since bulk diffusion is frozen in low temperature regions ($T < 0.5T_m$), as it is pointed out above, at these temperatures short-circuit diffusion processes play a crucial role in the phenomena involving some material transport, such as sintering, creep, grain growth, re-crystallization, segregation, eutectoid decomposition, etc. The wide spectra of short-circuit diffusion related phenomena underlines the importance of grain and/or interface boundary diffusion from the point of view of technology and basic research as well.

3.2. Classification of diffusion regimes

In phenomenological models describing grain or interphase boundary tracer diffusion, it is supposed that the boundary is a homogeneous slab of thickness of δ, with a high diffusivity D'. The grain boundary width, δ, is supposed to be a constant. The generally accepted value of the GB width for metals is $\delta=0.5$ nm. In principle, three types of kinetic behavior may be distinguished, as proposed first by Harrison [26]. The most important factors determining the type of kinetics are the volume diffusion length $\sqrt{Dt}$ - (D is the volume diffusion coefficient, t is the diffusion heat treatment time), the grain-boundary width, δ, and the grain size, d, of the investigated material.

Type-C kinetics

This regime refers to the limit when the bulk diffusion length is negligible, as compared to δ, i.e.

$$\sqrt{Dt} \langle\langle \delta .$$
(2)

In this case, the diffusion is restricted to grain/interphase boundaries only. In order to obtain the boundary diffusion coefficient from sectioning experiments, the Gaussian solution of the Fick-equation can be used:

$$c_B = \frac{M}{\sqrt{\pi D' t}} \exp\left(-\frac{y^2}{4D't} \right),$$
(3)

where c_B is the tracer concentration in the grain/interphase boundary, y is the penetration depth, t is the annealing time, M is the surface concentration of the tracer atoms. The D' diffusion coefficient can be obtained from the slope of the $\ln(c_B)$ vs. y^2 curve. In radiotracer experiments $c_B(y)$ is proportional to the specific activity of the section taken at the depth y.

The grain or phase boundary diffusion coefficient follows the Arrhenius law:

$$D' = D_0' \exp\left(-\frac{E_B}{RT} \right),$$
(4)

where D_0' is the pre-exponential factor, E_B is the activation energy of the diffusion along boundaries, R is the gas constant and T is the absolute temperature.

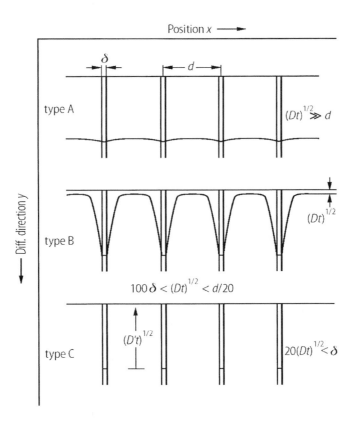

Fig. 2.　Classification of grain boundary diffusion kinetic regimes. Grain boundaries are perpendicular to the terminal surface of the samples. Concentration contours illustrate the distribution of the tracer after diffusion heat treatments. Tracer atoms initially are at the terminal surface of the sample at $y=0$.

Type-A kinetics

In this regime, the diffusion fields, developed around each boundary in the bulk, overlap since the condition

$$\sqrt{Dt} \rangle\rangle d \qquad (5)$$

holds (Fig. 2). Thus, the diffusion process can be characterized by an effective diffusivity:

$$D_{eff} = gD' + (1 - g)D.$$ (6)

This is the Hart's equation, which is obtained by assuming that the time fractions of the migrating atoms spent in the boundary and inside the grains are proportional to the volume fraction of the grain boundaries (g) and to the fraction of the bulk ($1-g$) [27].

For the determination of coefficients D_{eff}, the same methods can be used as for the bulk diffusion coefficient. The diffusion takes place according to type-A kinetics when any two of the following circumstances hold simultaneously:

- diffusion anneals are carried out at high temperature,
- annealing times are very long,
- microcrystalline samples are used.

Type-B kinetics

This is the usually realized situation in most of the experiments. The volume diffusion length is much higher than the grain boundary width, but the concentration profiles inside grains do not overlap (Fig.2):

$$100\delta \leq \sqrt{Dt} \leq d/20$$ (7)

The diffusion process in a bicrystal was first treated by Fisher [28]. In order to obtain the basic equations, one should consider the diffusion in a high diffusivity slab, which represents either a grain boundary or an interphase boundary (Fig. 3).

The slab is perpendicular to the free surface, where the tracer atoms enter the sample. In the coordinate system, shown on figure 3, the diffusion flux along the axis z is zero, J_y is the flux along the boundary, J_x is the flux from the boundary into the grains. Considering the fluxes in and out of an element of the slab, the rate of change of concentration inside the slab (c_B) can be given as (see Fig. 3):

G. Erdélyi, D.L. Beke and I.A. Szabó

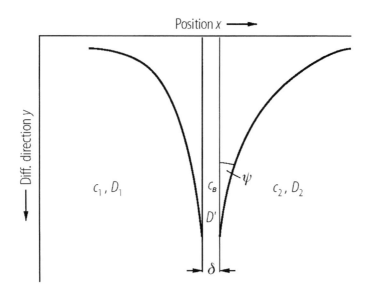

Fig. 3. Isoconcentration lines in a section perpendicular to the axis z. The concentration
contours are symmetric for grain boundary and asymmetric for interface diffusion. For
the latter case the volume diffusion coefficients in the grains are different [23]. The
parameter ψ denotes the angle at which the isoconcentration line crosses the interface.

$$\frac{\partial c_B}{\partial t} = \frac{1}{1 dy \delta}\left(\delta\left(J_y - \left(J_y + \frac{\partial J_y}{\partial y} dy\right)\right) - dy J_{\leftarrow x} - dy J_{\rightarrow x}\right) = -\frac{\partial J_y}{\partial y} - \frac{J_{\leftarrow x}}{\delta} - \frac{J_{\rightarrow x}}{\delta}. \quad (8)$$

The fluxes out of the boundary into the adjacent grains can be given as:

$$J_{\leftarrow x} = D_1 \frac{\partial c_1}{\partial x}\Big|_{x=-\delta/2} \quad \text{and } J_{\rightarrow x} = D_2 \frac{\partial c_2}{\partial x}\Big|_{x=\delta/2}. \quad (9)$$

Here, we supposed that the adjacent grains are characterized by different volume
diffusion coefficients D_1 and D_2. Moreover, the concentrations in the adjacent
grains can be different, i.e., interphase boundary diffusion is also included in the
consideration. Similarly, the flux J_y along the boundary can be given by

$$J_y = -D' \frac{\partial c_B}{\partial y}. \quad (10)$$

Substituting the fluxes into Eq. (7), we get the differential equation describing the transport in the slab:

$$\frac{\partial c_B}{\partial t} = D'\frac{\partial^2 c_B}{\partial y^2} + \frac{D_1}{\delta}\left(\frac{\partial c_1}{\partial x}\right)_{x=-\delta/2} + \frac{D_2}{\delta}\left(\frac{\partial c_2}{\partial x}\right)_{x=\delta/2}. \tag{11}$$

Outside the boundary, the well-known form of Fick-equations should be solved:

$$\frac{\partial c}{\partial t} = D_1 \Delta c_1; \quad \text{and} \quad \frac{\partial c}{\partial t} = D_2 \Delta c_2. \tag{12}$$

The rate of change in the concentration of an element of the boundary slab is caused by the divergence of the flux along the boundary and by the exchange of matter between the boundary and the grain ("leakage" of atoms into the bulk). If the relations $D_1 = D_2 = D$ and $c_1 = c_2 = c$ hold, we talk about *grain boundary diffusion*. The general case, i.e. $D_1 \neq D_2$ describes the diffusion along *phase boundaries*. In the following paragraphs when grain boundary diffusion is considered, D_{GB} stands for GB diffusion coefficient, while D_{IB} represents the coefficient of diffusion along phase boundaries. This latter process will be treated in section 6.

The solution $c(x,y,z,t)$ should simultaneously satisfy these differential equations and the appropriate initial and boundary conditions. For example, the so-called *constant surface concentration case* can be described by the following set of initial and boundary conditions:

$$c(x, y=0, t=0) = c_0; \; c(x, y, t=0) = 0;$$
$$c(x, y=\infty, t) = 0; \; c(x=\pm\delta/2, y, t) = f(c_B) \tag{13}$$

3.3. Grain boundary self- and heterodiffusion

3.3.1 Sectioning experiments, self-diffusion
For self-diffusion, the condition

$$c_{GB}(y,t) = c(x=\pm\delta/2, y, t) \tag{14}$$

holds, i.e. the concentration in the slab (c_{GB}) and at the interface of the grain and the slab (c) are equal.

The exact solution for the constant surface concentration case (see Eq. (13)) was given by Whipple [29] and for thin films by Suzuoka [30, 24]. Introducing the following dimensionless quantities:

$$\alpha = \frac{\delta}{\sqrt{Dt}}; \qquad \beta = \frac{\delta D_{GB}}{2D\sqrt{Dt}} = \frac{P}{2D\sqrt{Dt}}; \qquad \eta = \frac{y}{\sqrt{Dt}}, \qquad (15)$$

both solutions can be given as a function of these variables. However, in tracer experiments, we measure the averaged concentration (specific activity) in a thin section perpendicular to the grain boundary, i.e. in the x-z plane (see Fig. 3.), so we should calculate the averaged concentration $\bar{c}(y)$ at a depth y. The average tracer concentration in the section can be calculated by performing the integration in the x-z plane (for details, see for example Ref. 24) and the averaged concentration can be given as a function of dimensionless parameters α, β, η. Le Claire, using the results of Levine and MacCallum, showed that the exact analytical solutions of Whipple and Suzuoka could be given in a simplified form, which is convenient to evaluate sectioning experiments [31,32]. According to the results of LeClaire, the P parameter, the product of the grain boundary diffusion coefficient and the grain boundary width, can be given by

$$P = D_{GB}\delta = \left(\frac{\partial \ln \bar{c}}{\partial y^{6/5}}\right)^{-5/3} \left(\frac{4D}{t}\right)^{1/2} A^{5/3}, \qquad (16)$$

where $A=0.78$ (constant surface concentration at the surface, Whipple-solution), and $A = 0.72\beta^{0.008}$ (thin film- or Suzuoka-solution).

Eq. (16) can be used if the conditions $\alpha \langle 0.01; \beta \rangle 10$ hold. It is a considerable advantage that the form of the tracer penetration function, $\ln \bar{c}$ vs. $y^{6/5}$, is not sensitive to the boundary conditions. If the requirements of the type-B regime are fulfilled (Eq. (7)), the product $P = D_{GB}\delta$ can be determined from the slope of the linear part of the $\ln \bar{c}$ vs. $y^{6/5}$ plot, provided the bulk diffusion coefficient (D) is known.

The penetration function can be established by removing thin sections from the specimen by means of microtome sectioning or ion-sputtering. The radioactivity of the removed thin sections is usually measured by means of standard counters or detectors. The average tracer concentration $\bar{c}(y)$ in a section is proportional to the specific activity of the section, which is calculated as the ratio of the activity and the mass of the section (section activity method).

The tracer concentration profiles may also be obtained by measuring the activity at the surface of the sample left after each successive section removal. The latter technique, known as the residual activity method, is applied when the activity of the removed sections cannot simply be measured (grinding sectioning). In order to establish the relation between the measured residual activity and the average tracer concentration, the absorption characteristic of the sample material should also be taken into account (for details see Ref. 24). However, for the limiting case of very strongly absorbed radiation, the measured residual activity is proportional to the average concentration of the tracer at this surface, so the product $P = D_{GB}\delta$ can be determined from the slope of the linear part of the ln (*residual activity*) *vs.* $y^{6/5}$ plot (Fig.4).

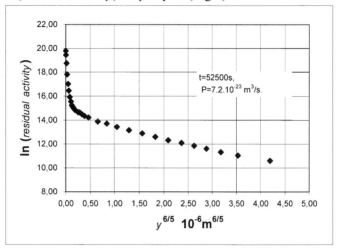

Fig. 4. Typical penetration profile measured in type-B regime using residual activity technique. The ^{63}Ni profile was measured in boron-doped Ni$_3$Al at 969K. Since the isotope emits low-energy β-particles, the residual activity of the sample at a given y depth is proportional to the tracer concentration in the surface layer. The GB diffusivity was evaluated from the linear part of the profile, using Eq. (16).

It should be stressed that from measurements carried out in type-B regime only the product P can be evaluated. The grain boundary diffusion coefficients can be determined from experiments carried out in type-C regime (see Eqs. (2) and (3)). However, performing experiments in type-C regime is rather difficult because confining tracer atoms in boundaries requires very short diffusion time or diffusion heat treatments at very low temperature. Furthermore, very low tracer activities need to be measured, and this is why these measurements became used commonly only from the middle of the eighties of the last century.

3.3.2. Contour angle measurements, autoradiography

Besides sectioning methods, where the averaged concentration (specific activity) of sections is measured, there are additional experimental techniques to reveal the equal concentration contours of the diffusing element in the diffusion zone. For example, by means of autoradiography, one can map the isoconcentration lines and the angle (ψ) at which they cross the grain boundary/grain interface (Fig.3). The latter parameter can be related to grain boundary diffusivity using the approximate solution of the problem (Fisher solution [28]):

$$P = 2\pi \sqrt{t} D^{\frac{3}{2}} \cot^2 \psi \tag{17}$$

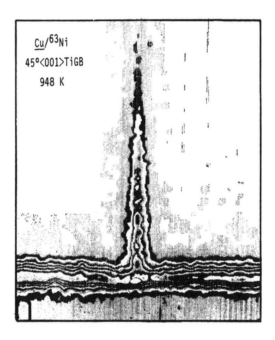

Fig. 5. Isoconcentration contours of ^{63}Ni, measured in Cu symmetric tilt GBs [33]. The grain boundary parameters are in the insert.

In this approximation, the contour angle is the same at any depth and the diffusivity can be deduced from Eq. (17).

Autoradiograms are usually taken on a metallographically prepared bicrystal surface along a plane, which is normal both to the grain boundary plane and the plane of the diffusion source (Fig. 3).

Using radiotracers emitting low-energy β-rays, the radiation-induced blackening of film can be considered to be proportional to the surface concentration of tracer atoms. The isoconcentration contours and the contour angle can be determined by standard densitometric measurements (Fig. 5). The relation between the contour angle and the diffusivity becomes more complicated, if one uses the accurate solution (Whipple) of the problem. For details see Ref. 24.

3.3.3 Grain boundary width

In the phenomenological models, grain boundaries are treated as homogeneous slabs with the thickness of δ. For self-diffusion the δ parameter can be experimentally measured, performing experiments both in type-B and type-C kinetic regimes:

$$\delta = \frac{\delta D_{GB}\big|_B}{D_{GB}\big|_C}. \tag{18}$$

The grain boundary width was first determined in a metal from the above relation for Ag self-diffusion by Gas et al. [34] using a rather sophisticated analysis of the tracer profiles in the transition range between type-B and type-C kinetic regimes [35]. According to their results, δ=0.5 nm is a good approximation for the grain boundary thickness (in Fig. 6, the two lines are shifted from each other by this factor). This result was confirmed by other authors [36] and today the value of 0.5 nm is generally accepted in the literature. In NiO, similar type of experiments proved earlier that δ is about 0.7 nm [37].

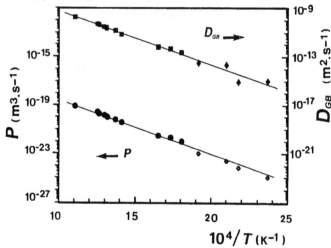

Fig. 6. Arrhenius plots of D_{GB} and P for silver GB self-diffusion [34]. Points in the transition region between type-B and C regimes (symbols ◆ and ●, respectively) were corrected by the method given in Ref. 34.

4. STRUCTURE AND THE MECHANISMS OF DIFFUSION

4.1. Structure and boundary diffusivity

There is a strong link between the structure of grain and phase boundaries and the diffusion along them. The boundary structure depends primarily on crystallographic parameters describing the orientation of the adjoining crystals and the interface between them.

The actual structure of a boundary depends not only on the lattice structure and the relative orientation of the adjacent grains, but also on the details of interatomic potentials. The relaxations lead to local rearrangements of the structure and, thus, the grain boundary provides a special environment for the creation and movement of point defects. These structural details can be studied theoretically only by computer simulations. Computer simulations have also shown that the GB structure can be modified by the segregation of impurities at the grain boundary and also by stresses and dislocations being present in the material.

Different classes of grain boundaries can be distinguished depending on their structure: small-angle, large-angle, special and general (random) boundaries.

The structure of small-angle boundaries can be visualized as an array of individual dislocations, separated by patches of a strained single crystal [38]. The dislocation model fails for misorientations above 15°, when the spacing becomes comparable with the dislocation pipe diameter. In early works, these "general" boundaries were described as an amorphous, liquid-like region between the two crystals. The experimentally observed anisotropies in these general boundaries led Mott [39] to introduce the island model. In this model, boundaries are composed of islands of disordered material (corresponding to a bad fit between the grains) separated by regions where the fit between the grains is relatively good.

The concept of special grain boundaries was introduced trough the coincident site lattice model. For certain relative orientations between the two crystals, a fraction of the total lattice sites are common to both lattices when they are formally extended to the whole space. These sites are forming a superlattice called the *coincidence site lattice* (CSL). The goodness of the match can be represented by an integer, Σ, which is the inverse of the relative density of the CSL sites. The boundaries, which pass through planes of the CSL, are called special grain boundaries. These boundaries have a periodic structure with a repeat unit length of only several atomic distances, which are the so-called structural units. Small deviations from the orientation of a special boundary are accommodated by an array of GB dislocations, thus special boundaries have smaller grain boundary energy than the neighboring orientations, leading to characteristic cusps in the misorientation dependence of various physical quantities.

On the experimental side, the radiotracer technique provides the necessary data to check these structural models. There are two kinds of experiments relevant in this respect: the measurement of the angle dependence for a set of grain boundaries and the determination of the anisotropy of matter transport in low-angle or some special boundaries. Such measurements can be performed only on well-oriented bicrystals. The production of oriented bicrystals is an extremely difficult problem, which is done by either bonding or directional crystallization. It is not an easy task to reach the desired misorientation with high accuracy and uniformly along the sample,

The structure of a small angle grain boundary can be represented as an array of uniformly spaced dislocations [40]. We can replace the dislocation array with a homogeneous slab of thickness, δ and diffusivity, D_{GB}:

$$\delta D_{GB} = \pi \cdot r_d^2 \cdot D_d / \lambda_d , \tag{19}$$

where r_d is the dislocation pipe radius, λ_d is the dislocation spacing, and D_d is the dislocation pipe diffusion coefficient. The dislocation spacing can be expressed by the misorientation angle. For example, for a tilt boundary in an fcc crystal,

$$\lambda_d = \frac{a}{2\sqrt{2} \sin(\theta / 2)}, \tag{20}$$

where θ is the tilt angle, and a is the lattice parameter.

According to this model, the diffusion current and, thus, the penetration depth is approximately proportional to the square of the misorientation angle. Furthermore, the activation energy for small angle boundaries is independent of the misorientation, and equal to the activation energy for dislocation pipe diffusion.

The orientation dependence of the grain boundary diffusion was measured by Upthegrove and Sinnot [41] for self-diffusion along <001> symmetrical tilt grain boundaries in Ni. The penetration depths are plotted on Fig. 7, which fits to the $\sin(\theta/2)$-type angular dependence curve very well.

Turnbull and Hoffman [40] found the activation energy for grain boundary self-diffusion in Ag tilt bicrystals to be almost independent of the misorientation angle in agreement with the dislocation model. Similar conclusions were also made by Canon and Stark [42].

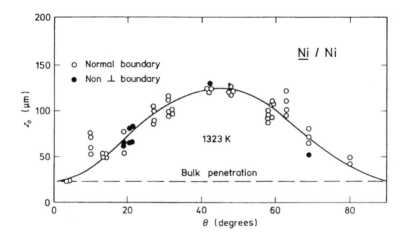

Fig. 7. Orientation dependence of the grain boundary penetration depth for self-diffusion along <001> symmetrical tilt grain boundaries in Ni, after Upthegrove and Sinnot [41].

According to the dislocation pipe model, diffusion is much faster in the direction parallel to the dislocation lines than that in the direction perpendicular to them. Let $D_{GB}{}^{//}$ and $D_{GB}{}^{\perp}$ represent the grain boundary diffusion coefficients parallel and perpendicular to the tilt direction, respectively. The results, published by Hoffman [43], show this anisotropy unambiguously (see Fig. 8.).

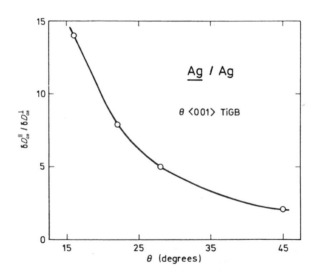

Fig. 8. Orientation dependence of the grain boundary diffusion anisotropy for self-diffusion along < 001> tilt grain boundaries (TiGB) in Ag after Hoffman [43].

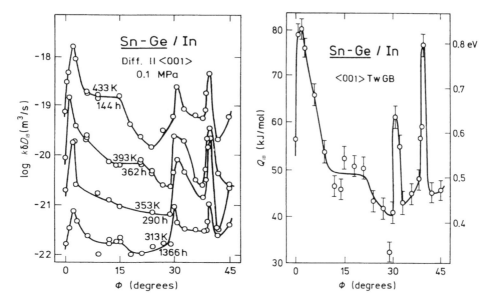

Fig. 9. Angular dependence of the interphase boundary diffusivity of In in Sn/Ge twist (TwGB) boundaries. On the right hand side, the orientation dependence of the activation energy is displayed, after Straumal et al. [44].

The anisotropy decreases smoothly with increasing tilt angle. The finite value of the anisotropy, even at the maximum tilt angle of 45°, indicates that also large-angle grain boundaries possess a structure.

For large-angle grain boundaries, there is convincing evidence that the activation energy shows a systematic dependence on the misorientation angle, with maxima occurring at CSL misorientations. On the other hand, the measured diffusivities show a less systematic picture. Some authors found smooth orientation dependence, while in other experiments, the CSL orientations coincided with local minima of the diffusivity. In a few cases, local maxima were observed, for example for the diffusion along the Sn/Ge interface by Straumal et al. [44]. Figure 9. shows that both the interface boundary diffusivity and the activation energy show maxima at CSL orientations.

Furthermore, Straumal et al. showed that a linear relation exists between the pre-exponential factor and the activation energy for grain boundary diffusion, as it is shown on Fig. 10. For boundaries, where the activation energy is large, the pre-exponential factor is large as well. These parameters affect the diffusivity in opposite ways at each temperature. It turns out that there is a temperature where they compensate each other, and the orientation dependence of the diffusivity disappears. Below this temperature, the special boundaries have lower diffusivities, while above this temperature they have higher diffusivities than the neighboring orientations.

The positive correlation between the pre-exponential factor, which contains the migration and formation entropy of the defects, and the activation energy was noticed in several other cases as well. These empirical rules can be justified with scaling arguments based on the similarity of the interatomic potentials [45].

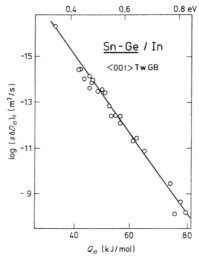

Fig. 10. Correlation between the activation energy and the pre-exponential factor for diffusion of In along <001 > Sn/Ge interphase twist boundaries, after Straumal et al. [44].

In contrast to the results presented above, several studies have found only a smooth orientation dependence of the grain boundary diffusivity [46,41,47]. This could be explained with the assumption that together with the change of the misorientation a gradual replacement of the structural units of the grain boundary takes place. The seeming contradiction can also be related to the difficulty of locating the minima. The minima in diffusivity can be localized to a narrow range around some special misorientations, as it is illustrated on Fig. 11. for Ge GB diffusion in Al, as observed by Surholt et al. [48]. These sharp minima could have been overlooked in the previous studies [46,41,47], where the measurements were carried out in steps of 5–7° orientation angle.

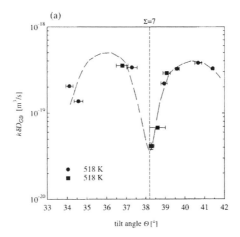

Fig. 11. Misorientation dependence of the diffusion parameter $k\delta D_{GB}$ of Ge in $\Sigma=7$, <111> tilt GBs in aluminum at $T=518$ K [48].

4.2. Heterodiffusion, segregation isotherms

Treating the problem of GB self-diffusion, we assumed that the concentrations of the diffusing atom at the two sides of the grain/grain boundary border are the same, i.e., there is a continuity of the concentration across the boundary (Eq. (13)).

If we would like to measure the hetero or impurity diffusion along grain boundaries, the above condition should be modified. As a consequence of the segregation, there is a discontinuity in the concentration of the tracer impurity atoms at the grain boundary and in the adjacent grains. If we take the distribution of the impurities into account by means of the simplest, the Henry-type segregation isotherm, the new matching condition is [49]:

$$c_{GB}(y,t) = c(y,t)k. \tag{21}$$

For the simplest, Henry-type segregation isotherm, the segregation coefficient k, depends on the temperature only:

$$k = \exp\left(-\frac{E_s}{RT}\right). \tag{22}$$

Here E_s is the segregation energy. The new coupling condition implies that there is a local thermodynamic equilibrium and the ratio of the impurity concentrations in the boundary and in the adjacent grain can be characterized by the same segregation factor at any depth. Taking the new matching condition into account, all the results obtained for self-diffusion remain valid, but one has to replace δ by $k\delta$ in the solutions [49]. For example, the modified dimensionless parameters for heterodiffusion are

$$\alpha = \frac{\delta k}{\sqrt{Dt}}; \qquad\qquad \beta = \frac{\delta k D_{GB}}{2D\sqrt{Dt}} = \frac{P}{2D\sqrt{Dt}} \qquad (23)$$

For heterodiffusion, the parameter $P = k\delta D_{GB}$, characterizing the diffusivity is called the *grain boundary triple product*. Sectioning experiments, carried out in type-B kinetic regime, provide the triple product, which can be deduced using Eq. (16).

According to the Henry-type matching condition, k is independent of the annealing time and the depth. Taking into account that in tracer experiments, the concentrations are extremely low, it is very likely that the Henry-type segregation isotherm can be applied at least at large penetration depths.

However, the validity range of the Henry-isotherm is rather limited: it holds only for very dilute ideal solutions. Surface saturation effect with increasing bulk concentrations requires more reasonable description of the segregation. The Langmuir-McLean isotherm [50] describes the saturation of the segregating component supposing ideal solid solution in the bulk:

$$c_{GB} = \frac{kc}{1 + (k-1)c} \qquad (24)$$

Note that Eq. (24), in the $k\rangle 1$ and $c\langle\langle 1$ limit, reduces to the Henry-isotherm (see Fig. 12). In Eq. (24) the c_{GB}/c ratio depends not only on the temperature, but on the c volume concentration as well. Since the volume concentration depends on the depth, the segregation distribution changes with the depth, too. Consequently, in such systems the segregation cannot be characterized by one depth independent segregation factor, i.e. the triple product changes with the depth. Accordingly, the penetration profiles are not linear any more, as it was pointed out by Bokhstein et al. [51] and Mishin and Herzig [53] (see Fig.13).

When the segregation tendency is strong, the interaction between the segregating atoms has to be taken into account. Such systems, far from the ideal solution, can be described by the Fowler-Guggenheim isotherm [54]:

$$c_{GB} = \frac{kc\exp(\gamma(2c_{GB}-1))}{1 + kc\exp(\gamma(2c_{GB}-1))}, \qquad (25)$$

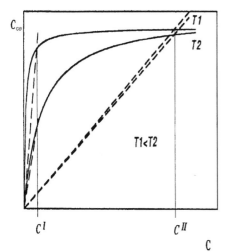

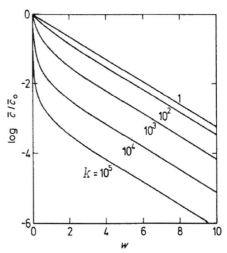

Fig. 12. McLean's isotherms at different temperatures. The broken lines determine the apparent segregation coefficients at volume concentrations c^I and c^{II}.

Around c^{II} small changes in c cause no detectable changes in the apparent value of k. Henry's isotherm can be represented by the initial slopes of the curves. T_1, T_2 are different temperature values, after Bernardini et al. [52].

Fig. 13. Effect of segregation on the penetration plot; $\bar{c}_0$ is the surface value of the average concentration, the horizontal axis represents the reduced penetration depths ($w = y4D^{1/4}t^{-1/4}P^{-1/2}$) after Mishin and Herzig [53].

where $\gamma = E_{mix}/RT$ and E_{mix} is the mixing energy in the boundary. The $c_{GB}(c)$ curves at some values of γ, become S-shaped, i.e. at some volume concentration two different surface phases coexist. The nonlinear segregation effect also disturbs the linearity of the penetration plot and causes an upward curvature [51].

The saturation of the $c_{GB}(c)$ function may occur in tracer GB diffusion experiments when a tracer B* diffuses in an AB alloy. The saturation equilibrium can be reached by means of a pre-diffusion anneal carried out at the planned diffusion temperature. During diffusion experiments the tracer atoms, B* are moving in B-saturated boundaries. The impact of strong segregation on the grain boundary diffusivity and on the shape of the penetration plots will be treated later.

4.3. Impact of the purity of the samples

The experimental data existing for grain boundary diffusion are of poorer reproducibility than the corresponding data for volume (bulk) diffusion. The pre-exponential factors ($P_0 = D_0'\delta$ for self-diffusion and $P_0 = kD_0'\delta$ for heterodiffusion), determined in different laboratories for the same materials, may differ by one or two orders of magnitude.

Gas and Bernardini demonstrated experimentally that self-diffusion is strongly influenced by solute segregation [55]. A strong solute-solvent binding causes a decrease in diffusivity, while a strong solute-solute binding does not affect the grain boundary mobility.

An analysis of the experimental data and systematic measurements revealed that the GB diffusivities are extremely sensitive to the purity of the samples.

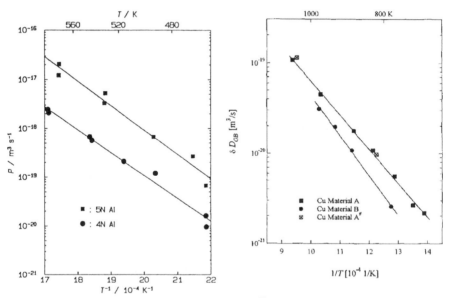

Fig.14. Sn GB diffusion in 4N and 5N purity aluminium [56].

Fig.15. ^{64}Cu tracer GB self-diffusion in Cu of different purity. Material A: 99.9998%, material B: 99.999%. , A<: samples from a second ingot of material A [57].

The triple product of Sn was found to be 10 times higher in 5N purity Al than in a less pure 4N Al material [56] (Fig. 14). The decrease of P in the less pure material was attributed to segregated impurities: the poorly soluble impurities - some of them tend to form compounds in the lattice - may block the Sn tracer movement by occupying boundary positions that could otherwise be involved in the tracer transport. Obviously, some co-segregation effects may also contribute to the reduction of the diffusivity.

Surholt demonstrated the dramatic effect of sulfur on grain-boundary self-diffusion in nominally high purity copper [57]. Two high purity (6N and 5N) copper materials were used, in which sulfur was the leading impurity with bulk concentrations of 0.11 and 1.1 wt ppm. The difference in sulfur content caused a dramatic change in the Arrhenius parameters: about 1 ppm increase of sulfur concentration reduced the diffusivity by a factor of about 15 at a temperature of 373 K (Fig.15)! The authors concluded that sulfur atoms find energetically favorable positions at the GBs at which the local surroundings are similar to those in an ordered (Cu_2S) compound. Because of these stable S-Cu complexes, Cu atoms cannot take an active part in the GB transport.

We recall that at low temperatures, GB diffusion is the rate–controlling mechanism of mass transport. The phenomenon is very important from a technological point of view: by means of micro-alloying, the GB transport can be reduced. This is a possible way to increase the lifetime of copper interconnects in microelectronic devices [57].

4.4. Different types of boundaries as different short circuit paths

The description of GB diffusion, outlined above briefly, concerns diffusion in a bicrystal sample with a single boundary. In a polycrystalline sample there is a distribution of grain boundaries, consequently the measured P triple product represents an average value of the diffusivity. The experimentally measured ln*(specific activity)* vs. $y^{6/5}$ plots cannot be approximated with single linear functions, they frequently show a definite upward curvature. The behavior of curved penetration plots were studied by Beke et al. [58] using polycrystalline samples with special multimodal grain structures: different large-angle grain boundaries, as well as sub-boundaries were present in the sample. Though they investigated grain boundary heterodiffusion (^{65}Zn in Al), segregation effects were supposed to be negligible. Varying the annealing parameters, they managed to separate the contribution of large-angle boundaries and sub-boundaries. They fitted the tracer penetration profiles by a sum of 2 or 3 exponential terms:

$$\bar{c} = c_{01}\exp\left(-m_1 y^{6/5}\right) + c_{02}\exp\left(-m_2 y^{6/5}\right),\qquad(26)$$

where $\bar{c}$ was the averaged tracer concentration (specific activity), m_1, m_2 were the slopes of the straight lines obtained from the decomposition of the

$\log \bar{c}$ $vs.y^{6/5}$ plots. The diffusivities, P_1 and P_2, characterizing large-angle boundaries and sub-boundaries, could be given using Eq.(15):

$$P_i = m_i^{-5/3} \left(\frac{4D}{t} \right)^{1/2} A^{5/3} \qquad i = 1,2 \qquad (27)$$

The penetration profiles and the Arrhenius plots they got are displayed in Figs. 16 and 17 respectively.

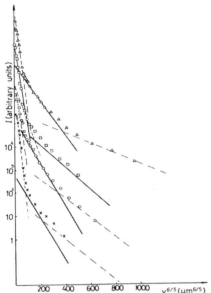

Fig.16. Zn GB penetration profiles measured in polycrystalline Al [58]. Contributions of large-angle and sub-boundaries are separated by the fit using Eq.(26).

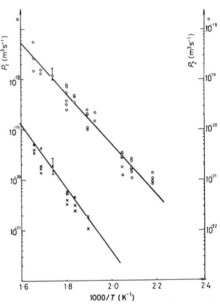

Fig.17. The Arrhenius diagrams of the P_1, P_2 triple products, [58].

It is obvious that in Fig. 16, the lines with higher slopes (lower diffusivity) are due to sub-boundaries, the lines at deep penetration correspond to the transport along large-angle boundaries. As it is expected, the activation energy of the diffusion along sub-boundaries was found to be higher than the activation energy of diffusivity along large-angle boundaries (Fig. (17)).

4.5. Mechanisms of grain-boundary diffusion

Until a decade ago, the prevailing paradigm was that GB diffusion is dominated by a simple vacancy mechanism, just like lattice diffusion, with the modification that the vacancy formation and migration energies are lower. A more detailed picture of the mechanism of GB diffusion is emerging from recent computer simulations [59]: a large variety of GB diffusion mechanisms may occur involving both vacancies and interstitials (see Fig.18). However, these defects can show interesting structural effects, such as delocalization and mechanical instability at certain GB sites.

Both vacancies and interstitials appear to be equally important in GBs with respect to their equilibrium concentrations and participation in the diffusion process.

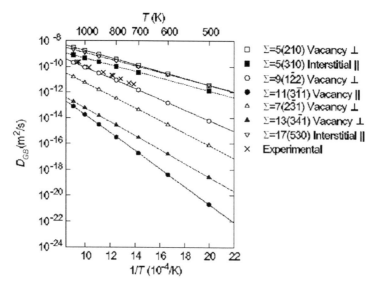

Fig.18. Arrhenius plot of GB diffusion coefficients in Cu obtained by simulations [59]. For each GB, only diffusion in the fast direction (⊥ - normal to the tilt axis, || - parallel to the tilt axis) is shown. The defect, dominating GB diffusion at a given boundary, is indicated in the legend.

The point defect formation energies in GBs are lower on average than in the bulk, but show dramatic variations from site to site, (see Fig.19).

Such variations, as well as the delocalization and instability effects, can be explained by large internal stress gradients that are invariably present in GBs [59].

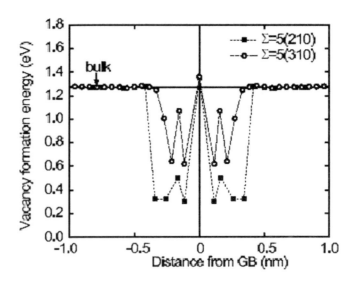

Fig. 19. Calculated vacancy formation energy as a function of distance to the GB core for
    selected GBs in Cu [59].

The experimental investigation of the pressure dependence of diffusion is a
good tool to reveal the role of point defects in the diffusion process. External
hydrostatic pressure affects the rate of solid state diffusion according to the
equation:

$$D = D_0 \exp\left(-\frac{\Delta V p}{RT}\right),$$ (28)

where ΔV is the activation volume of the diffusion process and p is the external
pressure. The activation volume can be determined from the $D(p)$ function by
diffusion heat treatments in a high pressure cell in the pressure range of at least
0-1 GPa. An activation volume close to the atomic volume is a strong evidence
for vacancy-assisted diffusion [60].

       High pressure effect on grain boundary diffusion was studied first by
Martin at al. [17]. They measured silver self-diffusion in random oriented Ag
bicrystals and their results of $\Delta V_{GB} = (1.1 \pm 0.2)\Omega$ (Ω is the atomic volume)
supported the assumption of vacancy mechanism.

       Erdélyi et al. measured the pressure effect on grain boundary diffusion in
microcrystalline Al material [18]. Separating the contribution of large-angle

boundaries and sub-boundaries, they deduced activation volumes for these types of boundaries (see Fig. 20).

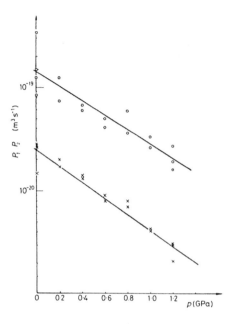

Fig. 20. Pressure dependence of the P_1, P_2 diffusivites for Zn GB diffusion in Al [18].

Their results (with less uncertainty than those of Ref. 17) confirmed the vacancy-mechanism and suggested that the activation volume does not change considerably with the type of the boundary as well. The GB activation volumes ($\Delta V_{GB1}, \Delta V_{GB2}$) were found to be a little bit smaller than the activation volume characterizing volume diffusion. This tendency is expected, since the relaxations around a vacancy situated at a boundary, are larger than those in the bulk.

Summarizing the results concerning the mechanism of diffusion: there are some indications from simulations that interstitials may contribute to the diffusion in GBs, however, experiments carried out till now, support the model of the vacancy-assisted diffusion unambiguously.

5. SEGREGATION EFFECTS

The measurement of the segregation factor is possible by means of Auger electron spectroscopy (AES), investigating the composition *in situ*, along grain boundaries on fractured surfaces. However, this method is suitable only for brittle materials such as ceramics and intermetallic phases. Pure metals can hardly be fractured, but the following indirect methods, based on GB heterodiffusion, can be applied:

   i) the segregation factor can be determined from grain boundary diffusion experiments carried out both in type-B and type-C kinetic regimes;

   ii) segregation parameters can be extracted from the shape of grain boundary diffusion profiles detected in type-B kinetic regimes.

5.1. Determination of segregation parameters from type-B and type-C measurements

The P triple product can be determined in a wide temperature range from B-type measurements, the grain boundary diffusion coefficient D_{GB}, as described above, can be extracted from C-type experiments. Combining the D_{GB} grain boundary diffusion coefficients, usually measured at rather low temperatures, and the $k\delta D_{GB}$ triple product values, extrapolated from type-B measurements, one can get the segregation factor [61,62]:

$$k = \frac{(\delta k D_{GB})\,|_B}{(\delta D_{GB})\,|_C} \ .$$

(29)

The triple product follows the Arrhenius-law

$$k\delta D_{GB} = (k\delta D_{GB})_0 \exp\left(-\frac{Q_{GB}}{kT}\right) \ ,$$

(30)

where Q_{GB} is the apparent activation energy. Since the segregation factor has similar temperature dependence (see Eq.(22)), the apparent activation energy is the sum of the grain boundary activation energy and the segregation energy:

$$Q_{GB} = E_{GB} + E_s$$

(31)

However, carrying out experiments in type-C regime is a rather hard task. Such series of measurements have been performed only in the last decade using carrier-free radioisotopes and detectors with large counting efficiencies and low background [34,61].

The condition for a type-C regime given by Eq.(2) can be rewritten for self- diffusion with the help of Eq.(14) in the following way:

$$\alpha = \frac{\delta}{\sqrt{Dt}} \rangle\rangle 1 \tag{32}$$

For heterodiffusion, the above condition should be corrected with the segregation factor:

$$\alpha = \frac{\delta k}{\sqrt{Dt}} \rangle\rangle 1 \tag{33}$$

Taking into account that the segregation factor of a strongly segregating impurity can be rather high, the condition of type-C regime becomes less restrictive. It means that the temperature range of type-C measurements for strongly segregated impurities can be extended towards higher temperatures.

Till now segregation factors from type-C tracer diffusion measurements have been determined for the following systems: Te, Ni and Se in Ag [61,62], Se in Cu [63], Au in Cu [64].

In Fig. 21 a typical penetration profile is shown measured in C-regimes [36]. Fig.22 presents the Arrhenius plots of $k\delta D_{GB}$ and δD_{GB} diffusivities measured for Te grain boundary diffusion in Ag [61].

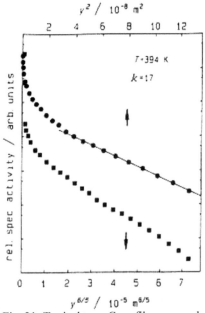

Fig. 21. Typical type-C profile measured for GB self-diffusion in Ag [36]. The tail of the profile shows a downward curvature when plotted the

$\ln \bar{c}$ vs. $y^{6/5}$ function. The

$\ln \bar{c}$ vs y^2 fits better, showing that the conditions of type-C regime are fulfilled.

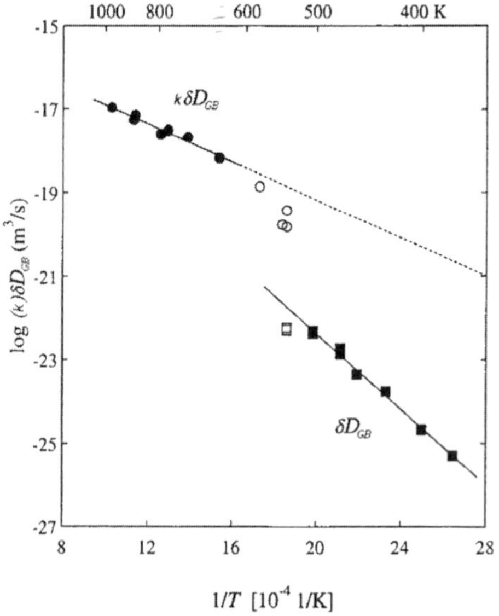

Fig. 22. Te impurity GB diffusion in Ag [61]. Arrhenius plot of the $k\delta D_{GB}$ (circles) deduced from type-B measurements and δD_{GB} (squares), evaluated from type-C experiments. The segregation factor can be deduced using Eq.(29). From the Arrhenius plot the relation $E_{GB} \rangle Q_{GB}$ is obvious.

From tracer measurements on the segregation factors the following trends can be seen [25]:
- Fast (or low) diffusers in the bulk tend to remain fast (or low) diffusers in the grain boundaries;
- Strong segregation tends to reduce the diffusivity in GBs. The stronger the segregation the larger the retardation effect.

Ni for example is a slow diffuser in the bulk and proved to be a strong segregant in Ag. These tendencies result in very low diffusivity in GBs, see Fig. 24.

As for Te and Se, the trends act in opposite directions. These elements are fast diffuser in an Ag lattice, but they are strong segregants in Ag (see Fig. 23). The competition between the two trends results in a mixed behavior: their GB heterodiffusion is faster than self-diffusion at high temperatures and slower than that in low temperature region (Fig.24).

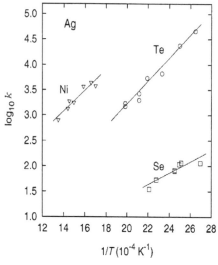

Fig. 23. Arrhenius plot of the GB segregation factors of Te, Se, and Ni in Ag, after [61,62]. The segregation factors increase with decreasing temperature, i.e. E_s <0, see also Eq.(22).

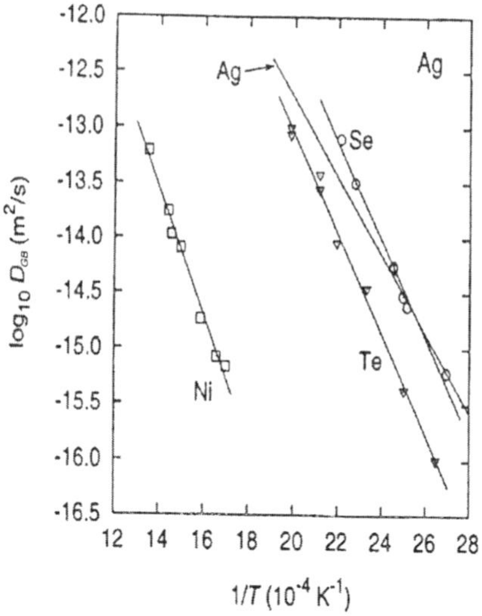

Fig. 24. Temperature dependence of P for GB heterodiffusion in Ag, [61,62,25].

5.2. Extraction of segregation parameters from penetration profiles, measured in B-kinetic regime

It is a general observation that for strongly segregating impurities, the first part of the penetration plot is curved; the triple product can be evaluated from the deeper region of the plot. Theoretical calculations attributed the curvature to a strong effect of solute segregation [51] (see Fig. 13.).

Supposing McLean's segregation isotherm, in the near surface region of the profile, the average bulk concentration drops rapidly, but the grain boundary concentration remains in saturation (see Fig. 12), i.e. the apparent segregation factor changes with the penetration depth. The idea of Bernardini et al. [52] was that the curvature should disappear when the c_{GB}/c ratio is held constant in the saturation regime in the whole volume of the sample. They studied tracer diffusion of Ag in pure Cu and in Cu-0.91 at% Ag alloy. The grain-grain boundary distribution was established by the pre-annealing of the sample. The experiments confirmed the expectations: the plots $\ln \bar{c}$ vs. $y^{6/5}$ measured on the alloy samples were linear and, on the other hand, the penetration plot measured on pure Cu samples showed an extended upward curvature (Fig. 25).

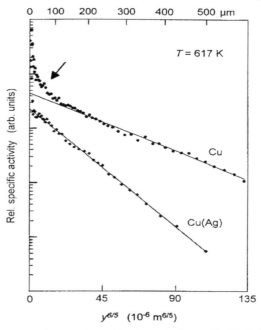

Fig. 25. Ag penetration plots in pure Cu and in Cu(Ag) alloy at $T=617$ K [52].

The migration of tracer atoms is not affected by segregation in Cu(Ag) alloys since grain boundaries are saturated with inactive silver atoms, i.e. the temperature dependence of P_{alloy} does not contain the segregation energy. On the other hand, tracer atoms diffusing in pure copper establish a distribution that is required by segregation. Consequently, the energy of segregation can be determined by subtracting the energy terms describing the temperature dependence of the triple products P_{pure}, P_{alloy}. For the initial slope of McLean's isotherm, Bernardini et al. [52] gave the following expression:

$$k \approx \left(\frac{D_{GB\,alloy}}{D_{GB\,pure}} \frac{P_{pure}}{P_{alloy}} + c - 1 \right) \frac{1}{c}. \tag{34}$$

The ratio of the grain boundary diffusion coefficients in the above expression can be estimated from the $\dfrac{P_{pure}}{P_{alloy}}$ ratio, measured at the highest temperature [52]. The temperature dependence of the silver segregation factor was given as

$$k = 1.2\exp\left(\frac{37\text{kJmol}^{-1}}{RT}\right). \tag{35}$$

The energy of segregation was in good agreement with the value obtained from AES- measurements [65].

An interesting radiotracer technique, Scanning Radiochemical Microanalysis, (SRM) was proposed by Cabane et al., which is suitable for studying GB diffusion and segregation simultaneously in a bicrystal [66]. The principle of the method is shown in Fig. 26. The tracer distribution around a GB is measured through a narrow (10-50 μm) slit, moving very slowly along the surface of the sample. Only the particles, which pass through the slit, are detected and so the *activity vs. x* function can be measured. The same measurement is repeated, after removing a section parallel to the original surface, see Fig. 26. A quantitative analysis of the *activity* vs. *distance* functions, detected at different depths, permits to deduce both the segregation coefficient and the triple product. By means of this method, the sulfur segregation coefficient in a Ni bicrystal was measured and the segregation factor at 973 K was found to be $k=7000$ [66].

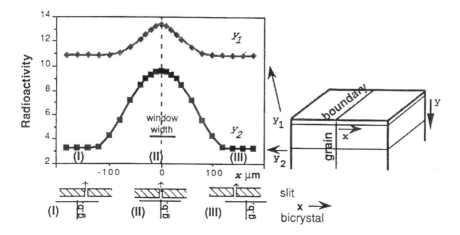

Fig. 26. Determination of the *activity vs. x* distance function measured at different y_1, y_2 depths moving a detector with a narrow slit [66].

Though the method was described more than 20 years ago, since then the technique has hardly ever been used. It is worth considering the revitalization of the method and with some technical improvement the technique could become competitive even with sophisticated ones.

Segregation and GB diffusion in thin films can be studied by the surface accumulation technique [67]: atoms A from a continuous layer diffuse along GBs through a nanocrystalline film (made of B atoms), to the opposite surface, where they spread out by surface diffusion and accumulate. If the surface diffusion is fast enough, the rate of accumulation is controlled by grain boundary diffusion. At low temperatures, the diffusion is restricted to the GBs, i.e. conditions of type-C regime hold. By means of Auger-electron-spectroscopy (AES) the rate of the accumulation can be detected, and from the kinetics, the parameter, $\omega' = P/\delta_s k_s$ can be obtained. In this expression, δ_s is the thickness of the segregated surface layer and k_s is the surface segregation factor. The comparison of the results of surface accumulation and tracer experiments is not simple because ω' contains the surface segregation factor as well. However, combining the results of tracer and surface accumulation measurements, one can determine the surface segregation factor. For example, the ω' parameter was determined for Ag GB diffusion in nanocrystalline copper film in Ref. 68. From the comparison of these data with the triple products determined by tracer technique in type-B kinetic regime in pure Cu, and in Cu/Ag alloy [52], the energy of the surface segregation k_s was determined (E_s=34±14kJ/mol) [68]. Fig. 27 shows the Arrhenius plots of the grain boundary triple products (P) determined in type-B and type-C regimes by means of the tracer method and the surface accumulation technique, respectively.

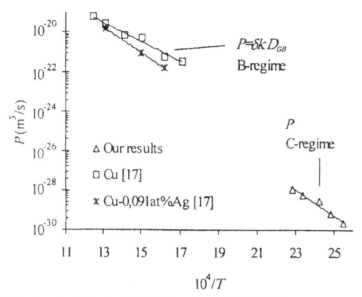

Fig. 27. Arrhenius plots for Ag GB diffusion in Cu. Diffusivities were measured at high
temperatures by tracer method [52], at low temperatures by surface accumulation
technique [68].

6. INTERPHASE BOUNDARY DIFFUSION

Two-phase alloys have important industrial applications. An important class of
them, those with lamellar structure, offers either good high temperature strength
(Ti-Al alloys) or they are indispensable in important technologies (Pb-Sn
solders). The mobility of atoms in them can be studied by an experimental
technique developed for studying the transport along grain boundaries.
Interphase boundary diffusion is the rate-controlling step in the kinetics of
degradation of such two-phase structural materials. However, data collected till
now are sparse and we are only at the beginning of establishing the relationship
between the structure of interphases and their diffusivity.

Bondy et al. [69] and later on Martin [70] developed phenomenological
models in order to describe the diffusion along a single interphase boundary
(IB). They generalized Fisher's model, which was developed originally for grain
boundary diffusion. A more rigorous solution of the problem was published later
by Mishin and Razumovskii [71]. According to their treatment, Eqs. (7-11)
describe the transport in an interphase boundary and the out-diffusion of the
atoms into the adjacent phases as well. The initial and boundary conditions
given by Eq.(12) for interphase boundary diffusion are modified in the
following way:

$$c_1(x, y = 0, t = 0) = c_0; \; c_1(x, y, t = 0) = 0;$$
$$c_1(x, y = \infty, t) = 0; \; c_1(x = -\delta/2, y, t) = c_{IB}(y,t)/k_1$$
$$c_2(x, y = 0, t = 0) = c_0 \; c_2(x, y, t = 0) = 0;$$
$$c_2(x, y = \infty, t) = 0 \; c_2(x = +\delta/2, y, t) = c_{IB}(y,t)/k_2 \tag{36}$$

Eqs.(36) represent the initial and boundary conditions on both sides of the boundary. The distribution of the tracer between the boundary and the adjoining volumes of phases is characterized by two segregation factors k_1, k_2. Correspondingly, two β parameters are necessary for the description:

$$\beta_i = \frac{k_i \delta D_{IB}}{2 D_i^{3/2} t^{1/2}}, \; i = 1,2, \tag{37}$$

where D_{IB} designates the phase boundary diffusion coefficient. The leakage of the tracer from the interphase boundary into the adjacent regions can be described by an effective volume diffusion coefficient, determined by

$$D_{eff} = \frac{1}{4} \left(\frac{D_1^{1/2}}{k_1} + \frac{D_2^{1/2}}{k_2} \right)^2. \tag{38}$$

This effective volume diffusion coefficient is used to define the dimensionless α parameter (see also Eq. (14)):

$$\alpha = \frac{\delta}{2\sqrt{D_{eff} t}}. \tag{39}$$

Taking into account the necessary modifications, as summarized above, the tracer phase boundary experiments can be evaluated in B-kinetic regime from the ln _(activity)_ vs. $y^{6/5}$ plots using the equation:

$$P = \left(\frac{\partial \ln \bar{c}}{\partial y^{6/5}} \right)^{-5/3} \left(\frac{4 D_{eff}}{t} \right)^{1/2} A^{5/3}. \tag{40}$$

In the following, some experimental results are reported. First, we review studies focused on the misorientation dependence of phase boundary diffusion using grown or diffusion bonded bicrystals. These investigations may serve as a test for different structure models.

Sommer et al. investigated the Ag tracer diffusion along (111) and (011) interfaces in diffusion bonded Ag/Cu bicrystals [72]. The following mutually perpendicular directions were chosen as diffusion directions: $[01\bar{1}]$ and $[\bar{2}11]$ for the boundary (111) and $[100]$ and $[01\bar{1}]$ for the boundary (011) (Figs. 28 and 29). Two different misfit dislocation networks were predicted by computer modeling for the investigated boundaries. Results on diffusivities seem to correlate with the boundary structures. Diffusion along the (011) interphase boundary was found to be anisotropic and the diffusivity was higher in the [100] direction than in the $[01\bar{1}]$ direction, with activation energies of 104 kJ/mol and 117 kJ/mol, respectively. These results are in agreement with the predicted anisotropic misfit dislocation structures, where channels of easy vacancy formation sites run along [100].

The significant scatter of the data measured in (011) boundaries (Fig. 29) was explained by facets formation.

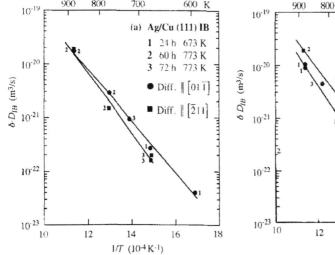

Fig. 28. Arrhenius plots for the diffusion of [110m]Ag along Ag/Cu (111) interphase boun-daries [72].

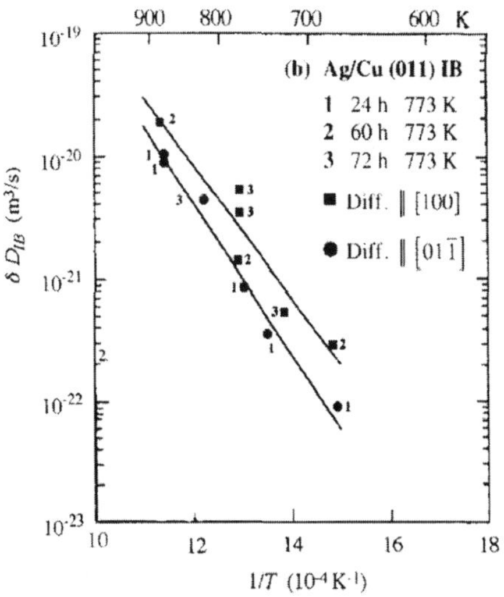

Fig. 29. Arrhenius plots for the diffusion of [110m]Ag along Ag/Cu (011) interphase boundaries [72]. The anisotropy is more pronounced in this structure.

The diffusion in the (111) boundaries was found to be less anisotropic than expected. The anisotropy was detectable only at low temperatures (Fig. 28). The diminishing anisotropy was explained by the influence of excess vacancies formed during the interdiffusion between the two parts of the Ag/Cu bicrystal.

The site-to-site variation of the vacancy formation energy is "washed out" when excess vacancies are annihilated at the interface.

Phase boundary diffusion of the constituents in Pb-62wt% Sn eutectic specimens showing lamellar structure was studied by Gupta et al. [73]. The phase boundary diffusivities were found to be definitely lower and the activation energy higher, (about 77-85 kJ/mol) than the corresponding data for GB diffusion measured in polycrystalline Pb-Sn alloys (Q_{GB}~40-50 kJ/mol). This result also reflects highly ordered, coincident phase boundary structure. However, the lamellar structure proved to be unstable above 400 K. Above that temperature, diffusivities increase radically and approach a range, which is characteristic for GB diffusion (see Fig. 30.).

Studies concerning diffusion along interphase boundaries may serve as indirect tests of structural models. Tracer investigations highlighted the unstable character of phase boundaries. Structural changes, such as coherent-incoherent transition, movements of misfit dislocations, faceting affect the anisotropy and the magnitude of the diffusivity.

Unfortunately, the traditional sectioning methods are not appropriate to detect fine details of the profiles, for example, their asymmetrical character (Fig.3). In the future, traditional experimental techniques, for instance, autoradiography and SRM method, may have an important role in the investigation of the interphase diffusion.

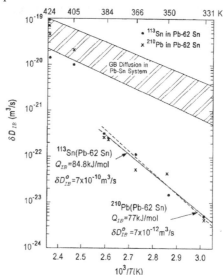

Fig. 30. Arrhenius plot of the interphase diffusivity of ^{210}Pb and ^{113}Sn tracers in Pb-62 wt% Sn eutectic specimens. Note the abrupt change above 400 K, after [73].

7. OPEN PROBLEMS, FUTURE CHALLENGES; INVESTIGATIONS IN NANOSTRUCTURED MATERIALS

7.1. New problems in the study of nanostructured materials

As we have seen, grain boundaries and interfaces are generally diffusion short circuits; consequently, the major part of material transport will occur by grain-boundary/interface diffusion in nanomaterials where a large amount of atoms can lie on grain or interphase boundaries (about 50% for $d=5$ nm and 20% for $d=10$ nm). The development of nanolayered materials and bulk nano-materials prepared, e.g., by compaction of nanocrystalline powders, by severe plastic deformation or by deposition of thin films and multilayers, is now raising the question of how the overall mass transport changes as the microstructural scale (at least in one direction) is reduced to nanometer dimensions [74].

It is an interesting question whether the GB diffusion coefficients measured in these alloys are identical to those obtained in microcrystalline state (e.g. from diffusion experiments made in type-B regime) or not. There is an increasing number of experimental evidence that the above diffusion coefficients agree well with each other.

Stress effects on diffusion in nanomaterials can also be much more important than in microcrystalline samples. Here, large stress gradients can exist, either on atomic scale in the interfaces and/or long-range stress fields along them. Furthermore, due to shorter diffusion distances (times), in many cases, the stress relaxation cannot take place during the diffusion anneal, and feedback effects can become important.

In this section, we will see that one of the interesting questions is related to the traditional classification [75, 23] of type-A, B and C regimes of GB diffusion. This classification fails in nanomaterials, even if they contain only one type of short circuits with a single diffusion coefficient only [76], because the B regime cannot be realized and a variety of possible sub-regimes should be introduced (C-C, B-B, AB-B, A-B sub-regimes) and investigated experimentally [77,78]. Similarly, segregation effects can also be different from the well-known microcrystalline case [54,76].

7.2. Diffusion mechanisms

There is experimental evidence that the diffusion coefficients measured in microcrystalline and nanocrystalline state agree with each other very well, i.e., in most of the cases the structure of relaxed grain-boundaries in nanocrystalline and polycrystalline samples are very similar. Nevertheless, in some cases, the diffusion coefficients and, thus, the nature of GB seems to depend on the technique applied for the synthesis of the material more than on the grain size. This can be caused either by non-relaxed structures of the boundaries and/or by contamination effects [78,79]. On the other hand, there is still a discussion in the literature on how large the difference in the diffusion coefficients measured along non-relaxed and well-relaxed boundaries can be. It is still an open question whether the time of structural relaxation is long enough to allow finding time dependent D values (similarly to that illustrated clearly for tracer diffusion in amorphous metals during structural relaxations [80]).

In subsection 4.5., our present knowledge on grain boundary diffusion mechanisms has been summarized on the basis of a very recent review by Suzuki and Mishin [59]. As we have seen, it was obtained from computer simulations that vacancies and interstitials are equally important even for GB self-diffusion and large internal stresses and stress-gradients exist in GB, resulting in an unusual behavior of point defects and a multiplicity of diffusion mechanisms. These statements are challenges for future experiments. Indeed, we have seen that experimental evidence exists only for vacancy mechanism (at least for the cases when the bulk diffusion is realized by a vacancy mechanism) and no direct evidence exists for the interstitial mechanism. The role of local stress fields (changing abruptly even at an atomic scale) is also very interesting and there are no experimental confirmations of results obtained from computer simulations.

7.3. Effect of stresses

Beside the role of local stresses, further difficulty can be related to the role of (either diffusion-induced or built-in) long-range stress fields. In many treatments of diffusion (at least in metallic systems), it is supposed that the relaxation of diffusion-induced stresses, proportional to stress free strains created by the differences in atomic volumes and intrinsic diffusion coefficients, is fast and complete (see e.g. Darken's treatment of interdiffusion [81]). However, for short diffusion distances, the time of relaxation of stresses can be longer than the time of diffusion. Thus feedback effects of stresses might become important, and cannot be ignored any more [76,78,82,83].

Furthermore, owing to the very short length of the internal interfaces in thin films and nanomaterials, large stress gradients can exist in them. This may explain the large differences in D_{GB} values obtained in different thin film systems under different stress conditions [78]. Indeed, Bokstein and his co-workers [78,84,85] illustrated that *the mechanical stress field* (both the stress gradient and its hydrostatic component) can have a considerable influence on the GB diffusion in thin films. The effect of the stress gradient and the effect of the hydrostatic component can be separated as "gradient effect' and "mobility effect". In Ref. [84] the authors argue that the mechanical stress can reach as high values as 1 GPa and the small film thickness can lead to large gradients as well. Furthermore, at low temperatures, the stress relaxation can be hindered and the hydrostatic component of the stress field can either enhance or suppress the grain boundary diffusion (through the exponential dependence of the mobility on the pressure). At the moment, there is no equivocal evidence for the correctness of the above explanation for the observed an order of magnitude high differences of the grain boundary diffusion coefficients measured in thin films prepared on different substrates (and thus being in different stress states), although the above explanation offers a plausible solution.

7.4. Refinements of Harrison's classification

We have seen in subsection 3.2. that in classical treatments of grain or interface diffusion, three different grain-boundary diffusion regimes are distinguished: type-A, B and C. In type-C regime the diffusion takes place only along grain boundaries, in type-B regime there is also diffusion into the bulk and in type-A regime the bulk diffusion fields are overlapping.

In nanomaterials, i.e. for grain sizes, d, in the order of 10 nm, type-A *or* C regimes will be more important than the type-B regime. In contrast to micro-crystalline samples, where it most frequently occurs, in nanomaterials the conditions for the type-B regime can not be fulfilled. It can be seen from the condition for type-B regime [23,75] ($100\delta < (Dt)^{1/2} < d/20$; see also Eqs. (7), that if $d\cong10$ nm this condition can not be fulfilled. In type-A regime – according to classical handbooks on diffusion [86], a significant enhancement of intermixing or solid state reactions is observed with an effective interdiffusion coefficient given by the Hart's equation (see Eq. (6)). However, if $d\cong10$ nm then the first term will be dominant, since usually $D_{GB}/D>10^4$. Very recently, Belova and Murch [87] have shown that in order to avoid an overestimation of the effective diffusivity, the generally accepted Hart's equation (Eq. 6) should be replaced by the following equation

$$D_{eff} = \frac{D_{GB}\left[(3 - 2g)D + 2gD_{GB}\right]}{\left[gD + (3 - g)D_{GB}\right]} . \tag{41}$$

Monte Carlo simulations confirmed that in nanomaterials the usual condition for type-A regime (Eq.(5)) is too stringent, the condition for a type-A regime should be modified: $\sqrt{Dt} \geq 0.2d$.

Accordingly, in type-A regime, a high effective diffusion coefficient will characterize the rate in any technically important interdiffusion or solid state reaction process (e.g. in surface alloying). On the other hand, in many cases, the process will take part dominantly along grain- or phase boundaries (type-C regime) leading to such phenomena as degradation of multilayers by grain-boundary grooving, pinhole formation and coarsening [88,89] or solid state phase transformations in thin films.

We have seen above that the type-B regime cannot be realized in a nanocrystalline material. Experiments are carried out either in type-C or type-A regime, depending on the annealing time. Till now it was supposed that GBs of our nanocrystalline material could be characterized by a single value of the GB diffusivity, D_{GB}. However, it became obvious from GB diffusion measurements carried out in microcrystalline materials that the presence of different types of GB's is manifested in a wide distribution of D_{GB}s. Furthermore, if there is a multi-level organization in the structure, then the above clear classification fails even in microcrystalline materials (see, e.g. [75,77,78]). For example, the measurements in type-B regimes can not be simply fitted by one single straight line on the lg *specific activity vs.* $y^{6/5}$ plots, but – because of the presence of different type of GBs with different diffusion coefficients – a non-linear penetration plot is usually observed (see also subsection 4.4.).

In nanomaterials it may occur that dominantly two different types of grain or interphase boundaries (e.g. closed free surfaces and GBs in vacuum condensed nanomaterials), exist with characteristically different D_{GB} values. In this simple (bimodal structure) limit, a variety of additional regimes can be treated (type-C-C, type-B-B, type-AB-B, type-A-B sub-regimes) and investigated experimentally [77,78].

In hetero-GB diffusion experiments with radiotracers (i.e. when the tracer is different from the matrix atoms), segregation coefficients describing the matching conditions between the diffusion source and the grain-boundary (as well as between the grain-boundary and the free surface, if there is a terminal free surface present in the experiment) should also be taken into account. Thus, segregation effects are also important in the evaluation of the diffusion kinetics or profiles (see section 5). In the following a possible solution for the problems exposed above will be illustrated by the recent work of Divinski et al. [77], where *intermediate regimes* between the type-A, type-B and type-C diffusion regimes were defined due to the presence of grain and inter-agglomerate boundaries in nanocrystalline γ-FeNi [77,90].

A stable grain structure is required for practical applications of nanostructured materials, which can be reached by proper heat treatments. The stabilization of the grain structure is also important for tracer diffusion studies (to reach a "relaxed" boundary structure). The two-step nature of the material preparation often leads to a duplex structure: the nano-sized grains are embedded in a network of coarser grains as shown on Fig. 31.

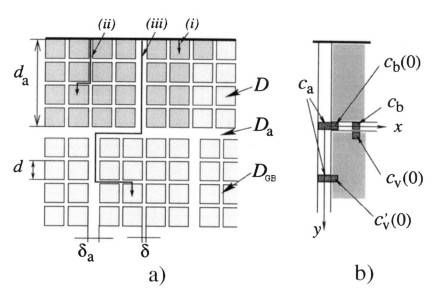

Fig. 31. Schematic representation of a bimodal microstructure. Small nano-grains (of size d) are clustered in agglomerates of the size d_a. δ and δ_a are the widths of nano-GBs and inter-agglomerate boundaries, respectively; D, D_{GB}, and D_a are the bulk, nano-GB, and inter-agglomerate boundary diffusivities, respectively. Possible diffusion walks of individual tracer atoms, which contribute to the fluxes *(i)*, *(ii)*, and *(iii)*, are schematically shown. In (b) the concentrations, which are relevant to the definitions of the segregation factors k and k_a, are indicated [77].

Table 1.
Parameters of kinetic regimes of GB diffusion in a nanocrystalline material with a bimodal structure. c is the layer concentration, y is the penetration depth, q_1 and q_2 are numerical factors, and $\lambda = 2\delta/d$ is the density of nano-GBs intersecting an inter-agglomerate boundary [77]. Other parameters are defined in the text.

Regime	Conditions	Measured parameters	Typical concentration dependence
C-C	$\alpha = \dfrac{k\delta}{2\sqrt{D\,t}} > 1$; $\alpha_a = \dfrac{k_a\delta_a}{2\lambda\sqrt{D_{GB}t}} > 1$	$D_a = \dfrac{1}{4t}q_2^{-1}$	$\bar{c} \sim \exp(-q_2 y^2)$
C-B	$\alpha > 1$; $\alpha_a < 0.1$, $\beta_a = \dfrac{P_a}{2D_{GB}\sqrt{D_{GB}t}} \geq 2$, $\sqrt{D_{GB}t} < d_a/4$	$D_{GB} = \dfrac{1}{4t}q_1^{-1}$ $P_a = \dfrac{k_a\delta_a D_a}{\lambda} = 1.31\sqrt{\dfrac{D_{GB}}{t}}\,q_2^{-5/3}$	$\bar{c} \sim \exp(-q_1 y^2)$ $+ \exp(-q_2 y^{6/5})$
B-B	$\alpha < 0.1$, $\beta \geq 2$, $\sqrt{D_v t} < d/4$; $\alpha_a' < 0.1$, $\beta_a' \geq 2$, $\sqrt{P}\left(\dfrac{\pi t}{4D}\right)^{1/4} < d_a/2$	$P = k\delta D_{GB} = 1.31\sqrt{\dfrac{D}{t}}\,q_1^{-5/3}$ $P_a = \dfrac{k_a\delta_a D_a}{\lambda} = 1.80\sqrt{\dfrac{D_{GB}}{k\delta}}\left(\dfrac{D}{t}\right)^{1/4}q_2^{-2}$	$\bar{c} \sim \exp(-q_1 y^{6/5})$ $+ \exp(-q_2 y)$
AB-B	$d/4 < \sqrt{D\,t} < 3d$; $\alpha_a'' < 0.1$, $\beta_a'' \geq 2$, $\sqrt{D_{eff}^a t} < d_a/4$	$kD_{GB} = 16.48\dfrac{D^{0.1}}{\delta^{0.2}t^{0.9}}q_1^{-4/3}$ $P_a' = k_a\delta_a D_a = 1.31\sqrt{\dfrac{D_{eff}^a}{t}}\,q_2^{-5/3}$	$\bar{c} \sim \exp(-q_1 y^{3/2})$ $+ \exp(-q_2 y^{6/5})$
A-B	$\sqrt{Dt} > 3d$, $\sqrt{D_{eff}^a t} < d_a/4$; $\alpha_a'' < 0.1$, $\beta_a'' \geq 2$	$D_{eff}^a \cong \dfrac{1}{2}\dfrac{kf_{GB}}{1+kf_{GB}}D_{GB} = \dfrac{1}{4t}q_1^{-1}$ $P_a' = k_a\delta_a D_a = 1.31\sqrt{\dfrac{D_{eff}^a}{t}}\,q_2^{-5/3}$	$\bar{c} \sim \exp(-q_1 y^2)$ $+ \exp(-q_2 y^{6/5})$
A	$\sqrt{D_{eff}^a t} > 3d_a$	$D_{eff}^M \cong \dfrac{kf_{GB}/2}{1+kf_{GB}}D_{gb} + \dfrac{kk_a f_a/2}{1+kk_a f_a}D_a = \dfrac{1}{4tq}$	$\bar{c} \sim \exp(-q_2 y^2)$

Using the technique of powder metallurgy and sintering, a nanocrystalline Fe–Ni alloy was produced in Ref. 90, which was stable up to 1000K. The authors could reach penetration depths up to several hundred nanometers, which allows tracer atoms to sample a huge number of grain boundaries. The systematic measurements were combined with new techniques of analysis, providing information on the diffusion and segregation properties of the grain boundaries, as well.

The analysis of the penetration profiles must also take the duplex structure of the material into account. Three potential diffusion paths should simultaneously be taken into account: diffusion along inter-agglomerate boundaries (D_a), diffusion along nanograin-boundaries (D_{GB}), and diffusion into the nanograins (D). One can assume that, in the temperature range of the tracer experiments, $D_a >> D_{GB} >> D$ segregation can take place both at the inter-agglomerate boundaries and at the nanograin-boundaries with segregation factors, and boundary widths k_a, k and δ_a, δ, respectively. Thus, regimes of interface diffusion for the nanograin and inter-agglomerate boundaries can be denoted by a two-letter notation. For example, C-B corresponds to type-C kinetics along the nanograin-boundaries (no out-diffusion into the bulk) and type-B kinetics for the diffusion along inter-agglomerate boundaries (fast diffusion along the inter-agglomerate boundaries and outdiffusion into the adjacent nanograin-boundaries). Note that in this case, there is no diffusion directly into the bulk. Table 1. summarizes the classification, the resulting penetration profiles, and the parameters that can be evaluated from tracer measurements.

In the nanocrystalline Fe-Ni alloy, the tracer self-diffusion of Fe and Ni, as well as, Ag heterodiffusion were investigated. The experimental penetration profiles were analyzed according to the above schema. In spite of the different mathematical treatments of different kinetic regimes the derived diffusion characteristics turned out to be quite consistent. A few examples of the fitted experimental profiles are shown in Figure 32.

The Arrhenius parameters could be determined for the diffusion along both nano-GB's and inter-agglomerate boundaries. From the comparison of the diffusion parameters measured in different kinetic regimes in coarse-grained ($P = k\delta D_{GB}$) and nanocrystalline (D_{GB}) alloys of the same chemical composition, the temperature dependence of Ag segregation coefficient, $k = c_{GB}/c\,(0)$, was established. It was found that Ag segregates very strongly at nanograin-boundaries in Fe–Ni alloys, with a segregation enthalpy of $H_s = 47$ kJ/mol. The diffusivity along nano-GBs in these nanocrystalline materials (grain size $d \sim 100$ nm) is similar to that of conventional coarse-grained materials. The presence of inter-agglomerate boundaries affected the diffusion processes considerably and altered the kinetic regimes of interface diffusion, since these inter-agglomerate boundaries present the fastest short-circuit diffusion paths. The Arrhenius functions for Fe Ni and Ag diffusion are shown in Fig. 32.

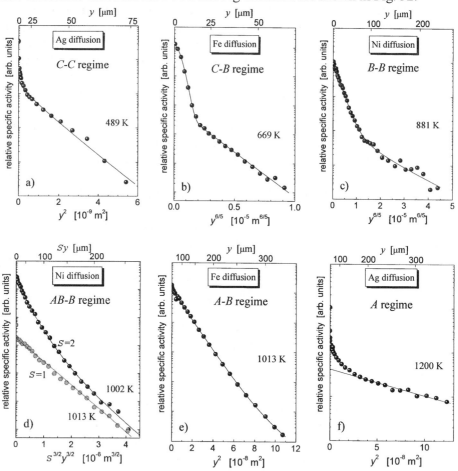

Fig. 32. Examples of penetration profiles measured for Ag, Fe, or Ni diffusion in nanocrystalline Fe-40wt.%Ni alloy in the C-C (a), C-B (b), B-B (c), AB-B (d), A-B (e), and A (f) kinetic regimes. s is a numerical factor; y is the penetration depth [77]. The solid lines represent the relevant fits used to determine interface diffusion parameters (see Table 1).

The work summarized above also demonstrates that a nanocrystalline material with stable and relaxed grain structure after a certain grain growth, presents a suitable object for GB diffusion measurements in type-C regime. The huge number of short-circuit diffusion paths can help to overcome the experimental limitations of tracer measurements in coarse-grained materials.

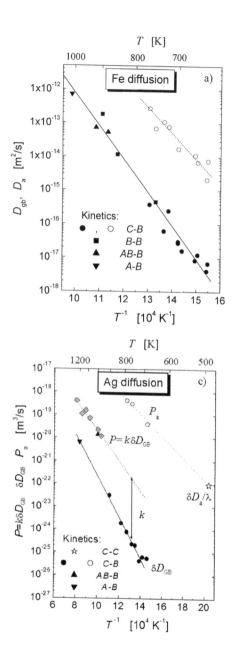
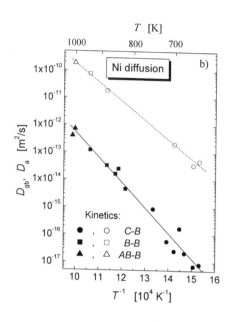

Fig. 33. Arrhenius diagram for Fe (a), Ni (b), and Ag (c) diffusion in nanocrystalline γ-Fe-40wt.%Ni alloy [77]. Filled symbols represent the diffusivity of nanocrystalline GBs and open symbols correspond to inter-agglomerate boundary diffusion. The conditions δ=0.5 nm and δ_a=1 nm were used to recalculate measured P_a and P values to the relevant diffusivities D_a and D_{GB} for Fe (a) and Ni (b) diffusion. In (c) the triple product P for Ag diffusion in coarse-grained FeNi alloy is shown (diamonds). The method of calculation of the segregation factor k for Ag in FeNi is illustrated (c).

Further work in this direction can reveal inherent nanoscale effects using the tracer technique. These effects are related to the break-down of the continuum limit, when the diffusion distance is comparable to the lattice spacing. It was shown theoretically that nonlinear effects can enhance these nanoscale phenomena considerably [82], but direct proof awaits further experiments.

ACKNOWLEDGMENT

This work was supported by the OTKA Board of Hungary (No. T038125).

LIST OF ABBREVIATIONS

AES   Auger Electron Spectroscopy
CSL   Coincident Site Lattice
DIGM   Diffusion-Induced Grain Boundary Migration
GB   Grain Boundary
IB   Interphase Boundary
SRM   Scanning Radiochemical Microanalysis
TiGB   Tilt Grain Boundary
TwGB   Twist Grain Boundary

REFERENCES

[1]   W.C. Roberts-Austen, Phil. Trans. Roy. Soc. (London) A, 187 (1896) 404.
[2]   J. Gróh and G. v Hevesy, Annalen der Physik, 63 (1920) 85.
[3]   J. Gróh and G. v Hevesy, Annalen der Physik, 65 (1921) 216.
[4]   G. Hevesy and A. Obrutsheva, Nature, 115 (1925) 674.
[5]   G. v. Hevesy, Zeitschrift für physikalische Chemie, 101 (1922) 337.
[6]   G. v. Hevesy, Zeitschrift für Physik, 10 (1922) 80.
[7]   R. S. Barnes, Nature (London), 166 (1950) 1032.
[8]   P. A. Beck, P. R. Sperry and H. Hu, J. Appl. Phys., 21 (1950) 420.
[9]   M. R. Achter and R. Smoluchowski, J. Appl. Phys., 22 (1951) 1260.
[10] S. R. L. Couling and R. Smoluchowski, J. Appl. Phys., 25 (1954) 1538.
[11] D. Turnbull and R. E. Hoffman, Acta Met., 2 (1954) 419.
[12] D. Gupta, J. Appl. Phys., 44 (1973) 4455.
[13] A. Atkinson and R. 1. Taylor, Philos. Mag. A, 43 (1981) 979.
[14] H. Mehrer, K. Maier, G. Hettich, H. J. Mayer and G. Rein, J. Nucl Mater., 69 (1978) 545.
[15] Y. Ishida, T. Inoue, T. Yamamoto and M. Mori, Metal. Sci., 224 (1976).
[16] H. G. Tompkins and M. R. Pinnel, J. Appl. Phys., 47 (1976) 3804.
[17] G. Martin, D. A. Blackburn and Y. Adda, Phys. Stat. Sol., 23 (1967) 223.

[18] G. Erdélyi, W. Lojkowski, D.L. Beke, I. Gödény and F.J. Kedves, Phil Mag. A, 56 (1987) 673.
[19] J. T. Robinson and N. L. Peterson, Surf Sci., 31 (1972) 586.
[20] J. Horvath, R. Birringer and H. Gleiter, Sol. State Commun., 62,319 (1987).
[21] D. Turnbull, in: Atom Movements, ASM, Cleveland, 1951 p.129.
[22] P. Shewmon, in: Diffusion in Solids, The Minerals, Metals& Materials Society, 1989 p.189.
[23] D.L. Beke (ed) Difusion in Semiconductors and Non-metallic Solids , Landolt-Börnstein New Series, 1998Vol. 33, subvolume A, Springer, Berlin.
[24] I. Kaur, W. Gust, Fundamentals of Grain and Interphase Boundary Diffusion, Ziegler Press Stuttgart, 1988.
[25] Y. Mishin, Chr. Herzig, Mater. Sci. Eng., A260 (1999) 55.
[26] L.G. Harrison, Trans. Faraday Soc., 57 (1961) 1191.
[27] E.W. Hart, acta Met., 5 (1957) 597.
[28] J.C. Fisher, J. Appl. Phys., 22 (1951) 74.
[29] R.T.P. Whipple, Phil. Mag., 45, (1954) 1225.
[30] T. Suzuoka, Trans. Jap. Inst. Metals, 2 (1961) 25.
[31] H.S. Levine, C.J. McCallum, J. Appl. Phys., 31 (1960) 595.
[32] A.D. LeClaire, Brit. J. Appl. Phys., 14 (1963) 351.
[33] T.J. Renaouf, Phil.Mag., 22 (1970) 359.
[34] P. Gas, D.L. Beke, J. Bernardini, Phil. Mag. Letters, 65 (1992) 133.
[35] I.A. Szabo, D.L. Beke, F.J. Kedves, Phil. Mag. A, 62 (1990) 227.
[36] J. Sommer, Chr. Herzig, J. Appl. Phys., 72 (1992) 2758.
[37] A. Atkinson, R.I. Taylor, Phil. Mag., A39 (1979) 581.
[38] W.T Read, W. Shockley, Phys. Rev., 78 (1950) 275.
[39] N.F. Mott. Proc. Phys. Soc., 60 (1948) 391.
[40] D. Turnbull, R.E: Hoffman, Acta Metall., 2 (1954) 419.
[41] W.R. Upthegrove, M.J. Sinnot, Trans ASM 50 (1958) 1011.
[42] R.F: Cannon, J.P. Stark, J. Appl. Phys. 40 (1969) 4361.
[43] R.E. Hoffman, Acta Metall. 4 (1956) 97.
[44] B.B. Straumal, B. Bokstein, I.M. Klinger, L.S. Shvindlerman, Scripta Metall. 17 (1983) 275.
[45] D.L. Beke, in „Diffusion in Solids- Unsolved problems" (Ed. G. E. Murch), Trans. Tech. Publ. Zürich, 1992 p. 31.
[46] J. Sommer, Chr. Herzig, S. Mayer, W. Gust, Defect Diff. Forum 66–69 (1989) 843.
[47] Q. Ma, R.W. Balluffi, Acta Metall. Mater. 41 (1993) 133.
[48] T. Surholt, D.A. Molodov, Chr. Herzig, Acta Mater. 46 (1998) 5345.
[49] G.B. Gibbs, phys. stat. sol., 16 (1966) K27.
[50] D. McLean, Grain-Boundaries in Metals, Clarendon Press, Oxford (1957).
[51] B.S. Bokstein, V.E. Fradkov, D.L. Beke, Phil. Mag. A65 (1992) 277.
[52] J. Bernardini, Zs. Tőkei, D.L.Beke, Phil. Mag., A73 (1996) 237.
[53] Y. Mishin, Chr. Herzig, J. Appl. Phys., 73 (1993) 8206.
[54] D.L. Beke, C. Cserháti, Z. Erdélyi and I.A. Szabó "Segregation in Nanostructures" in Nanoclusters and Nanocrystals" (Ed. H.S. Nalwa) American Scientific Publishers, North Lewis Way, Stevenson Ranch, California, 2003 p. 211.
[55] P. Gas, J. Bernardini, Surf. Sci., 72 (1978) 365.
[56] G. Erdélyi, K. Freitag, G. Rummel, H. Mehrer, Appl. Phys., A53 (1991).
[57] T. Surholt, Chr. Herzig, Acta Mater., 45 (1997) 3817.
[58] D.L. Beke, I. Gődény, G. Erdélyi, F.J.Kedves, Phil. Mag., A56 (1987) 659.

[59] A. Suzuki and Y. Mishin in „Nanodiffusion" (ed. D.L. Beke) Trans Tec., J of Metstabl. and Nanocyst. Mat. 19 (2004) p. 1.

[60] F.J. Kedves, G. Erdélyi, Defect and Diffusion Forum 66-68 (1989)175-188.

[61] Chr. Herzig, J. Geise, Y. Mishin, Acta Met., 41 (1993) 1683.

[62] T. Surholt, Chr. Minkwitz, Chr. Herzig, Acta Mater. 37 (1997) 119.

[63] Chr. Herzig, Mater. Sci. Forum, 207-209 (1996) 481.

[64] T. Surholt, Y. Mishin, Chr. Herzig, Phys. Rev., B50 (1994) 3577.

[65] M. Menyhard, Mater. Sci. Forum, 126-128 (1993) 205.

[66] M. Pierantoni, B. Aufray, F. Cabane, Acta Metall., 33 (1985) 1625.

[67] J.C.M. Hwang, J.D. Pan, R.W. Baluffi, J. Appl. Phys., 50 (1979) 1339.

[68] Z. Erdelyi, Ch. Girardeaux, G.A. Langer, D.L. Beke, A. Rolland, J. Bernardini, J. Appl. Phys. 89 (2001) 3971.

[69] A. Bondy, P. Regnier, V. Levy, Scripta Met., 5 ((1971) 345.

[70] G. Martin, Acta Metall., 23 (1975) 967.

[71] Y. Mishin, I.M. Razumovski, Acta Met., 40 (1992) 597.

[72] J. Sommer, T. Muschik, Chr. Herzig, W. Gust, Acta Mater., 44 (1996) 327.

[73] D. Gupta, K. Vieregge, W. Gust, Acta Mater., 47 (1999) 5.

[74] D.L Beke (editor) „Nanodiffusion" Trans Tec., J of Metstabl. and Nanocyst. Mat. 19 (2004).

[75] I. Kaur, Y. Mishin, W. Gust, Fundamentals of Grain and Interphase Boundary Diffusion, London, J. Wiley & Sons, 1995.

[76] J. Bernardini and D.L. Beke, "Diffusion in Nanomaterials" in " Nanocrystalline Metals and Oxides: Selected Properties and Applications"; Eds. P. Knauth and J. Schoonman; Kluwer Publ. Boston (2001).

[77] S. Divinski, J.-S. Lee and Chr. Herzig, in „Nanodiffusion" (ed. D.L. Beke) Trans Tec., J of Metstabl. and Nanocyst. Mat. 19 (2004) p 55.

[78] B. Bokstein, M. Ivanov, Yu. Kolobov and A. Ostrovsky, in „Nanodiffusion" (ed. D.L. Beke) Trans Tec., J of Metstabl. and Nanocyst. Mat. 19 (2004) p 69.

[79] J. Bernardini, C. Girardeaux, Z. Erdélyi and C. Lexcellent, in „Nanodiffusion" (ed. D.L. Beke) Trans Tec., J of Metstabl. and Nanocyst. Mat. 19 (2004) p 35.

[80] J. Horváth and H. Mehrer, Cryst. Lattice Defects Amorphous Mater., 13 (1986) 1.

[81] D.L. Beke and I.A. Szabó (Eds.) „Diffusion and Stresses" Defect and Diffusion Forum 129-130 (1996).

[82] D.L. Beke, Z. Erdélyi, I.A. Szabó and C. Cserháti, in „Nanodiffusion" (ed. D.L. Beke) Trans Tec., J of Metstabl. and Nanocyst. Mat. 19 (2004) p 107.

[83] O. Thomas, S. Labat, T. Bigault, P. Gergaud and F. Bocquet, in „Nanodiffusion" (ed. D.L. Beke) Trans Tec., J of Metstabl. and Nanocyst. Mat. 19 (2004) p 129.

[84] A. S. Ostrovsky, B. S. Bokstein, Appl. Surf. Sci., 173, (2001) 312.

[85] N. Balandina, B.S. Bokstein, A. Ostrovsky, Def. and Dif. Forum, 143-147 (1997) 1499.

[86] J. Philibert: „Atom Movements. Diffusion and Mass Transport in Solids " Les Ulms, France, Les Editions des Physique, Paris, (1991).

[87] I..V. Belova and G.E. Murch, in „Nanodiffusion" (ed. D.L. Beke) Trans Tec., J of Metstabl. and Nanocyst. Mat. 19 (2004) p 25.

[88] D.L. Beke, G.A. Langer, A. Csik, Z. Erdélyi, M. Kis-Varga, I.A. Szabó, Z, Papp, Defect and Diffusion Forum, 194-199 (2001) 1403.

[89] M. Bobeth, A. Ullrich, W. Pompe, in „Nanodiffusion" (ed. D.L. Beke) Trans Tec., J of Metstabl. and Nanocyst. Mat. 19 (2004) p 153.

[90] S.V. Divinski, F. Hisker, Y.-S. Rang, J.-S. Lee, Chr. Herzig, Acta Mater. (2003) in press.

Radiotracer Studies of Interfaces
G. Horányi (editor)

Chapter 8

Sorption (binding) and transport phenomena in biomembranes

J. Kardos

Department of Neurochemistry, Chemical Research Center, the Hungarian Academy of Sciences, P.O. Box 17, 1525, Budapest, Hungary

1. BIOMEMBRANES

Responsiveness within the animal kingdom translates in molecular sorption (binding) and transport phenomena at interfaces of semi-permeable membranes that isolate the cell compartment. The function of this chapter in the book is to focus on some of the principles, which have applicability to a great variety of biological membranes.

In the second half of the last century, the study of the movement of solutes across biological membranes has progressed from the collection of phenomenological observations on cellular transport and electrical behavior to the identification, purification, characterization and reconstitution of some of the integral membrane proteins responsible for the phenomena [1, 2]. With the objective of reconstituting the very same binding and transport functions seen in the living cell, reconstitution requires the imposition of asymmetry of solute binding and protein's transmembrane orientation upon the system so as to mimic the general asymmetry of biological membranes, a clue for understanding responsiveness. It is not a quaternary structure of integral membrane protein that is reconstituted, but rather a transmembrane function, or permeability, for the systems considered here. We are now beginning to witness a further step in the development of the field: a combination of high-resolution structure determination [3] and bio-informatics [4], in which relationships between stereo-structure and function of integral membrane proteins are being disclosed.

As early as 1926 [5, 6] Clark estimated the number of specific drug receptors present on responsive cells, which appeared remarkably small, e.g., about 10^5 per cell. These estimates - based on pharmacological data - were in excellent agreement with the first direct measurement of the uptake of radioactive atropine by intestinal muscle, which appeared to involve acetylcholine receptors [7], e.g., about 1.5×10^5 per smooth muscle cell. Ranging 10^3-10^4 per cells, hormone receptors, such as insulin, epidermal growth factor and beta-adrenergic, exhibited high affinities with dissociation constants of the order of 10^{-8} to 10^{-10} M [8]. In addition to the small numbers of sites and the low concentrations over which range measurements must be made, the practical requirement to use small quantities of cells or biomembranes necessitated the use of highly sensitive radioisotope techniques [9, 10]. In what follows, the molecular mechanisms of processes occurring at excitable cell, e.g., neuronal membrane interfaces will be addressed as constituting a phenotype whereby radiotracer sensitivity and versatility has been explored most.

Excitable cells respond to external stimuli by changes in permeability and conductivity of monovalent ions. The equilibrium potential (V_m) generated from the difference in concentration of an ion across a permeable membrane is given by the Nernst equation

$$V_m = -2.3 \; RT/nF \; \log_{10} [X^{ne}]_i/[X^{ne}]_o \tag{1}$$

where R is the universal gas constant (1.99 cal/deg Kelvin); T is the temperature (degrees Kelvin), n is valence, F is the Faraday constant (23,500 cal/volt/mol), $[X^{ne}]_i$ and $[X^{ne}]_o$ denote concentrations of the ion on *in* and *out* sides of a permeable membrane, respectively. The presence of multiple permeant ions, primarily monovalent Na^+, K^+ and Cl^- ions gives rise the Goldman-Hodgkin-Katz equation for estimation of V_m in neuronal cells (for full review see [11]),

$$V_m = 2.3 \; RT/F \; (P_K[K^+]_o + P_{Na}[Na^+]_o + P_{Cl}[Cl^-]_i)/(P_K[K^+]_i + P_{Na}[Na^+]_i + P_{Cl}[Cl^-]_o) \tag{2}$$

where P_K, P_{Na} and P_{Cl} are the permeability values for Na^+, K^+ and Cl^- ions, respectively. Neuronal membranes are excitable, which means that membrane depolarization can be sustained and propagated along the membrane *via* the action of voltage-activated Na^+ ion- and K^+ ion-channels. Depolarization of membrane induces the *in* → *out* release of intracellularly stored bioactive ligands (chemical signal) and their binding to their specific membrane targets like ligand-gated ion channel (ion channel) and G-protein-coupled seven transmembrane helices (7TM) receptors embedded in the membrane of the next cell thereupon [11-13]. Molecular recognition does in turn trigger a series of processes involving transmembrane flux of ions involving voltage- and ligand-gated ion channel and 7TM receptors, ionic pumps and enzymatic and/or carrier-mediated *out* → *in* clearance of bioactive ligands, that constitute the primary messenger system of functioning neural cells [11-13].

2. BINDING OF BIOACTIVE LIGANDS AND DRUGS TO THEIR SPECIFIC MEMBRANE RECEPTORS

In general, the binding of a ligand (L) to its receptor site (R) can be described in the following reversible reaction scheme

$$L + R \underset{k_{-1}}{\overset{k_{+1}}{\rightleftharpoons}} LR \tag{3}$$

The kinetic rate constants, k_{+1} and k_{-1}, and the equilibrium dissociation constant, K_d are related such that

$$K_d = k_{+1}/k_{-1} = [L][R]/[LR] \tag{4}$$

2.1. Estimation of affinity and binding site concentrations with radioligands: Langmuir-type saturation isotherms

If we can assume i) that the reaction is reversible; ii) that the ligand concentration is much higher than the binding site concentration; and iii) that site-site interactions do not occur, than the saturation formalism of enzyme kinetics, i.e. a modified Michaelis-Menten equation can be derived [14] that yields

$$[LR] = [LR]_{max}[L]/(K_d + [L]) \tag{5}$$

where $[LR]$, $[LR]_{max}$ and $[L]$ denote the bound (B) ligand concentration, the maximal (B_{max}) concentration of the receptor-radioligand complex and the free (F) ligand concentrations, respectively. When $K_d = F$, then Eq. (5) indicates that $B = 0.5 B_{max}$. Thus, K_d is the concentration of a ligand, which produces a half-maximal occupancy of the binding sites. The linear representation of Eq. (5), e.g., Scatchard analysis [15] gives

$$B/F = (-1/K_d)B + B_{max}/K_d \tag{6}$$

Thus, by plotting the ratio of bound to free ligand concentrations against the concentration of bound ligand yields a straight line that has a slope and an abscissa intercept equal to the negative reciprocal of K_d and the total concentration of binding sites, respectively. As for all transformed plots, Scatchard plot may require extrapolation of the fitted line to obtain B_{max} as the x-intercept. Aside from errors accompanying extrapolation [16], Scatchard plots are mainly useful for visualizing data indicating departures from linearity that appear as curves that are concave downward (positive cooperativity) or concave upward (negative cooperativity).

2.1.1. Binding site multiplicity: identification of receptor subtypes

An example of negative cooperativity occurs in the binding of tritiated 1-(3-chlorophenyl)-4-methyl-7,8-dimethoxy-5H-homotphthalazine ([^{3}H]girisopam, EGIS-5810; Fig. 1.) binding sites in rat brain striatal membrane suspensions.

Ethanolic solution of [^{3}H]girisopam with 22.06 Ci/mmol (0.853 TBq/mmol) specific activity has been prepared by Dr. Géza Tóth (Biological Center, HAS, Szeged, Hungary). Rat brain striatal membrane suspension was prepared and the protein concentration was assayed according to [17] and [18], respectively. For equilibrium binding, aliquots (0.8 mg protein/ml) of rat brain striatal membrane suspension in Hepes buffer (pH 7.4) were incubated with 1-40 nM [^{3}H]girisopam for 60 min at 4°C in. The non-specific binding was determined in the presence of 10^{-5} M girisopam, diluted from a 10 mM stock solution in aqueous ethanol. Final ethanol concentration, 0.455 v/v %, had no effect on specific binding. After equilibration, bound [^{3}H]girisopam was separated from the free by centrifugation (10000x*g* for 5 min) in Eppendorf tubes pre-coated with polyethyleneimine to reduce surface-binding of [^{3}H]girisopam. The pellet was rinsed twice, resuspended in 0.3 ml Hepes buffer and the suspension was put in the scintillation vial. Eppendorf tube was rinsed one more time with 0.3 ml buffer. This volume was added to the suspension in the scintillation vial, and after addition of 10 ml benzylalcohol-toluene base scintillation fluid, the vials were left overnight at 4°C in the dark, and the radioactivity was measured by a liquid scintillation counter at 20 % efficiency. In the present example (Fig. 2.), comparisons of one-site and two-site models for equilibrium binding of [^{3}H]girisopam to recognition sites in rat brain striatal membrane suspensions indicated a better fit with the two-site model [19]. Different models (one-site vs. two-site) were compared for statistical differences in goodness of fit by the F-test for incremental improvement with addition of another term [20]. Goodness of fit (P) was calculated as the weighted sum of squares divided by the number of df (reduced χ^2 value: χ^2_r). Statistical analysis was done on untransformed data. The two-site model characterised by an F value of 10.2 greater than the critical level (F=6.7) gave a significantly better fit (P=0.001, χ^2_r=0.23) than the one-site model (χ^2_r=0.39), which fitted outside experimental error. Results from saturation binding experiments suggested an approximately 30 fold difference in binding affinities (K$_{d1}$=4.00 ± 0.4 nM, K_{d2}=130 ± 30 nM) and an about 5 fold difference in the number of these sites (B_{max1}=1.7 ± 0.2 pmol/mg protein, B_{max2}=9.5 ± 1.1 pmol/mg protein).

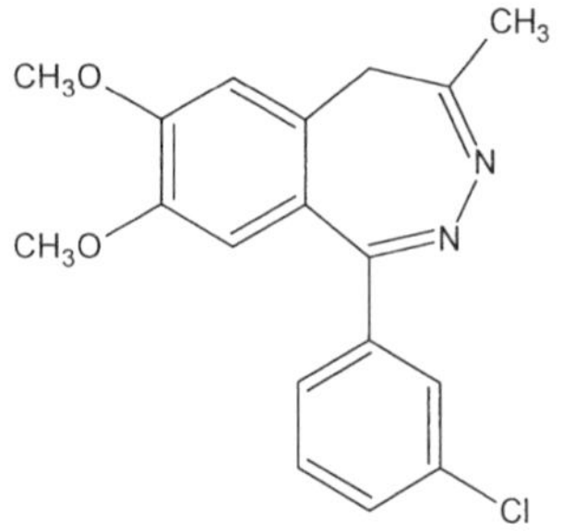

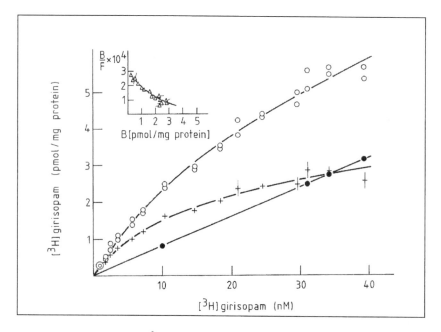

Fig. 1. Structure of the anxiolyitic 2,3-benzodiazepine derivative, 1-(3-chlorophenyl)-4-methyl-7,8-dimethoxy-5H-homotphthalazine (Girisopam, EGIS-5810 formerly GYKI-51189).

Fig. 2. Equilibrium binding of [^{3}H]girisopam to sites in rat brain striatal membranes at 4°C. Inset: Scatchard analysis of specific binding data. Symbols: ○ total binding; ● non-specific binding; + specific binding (vertical bars denote S.E. of the mean for duplicate determinations at each concentrations). All incubations were done in duplicate measurements. I. Kovács and J. Kardos, unpublished data.

Further support for the existence of multiple populations of binding sites can be assessed by dissociation kinetic measurements. The mean lifetime of the bound complex ($t_{1/2}$) can be determined using equations derived from the general solution obtained for the reversible reaction Eq. (3) as follows [21]

$$\ln\left[(x_o - x_\infty)/(x - x_\infty)\right] = k_{+1} Q t \tag{7}$$

$$Q = (c_0 B_{max})/x_\infty \approx K_d \tag{8}$$

$$\ln 2/k_{-1} = t_{1/2} \tag{9}$$

where c_0 is the total concentration of the radioligand. Quantities x_o, x_∞ and x denote concentration of the complex at $t = 0$, $t \to \infty$ and $t = t$, respectively.

To reveal fast dissociating components requires the techniques of rapid chemical kinetics as introduced by Cash and Hess [22] for measuring ion translocation through the plasma membrane (Fig. 3.). Determination of x_o in Eq. (6): rat brain striatal membrane suspension equilibrated with saturating concentration of [³H]girisopam was loaded into syringes 1 and 2 (Fig. 3.). After 3 minutes, aliquots (0.34 ml) were mixed (block 9) and incubated (block 10). After 130 ms of incubation, the suspension was displaced by 0.68 ml of buffer. The mixture was filtered applying reduced pressure. For filtration 2.5 cm diameter glass microporous filters (Whatman GF/B) were used. Each filter was previously treated with polyethyleneimine to reduce binding of [³H]girisopam to the filter. After the sample was filtered, the filter was washed twice with 5 ml of ice-cold Hepes buffer pH 7.4, placed in scintillation vials and counted as above. Determination of bound [³H]girisopam after complete dissociation (quantity x_∞ in Eq. (7) was performed the same way using rat brain striatal membrane suspension equilibrated with [³H]girisopam in the presence of 10^{-5} M girisopam. Determination of the loss of bound [³H]girisopam during incubation with 10^{-5} M girisopam for different times {quantity x in Eq. (7)} was performed as above loading rat brain striatal membrane suspension equilibrated with [³H]girisopam in the absence or presence of 2×10^{-5} M girisopam into syringe 1 and 2, respectively and incubated for different periods of time (130 ms – 30 s). Progress of dissociation with time indicated two phases (Fig. 4.). The fast phase, representing the major component of bound [³H]girisopam was completed within a few seconds with a $t_{1/2}$ value of 0.55 s. A minor component of binding activity was decreased much slower. This phase was characterized by $t_{1/2} = 24$ s. The two components of bound [³H]girisopam, that corresponded different rates of dissociation were in the ratio of 5:1. We can correlate equilibrium binding (K_{d1} and K_{d2}) and dissociation rate constants (k_{-11} and k_{-12}) assuming rate constants for association were unchanged. On this basis an about 40-fold difference in the affinity of sites can be estimated which is in agreement with the results of the equilibrium binding data.

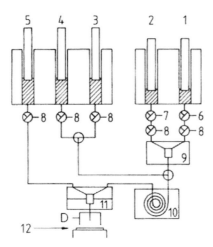

Fig. 3. Plumbing arrangement for the pulsed-mode technique of the quench flow apparatus. Reactant from syringes 1 and 2 are mixed (block 9) and incubated in a thermostatted tube (block 10) before being displaced by buffer from syringes 3 and 4, mixed with the quench solution (block 11) from syringe 5 and filtered (12) with the use of a filter-guide (decelerator, D). 1-5 Syringes are thermostatted. Numbers 6, 7 and 8 indicate outlets to reservoirs for the reactants and the buffer, respectively. After [22].

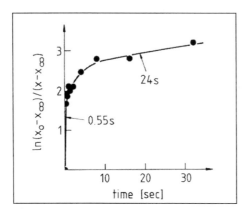

Fig. 4. Progress of dissociation of [³H]girisopam bound to recognition sites in rat brain striatal membranes at 4°C by 10-5 M girisopam. Data were plotted according to the Eq. (7). The numbers are mean lifetimes for the fast and slow phases (arrows) evaluated using Eq. (8) and Eq. (9). Data presented (filled circles) are from 2-6 separate determinations with an S.D. of 3 %. I. Kovács and J. Kardos, unpublished data.

2.1.2. Diagnostic use of a Hill plot

When there is a cooperative effect for ligand binding, Eq. (5) should be modified as follows [23]:

$$B = B_{max}F^{nH}/(K_d + F^{nH}) \tag{10}$$

$$\log(B/B_{max} - B) = \log K_d + nH\log F \tag{11}$$

Concave upward Scatchard plot was seen for example in the binding of [^{3}H]gamma-amino butyric acid ([^{3}H]GABA) to its recognition sites in rat brain membrane suspensions in a physiological buffer [24] (for a general discussion *see* [23]) suggesting positive cooperativity (Fig. 5.). Eq. (11) is known as a Hill plot [25], and the slope of the straight line drawn through the data, nH is often called the Hill coefficient (Fig. 6.). This value is often confused with the number of ligand binding sites, but the two quantities are related only by the restriction that nH cannot exceed the number of sites [26].

A more general approach in ligand binding is to treat each step in the binding process separately [27]. Accordingly, the shape of the Hill plot will depend on the ratio of the binding constants involved. In addition to its slope at the midpoint, a more analytical examination of the actual curve (and asymptotes) on either side of half-saturation can in some cases indicate relationships between the intrinsic binding constants. In the special case in which all the equilibrium constants are equal does it simplify to Eq. (11) [27]. Thus, the Hill plot has a diagnostic value: an almost straight Hill plot with an average nH appreciable different from unity will indicate whether the binding processes are positively ($nH>1$), negatively ($nH<1$), or non- ($nH=1$) cooperative.

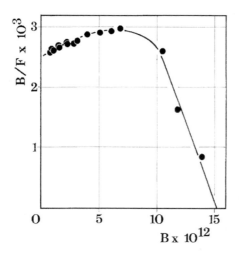

Fig. 5. Scatchard analysis of specific [³H]GABA binding to rat brain synaptic membrane in physiological buffer. Incubations were made with 0.2-16 nM [³H]GABA for 20 min at 4°C. Samples were filtered and bound radioactivity was assayed by liquid scintillation spectrometry. Specific [³H]GABA binding was determined by subtracting non-specific binding (obtained in the presence of 0.1 mM GABA) from the total binding. Data were taken from ref. [24].

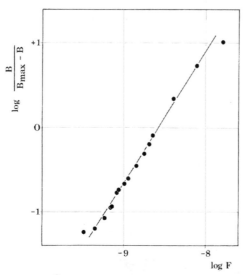

Fig. 6. Hill analysis of specific [³H]GABA binding data shown in Fig. 4. In accordance with the concave upward character of the Scatchard plot, analysis according to Eq. (11) yields n_H=1.55 (r^2=0.99). Data were taken from [24].

2.1.3. Indirect-binding

The earliest reported direct radioligand-binding assay for neurotransmitters measured nicotinic cholinergic receptors in fish and eels [28-29] and subsequently opioid receptors [30-32]. Recent years have seen the development of an arsenal of radiolabeled ligands and methodologies, that allow determination of the equilibrium binding parameters and the concentration of receptors in brain tissue *via* direct binding assays.

In addition to direct binding, one can estimate the binding affinity indirectly by using unlabeled ligand that interacts with the same receptor as the labeled one. Such 'displacement' studies are common and practical in dug design and development. A common procedure is to determine the bound concentration of the radioligand (B for the 'hot' ligand) at a fixed concentration (c_0) in the absence and presence of the increasing concentration of the 'cold' ligand. Concentration of the 'cold' ligand (c) that is necessary to reduce B to one-half (IC_{50}) is related to the equilibrium inhibition constant of the 'cold' ligand, K_I as follows:

$$K_I = IC_{50} / (1 + c_0/K_d) \qquad (12)$$

Eq. (12) is known as a Cheng-Prusoff equation [33]. A displacement graph (B in percent of control *vs.* log c/M) may not approach zero as the concentration of the 'cold' ligand increases, because there is binding to sites other than receptors. Such non-specific binding is often determined from this kind of plot. Since binding interaction between the ligand and receptor shows enantioselectivity (for examples see [34-41]), displacement studies using enantiomeric pairs of compounds like in Fig. 7. are often applied as a critical test to distinguish receptors from other (functionally 'silent') sites. Experiments using non- or slightly selective radioligands in combination with highly selective 'cold' ligands have been successfully applied to disclose multiple subtypes of receptors [42-44].

2.1.4. Radiolabeling for direct and indirect binding studies in vitro

The average concentration of receptors in brain tissue goes from some 10^{-12} M (peptide-receptors) to 10^{-9} M (neurotransmitter receptors). As a consequence, saturation and displacement binding studies require that the radioligand be labeled to a high specific activity while conserving the chemical structure which is critical for biological function. High specific activity (> 10 Ci/mmol) can be achieved by incorporation of ^{3}H ([45-52]), ^{35}S ([53-54]) and ^{125}I ([for a review see [55], [56-61]).

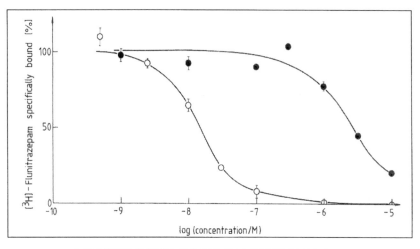

Fig. 7. Displacement of [³H]flunitrazepam specifically bound to cerebellar granule neurons in culture [41] by (*S*) and (*R*) 1,4-benzodiazepine derivatives. Symbols: ○ (*S*)Ro11-3128/002 ● (*R*)Ro11-3624/001. Vertical bars denote S.E. of the mean for duplicate determinations at each concentrations.

Previous analysis has shown that ^{125}I disintegration in radioiodinated peptide leads to the concomitant destruction of the carrier peptides such as insulin and alpha-bungarotoxin, in which the radiolabel was directly attached to tyrosine or histidine residues within the peptide [62-63]. The radiation induced 'decay-catastrophe' of the carrier molecule is explained by a combination of coulombic repulsion during the atomic delocalisation of large positive charge and the ionising energy of Auger cascade induced by electron capture of ^{125}I ([62], [64-65]). In contrast, the presence of an extra-aromatic group in Boltor-Hunter reagent conjugated [66] radioligands that spatially separates ^{125}I moiety from the peptide itself may protect the latter from being damaged during the decay process [67].

2.1.5. Scintillation proximity assays

To overcome the time-consuming separation of bound from free ligands, that cannot be automated, and that often can lead to increased non-specific binding, a scintillation proximity assay has been developed (for a review see [68]). The principle of the scintillation proximity assay is that only radiolabeled ligands, which are bound to the receptor molecule generate a signal on the fluomicrosphere coated with receptors. The technique, which is carried out in aqueous media (buffer) eliminates the use of liquid scintillants required for tritium-based assays. The scintillation proximity assay requires radiolabels that emit low-energy radiation, which is dissipated easily into an aqueous media. Because the mean path length for ^{125}I Auger electrons and ^{3}H β-particles is 35 µm and 4 µm, respectively, the majority of free radioligands are too far from the fluomicrospheres allowing the bound ligands to activate the fluor and produce light [68]. Beyond the unique and significant advantages as well as improved precision, the technology lends itself to high-throughput screening. Since its introduction in combination with the use of ^{125}I labeled ligands [68], other assays using ^{35}S [69-70] and ^{3}H [71-73] have been developed, also.

The biggest drawback of this methodology is that to attach the target molecule (receptor) to a fluomicrosphere it is necessary to isolate and purify. Membrane-bound receptors are difficult to purify and lose activity when they are removed from the hydrophobic environment of the membrane. For membrane bound proteins with oligomeric structures, isolation of the protein may result in the loss of some or more of the subunits thereby inactivating the receptor. To overcome these difficulties cells were cultured in multi-well plates containing scintillant molecules impregnated in the plastic and used in cellular uptake assays [74-77]. For studying integral membrane proteins only functional in their membrane environment, a novel, non-invasive scintillation proximity assay has been developed using membrane-soluble scintillants, having a scintillant 'head'-group (2,5-diphenyloxazole) attached to a lipophilic 'tail' [77]. These membrane-soluble scintillants can be incorporated into liposomes, which are fused with the plasma membrane of cultured cells. This system has been used to monitor [^{14}C]methionine uptake and may be used in radioligand binding assays, also [78].

2.1.6. Fluorescent ligands vs. radioligands

Since the use of fluorescence label is relatively safe and inexpensive compared with radiolabels, the history of fluorescent receptor ligands goes back to the early seventies (for review see [79]). However, the objective of these studies was the imaging of binding sites rather than characterization of their binding parameters. The most important disadvantages of any fluorescent ligand is that *i)* the large fluorescence group can affect the affinity of some ligands, *ii)* fluorescence change on binding may be different for different fluorophores, which complicates estimation of B_{max}, *iii)* very sensitive fluorescence detectors are required to detect low concentration of receptor-bound ligands, *iv)* at low [L] bleaching will occur, *v)* tissue autofluorescence can increase background fluorescence, *iv)* when internalised (*cf.* the next paragraph) fluorescent ligand may be subject to degradation. It seems that the fluorescent label, borate-dipyrimidomethene (BODIPY) overcomes these disadvantages and provides a reliably method for studying receptor-mediated mechanisms in living tissues mainly in combination with confocal microscopy. Most recent applications in complement to classical methods such as radioligand binding and site-directed mutagenesis [80], however, offer a multiplicity of information such as the mechanism of ligand binding [81-82], the movement and internalisation of receptors in living tissue [83-84], the distances between ligands and fluorescently labeled amino acids [85-86], the physical nature of the binding pocket [87-88], and the visualization of labeled receptors [89-91].

2.1.7. Determination of [R], [L] and $t_{1/2}$ called into question

The models discussed and all graphical approaches derived from these models have assumed, that the number of receptors is constant. There is evidence (for a review see [92], [93-98]), that the number of cell surface binding targets may be varied by activation. As a consequence, the receptor concentration would have to be treated as a function of receptor occupancy. Some of the consequences of receptor endocytosis (downregulation) and exocytosis (upregulation) were considered in a model [99] in which receptor concentration is replaced by a linear function of bound receptor (see also *paragraph 2.2.2.*).

Binding processes, whether simple or more complex, are presumed to depend on the physicochemical principles that underlie the mass action law and chemical kinetics that are the basis for the equation given above. However, in all organisms there is a fundamental distinction that must be made between endogenous and foreign substances. Drugs and other xenobiotics are eliminated from the living organism, whereas endogenous substances are strictly regulated within specific limits. The tendency toward stability of concentration of a few inorganic and organic solutes, such as Na^+, K^+, Cl^-, Ca^{2+} and Mg^{2+} ions, neurotransmitters, substrates and ATP in normal states (homeostasis) constituting a biochemical network [100-103], indicates a control *via* negative feedback mechanisms (see also *paragraph 2.2.2.*). Numerous negative feedback mechanisms do exist in the central nervous system (CNS), most importantly the action-potential evoked release of neurotransmitters [11].

The ligand-receptor interaction may have complex kinetics governing signal transduction such as interaction of ligands with subunits of receptors and subsequent interactions with membrane proteins or intracellular binding sites to form different reactive kinetic states (for a detailed discussion of the nature of the problem see section 3. of this chapter). Also, there is evidence that the rate parameter estimated for neurotransmitter-gated receptor channel opening depends on the temporal resolution of the techniques applied in the measurements (for a review see [104]).

As a consequence binding mechanisms are seldom so simple as the Michaelis-Menten, Hill and Cheng-Prusoff models describes. It is more likely that the functionally efficient interaction of a cell surface receptor and its ligand is a summation of the binding interaction and any cooperativity that may have been involved [105]. The general approach incorporates the slope function (K) of a displacement curve, which often deviates from unity. The inhibition constant is calculated according to the following power equations [105],

$$K_1 = IC_{50} / \{1 + (c_o)^K / K_d\} \tag{13}$$

Eq. (13) is the same as Eq. (12) when the slope function is exactly unity.

2.2. Visualization of receptors with radioligands *in vitro* and *in vivo*

Material elsewhere in this chapter provides ample documentation of the existence of a specific structure in the cell membrane to which the external signaling molecule in question binds in order to elicit the response. Usually this structure is an integral membrane protein, although exceptions to this statement can occur if we consider bacterial protein toxins that attack animal cells by binding to particular gangliosides [106, 107]. The binding sites on the receptor protein for a physiological ligand may be a carbohydrate attachment, as in the binding of plant lectins to membrane proteins [108]. For all membrane structures acting specifically at the cell membrane, new insight into their function will be obtained if we can view them by some non-destructive means spatially. Basic spatial parameters are the following: *i)* The identity of cells that bear the receptor, *ii)* The actual location of the receptor (enzyme, transporter) in question, *iii)* The total number of receptor on one single cell or at one single synapse, *iv)* The membrane density of receptors, i.e. the number per square micrometer of membrane surface and their arrangement over the active zone, *v)* Distinction between subtypes of receptors. Among methodologies of labeling and visualizing receptors available, such as autoradiography, electron-dense label attachment, enzymatic reaction product markers, fluorescent markers and X-ray microanalysis, the ones exploring radioisotopic labeling permit quantitative measurement of the number of visualized sites [109].

2.2.1. Cell and tissue autoradiography in vitro

Counting receptors in cell and tissue must make use of the particular binding characteristics of the receptor to achieve specificity, since ligands for neurotransmitter receptors often bind well to more than one type of receptor (enzymes, transporters). Ligands used for receptor autoradiography should be specific in terms of being displaced by known, receptor-specific agents at pharmacologically effective concentrations and, enantiospecific if applicable (*cf.* paragraph *2.1.3*). Likewise, the binding should have a major, saturable component, and a zero (covalent binding) or low dissociation rate, otherwise labeling would be loss during the removal of unbound ligand. While resolution of autoradiography is not at the highest limits, it provides quantitative data, both relative and absolute. In finding appropriate ligands for receptor autoradiography the following possibilities are exemplified: *i)* Direct use of a labeled receptor blocking agent, such as, [^{3}H]alpha-bungarotoxin, that is bound firmly enough to the nicotinic acetylcholine (ACh) receptor with proven selectivity [110-112], *ii)* Use of an affinity labeling reagent. The structural prerequisites for an appropriate affinity reagent are strict, since such a reagent must react covalently with a suitable acceptor group adjacent to the binding site. An alkylating agent, [^{3}H]propyl benzylcholine mustard has been successfully applied for muscarinic ACh receptor visualization [113], *iii)* Direct use of reversible antagonist or agonist, binding reversibly but with a dissociation rate constant sufficient to diminish risk of diffusion, such as [^{3}H]quinuclidinyl benzilate [114, 115].

When the ligand appropriate for autoradiography has been selected, the following additional variables must be considered. For tissues to be taken, the usual procedure is to inject the ligand in vivo and rapid freezing of the tissue at death for autoradiography by dry-mount methods [116]. However, with many appropriate ligand, the blood-brain barrier may prevent circulatory delivery to the brain. Thus, after intravenous injection of [^{3}H]diprenorphine, 98% of the brain radioactivity was at sites showing the enantiospecific binding characteristics of the opiate receptor, but with [^{3}H]naloxone applied similarly only 30-60% was, the rest being non-specific [117]. With covalent or pseudo-irreversible ligand, the best washing procedure is a brief postlabeling application of the same, 'cold' ligand, that displace the adsorbed excess isotope, followed by washing with the medium alone. In several cases, however, non-specific binding can persist through the normal washing procedures [110], which appears to be due to the entry of the ligand by endocytosis, characterised by its strong temperature dependence, its slowness, its non-saturability, its stimulation by polycations and by the failure of specific receptor ligands to inhibit it [118]. For a reversible ligand, it is necessary to assess the extent of non-specific binding by competition with unlabeled ligands, such as biologically inactive enantiomers [119].

Conventional autoradiographic methods for electron- and light-microscopic observation and interpretation of data are detailed in [109] and the references cited. Because of its simplicity and low investment costs these 'contact' autoradiographic methods have been used since the early seventies in receptor research. Among drawbacks, the narrow concentration range of quantitation and the long period required for radioactivity detection must be listed [120-122].

Based on Georges Charpak's invention of multi-wire proportional chamber awarded the Nobel Prize for Physics in 1992, digital autoradiography (DAR, [123, 124]) has been developed. DAR may offer a new and extremely fast technique in biological research, including the possibility of visualizing receptors in brain tissue sections [125]. DAR is more sensitive than contact autoradiography, and thus reduces the time and quantity required for detection of ^{3}H, ^{14}C, ^{32}P, ^{33}P, ^{35}S, ^{99m}Tc, ^{125}I and ^{131}I [125, 126].

2.2.2. In vivo observation of receptor binding by external imaging

In recent years, positron imaging in combination with the tomographic technique (Positron Emission Tomography, *PET* [127]*)* has become a clinical research tool and a routine diagnostic technique (for a review see [128] and references cited as well as [129-133]), however in this paragraph the potential of PET to understand receptor binding and activation phenomena *in vivo* will mainly be addressed.

In vivo observation of human brain function was reported first in 1983 [134, 135] using positron emitting radionuclide labeled receptor ligand derivatives, such as [^{18}F]-6-fluoro-L-3,4-dihydroxyphenylalanine ([^{18}F]-L-DOPA) and 3-N-[^{11}C]methylspiperone. Unstable ^{18}F and ^{11}C are produced in a cyclotron by bombarding ^{18}O and ^{14}N with protons, respectively. Since that time more positron emitting receptor ligands have been developed to monitor changes in brain activity resulting from drugs or disease (Table 1). Success of studies on receptors and transporter quoted in Table 1 as well as other human (for a review see [140]) and rat (for a review see [148]) brain targets based on the appropriate sensitivity of radiopharmaceuticals (for a review see [149] and references cited). It is to note, however, that consistency with the current occupancy model as described in Eq. (6) and Eq. (12) cannot be found in general [148]. Instead, as pointed out above (*c.f.* paragraph *2.1.7*), the agonist-mediated receptor internalization and fluctuations in concentration of endogenous ligands may occur *in vivo* explaining why changes in radiotracer binding potential are not directly related to changes in the occupancy of receptors. After over two decades of PET studies with radiotracers specific for different neurotransmitter systems, we can see now how neurons can influence their own activity, or how they can contribute to changes in the activity of other neurons to serve the dual process of stability and adaptability. Moreover, PET studies are drawing more attention to the possible multitransmitter genesis of psychiatric disorders [150]. In accordance, instead of the image analysis based on the assumptions of functional integration in certain brain areas selected as the 'region of interest' (ROI), several alternative image analysis strategies have been developed that have better resolution and use all of the information present in the image data. One such approach is the principal component analysis [151] applied to cross-sectional dataset [152] or to time-series of PET scans [153]. With the use of the two-dimensional Fourier transform, a method was developed for the analysis of PET metabolic brain images without the use of predefined anatomic ROIs [154].

Table 1
Imaging brain receptors and transporters with in vivo binding competition techniques

Brain target	Radiopharmaceutical	Reference
Human Muscarinic Acetylcholine Receptor	4-[^{123}I]quinuclidyl benzilate	[136]
Baboon Neuroleptic Receptor	[^{18}F]benperidol	[137]
	[^{18}F]spiroperidol	[137]
	[^{18}F]haloperidol	[137]
Rat Dopamine D2 Receptor	N-([^{11}C]methyl)nor-apomorphine	[138]
Human Dopamine D2 Receptor	[^{11}C]raclopride	[139]
	[^{76}Br]bromospiperone	[140]
	[^{76}Br]bromolisuride	[140]
	[^{11}C]pimozide	[140]
Rat Dopamine D1 Receptor	[^{11}C]SKF 82957	[141]
Human Dopamine D1 Receptor	[^{11}C]SCH 23390	[142]
Human Dopamin Reuptake Site	[^{11}C]beta-CIT	[143]
Human 5-HT2 Receptor	[^{18}F]altanserine	[140]
Human Gamma Aminobutyric Acid A Receptor	[^{11}C]flumazenil	[144-145]
Opiate Receptor	[^{11}C]carfentanil	[146-147]

3. TRANSPORT PHENOMENA IN BIOMEMBRANES

Success in measuring drug and neurotransmitter receptor recognition sites in the brain by radiotracer receptor binding studies were, very rightly, enormously influential for drug discovery programs. The recognition of receptor subtypes in simple test tube systems and by tissue autoradiography *in vitro* has suggested multiple ways for neurotransmitters to modulate neuronal functioning. It is believed, that a close correlation between pharmacological and binding potencies of these agents ensured that the binding sites represented pharmacologically relevant receptor. However, as early as the beginning of the eighties it was clearly pointed out the very important fact that "to establish that one is dealing with a biologically relevant receptor sites, it is important to show that relative potencies of numerous agents at the binding sites correlate with their biological effect" [155]. In binding studies the term 'receptor' is used in a

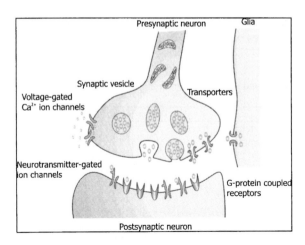

Fig. 8. Simplified scheme of major processes involved in chemical neurotransmission: Reaching the 'presynaptic' nerve terminal, the action potential activates transmembrane Ca^{2+} ion influx through voltage-gated Ca^{2+} ion channels and the release of neurotransmitter stored in synaptic vesicles thereupon. Hence, the concentration of the neurotransmitter jumps from the resting level to some two orders of magnitude higher value in the synapse. As a consequence, neurotransmitter occupancy of receptors embedded in the membrane of the 'postsynaptic' neuron, such as neurotransmitter-gated ion channels or G-protein coupled receptors increases. Change in receptor occupancy induces their functional response *via* either trans-channel flux of ions or activation of G proteins and metabolic changes downstream. Receptors are inactivated rapidly by 'desensitization' that allows relaxation of response even in the continuous presence of the neurotransmitter. Furthermore, relaxation is facilitated by specialized mechanisms like enzymatic degradation or pump like accumulation (uptake) into the presynaptic neuron or into glia or other neurons.

narrow sense, referring to the specific membrane protein that recognizes and binds ligands. This section deals primarily with receptor function to permit insights into how ligand recognition and binding in biomembranes perform chemical signaling and alters cellular response thereupon.

In general, information between neurons is processed as a cascade of consecutive *in → out* and *out → in* transmembrane flux of ions and molecules at the intracellular/membrane and extracellular/membrane interfaces, respectively. This type of information processing, called chemical signaling is triggered by recognition and binding of neurotransmitters (ligands) to specialized structures of neural membranes like ion channel and G protein-coupled receptors [11-13]. Sometimes, direct flux of cytosolic solutes between adjacent neurons triggered by membrane depolarization can also occur. When voltage-activated, gap junctions formed by coupling of apposed hexameric hemi-channels (connexons) open allowing direct (*in* → *in*) flux of bioactive molecules from one cell into the other [156].

Neurotransmitter (ligand)-triggered chemical signaling in neural tissue are represented by major processes that occur in a synapse (Fig. 8.). It involves the passage of chemicals (ligands), called neurotransmitters from nerve terminal of one (presynaptic) neuron to an adjacent (postsynaptic) neuron. Recognition of the neurotransmitter by the receptor embedded in the postsynaptic neuronal membrane surface alters ion permeability and/or intracellular metabolism and hence neuronal firing rates.

3.1. Link between ligand binding and receptor response

Most of the basic formalism outlined above supposes that the binding sites of membrane receptors are structurally well-defined sets of amino acid residues that recognize ligands eliciting receptor response (agonists). This is the general thesis underlying (stereo)structure-activity relationships of agonists (illustrated in Fig. 9.), or the effect of mutations on receptors. If any ligand interact directly with chiral amino acid residues constituting the binding site of membrane bound channel proteins, then its effect is expected to be enantioselective, also [157]. Thus the fact that the volatile anesthetic, (+)isoflurane is more potent than the (-) isomer in augmenting flunitrazepam binding to the GABA receptor complex in rat brain homogenates [158] was considered as evidence for the chloride channel forming GABA receptor being a potential anesthetic target site. However, the isomers equipotently reduced binding of the GABA receptor channel ligand, *t*-butylbicyclophosphorothionate, suggesting that in contrast to the benzodiazepine binding site of GABA receptor (see Fig. 7. and Fig. 9.), the GABA-gated anionic pore through which ions pass failed to distinguish between isomers [158]. To test directly whether isoflurane stereoselectively enhances GABA receptor channel function, the effect of isoflurane enantiomers on GABA-mediated transmembrane $^{36}Cl^-$ ion flux through rat brain synaptosomal membranes was measured [159]. Both isomers reduced the effective concentration of GABA to enhance transmembrane $^{36}Cl^-$ ion flux, however the (+) enantiomer was significantly more potent. In addition, the (+) enantiomer produced greater maximal enhancement of flux [159].

Matching agonist binding potency and receptor response efficacy, however, is not the rule, but the exception. The nature of the problem is illustrated below. Resolution of a thiadiazole analogue of glutamic acid, the major excitatory neurotransmitter in the brain, disclosed that (*R*)- and (*S*) enantiomers are approximately equipotent at the (+)[*S*]-α-amino-3-hydroxy-5-methylisoxazole-4-propionic acid (AMPA)-selective glutamic acid receptor subtype, however the (*S*) enantiomer shows somewhat higher affinity for [^{3}H]AMPA-labeled binding site [160]. By contrast, in response measurements, the (*R*) isomer is a selective AMPA

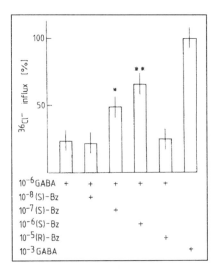

Fig. 9. Effects of (S) and (R) 1,4-benzodiazepine derivatives on GABA-responsive transmembrane $^{36}Cl^-$ ion flux into cerebellar granule neurons in culture [41]. Alterations of GABA receptor function to benzodiazepine enantiomers corroborate the results of displacement of [^{3}H]flunitrazepam specifically bound to sites in GABA receptor by (S) and (R) 1,4-benzodiazepine derivatives (*see* Fig. 7.). Abbreviations: (S)Ro11-3128/002 ● (R)Ro11-3624/001. Vertical bars denote S.E. of the mean for triplicate determinations.

agonist, more potent than the (S) isomer. This discrepancy between binding affinity and response efficacy can be explained by suggesting that the (S) isomer desensitizes AMPA receptor faster than the (R) isomer. Indeed, native ion channels show desensitization which only a few ms for some glutamate receptor [161, 162]. Also, binding constants of different nicotinic agonists in competition binding assays targeting different subtypes of nicotinic acetylcholine (ACh) receptor, such as α4β2, α7, α3β4 and α3β2* could not explain the differential behavioral effects observed [163]. It is suggested, that differences in their functional efficacy at nicotinic ACh receptor subtypes may instead be responsible.

The approach to the problem of discrepancy between affinity and efficacy roots in the realization that proteins could undergo global conformational changes, and that the ligand binding properties of the different conformations might be substantially different. In the early 1950s Wyman and Allen suggested that the affinity change during oxygen binding could be explained if all four subunits flip together from one conformation to the other [164]. In the neurotransmitter-gated ion channel class of receptors, equivalent idea was substantiated by he three-dimensional appearance of the cation-selective nicotinic ACh receptor, disclosed by Unwin and co-workers [165-169]. The structure determined at 9Å resolution by cryo-electron microscopy of tubular crystals of *Torpedo* electric fish postsynaptic membranes embedded in amorphous ice [for a review see 168]. ACh receptors embedded in tubular crystals were exposed to ACh for short (5 s for channel opening) and long (20 s for desensitization) time. Multiple, sequential conformations were visualized during channel opening that were different from receptor desensitization. These seminal series of studies allowed relating binding and response structurally, *via* the existence of multiple conformations for ion channels in membranes of excitable cells [169].

3.1.1. Affinity and efficacy

As discussed by Colquhoun [170], for any ligand recognized by ligand-gated ion channel and G-protein-coupled receptors (neurotransmitters, agonist, partial agonists and antagonists) that cause a conformational change binding experiments do not measure affinity. A conformational change sequential to ligand recognition and binding can be taken into account by Eq. (14)

$$L + A \leftrightarrows LA \leftrightarrows L\bar{A} \qquad \overset{K_1 \quad \Phi}{}$$

$$(14)$$

known as del Castillo-Katz mechanism [171], where A and $\bar{A}$ stands for the active and the open conformational isomer of the ligand-bound receptor, respectively. The equilibrium dissociation, K_1 and isomerization, Φ constants for individual reaction steps, are known as microscopic equilibrium (dissociation and isomerization) constants. Hence, if we want to learn about the genuine equilibrium dissociation constant of the recognition site of the active receptor than it should be related to the ratio of closed over open channel forms, $\Phi=[LA]/[L\bar{A}]$, that is the ability of ligands eliciting response. K_1 and Φ for individual reaction steps can be derived from measurements of receptor response as a function of time and ligand concentration (for K_1 and Φ values of ACh and GABA receptors *see* paragraphs *3.1.5.* and *3.1.7.*).

Conceptually, multiple conformational changes (pathways) that cause an ion channel to open, *i.e.* the molecular mechanism are defined by subunit structure and channel architecture, and should be independent of the agonist that caused it to open. Agonists will, however, determine the length of time the channel is open for, and can be specified accordingly as more or less desensitizing ones. Thus, relationships between channel opening and desensitization are crucial to the understanding of the mechanisms that determine ligand specificity. In what follows, an analysis of published data for two channel-forming neurotransmitter receptors, ACh and GABA, mediating transmembrane cation and anion fluxes, respectively, will be provided. The implication of these relationships are discussed in general, by focusing more specifically on problems associated with the measurement of ionic fluxes to and from rapidly exchanging pools, particularly the cytosolic sodium and chloride ion pool as an example.

3.1.2. Measurements of neurotransmitter receptor-mediated transmembrane tracer ion flux in the minute to hour time region

Up to the early seventies, excitability of biological membranes has been almost exclusively studied with whole cell electrophysiological recording techniques. To yield information on the molecular basis of chemical excitation a new, radiotracer method in combination with the use of a cell-free preparation has been introduced [172-175]. Membrane fragments rich in the enzyme, ACh esterase were isolated from the electric organ (electroplax) of the electric eel, *Electrophorus electricus* and are observed under the electron microscope. Membrane preparations make closed, approximately spherical vesicles. The preparations showing subunit structure and globular repeating units appear to be free of intracellular elements, such as mitochondria, nuclear envelopes or other cytoplasmic constituent [175]. Excitation of resealed membrane fragments by cholinergic agonists is measured after the vesicles are equilibrated with a medium containing $^{22}Na^+$ ion. Thereafter, the suspension is diluted into a nonradioactive medium and the $^{22}Na^+$ ion content of the vesicles as a function of time is followed by rapid filtration on Millipore filters [172]. Pharmacological similarities between ligand-induced ion flux in the vesicles and the electrophysiological response of electroplax cells were reported [172, 173]. Receptor-controlled flux of Na^+, K^+ and Ca^{2+} ions were demonstrated. In the presence of 10^{-4} M carbamoylcholine, an ACh analogue that is not hydrolized by ACh esterase, the time course of $^{22}Na^+$ ion efflux increases three- to fourfold, which is blocked by the cholinergic antagonist, 10^{-5} M *d*-tubocurarine and snake neurotoxins (for instance α-bungarotoxin). In the presence of carbamoylcholine, the time for half-equilibration is approximately 8 min. Under the conditions, the dose-response curves to cholinergic agents obtained *in vitro* agree, with the dose-response curves recorded by the electrical potential measurements on the whole cell *in vivo*. A significant difference between the absolute values of the maximal responses to carbamoylcholine recorded *in vivo* and *in vitro* is observed, the latter being 20 % to 40 % larger than that measured *in vivo* [172]. Quantitative comparison of the agonist-selective increase of permeability and the amount of cholinergic agonist ([^{14}C]decamethonium) specifically bound to the nicotinic ACh receptor in the membrane of vesicles showed that the dose-response curve and the binding curve superimpose almost exactly [174].

When the rate of receptor-controlled transfer of ions across membranes was calculated from measurements made with vesicles, the results were lower, by a factor 10^5, than those determined by electrophysiological techniques in intact cell [176-178]. In addition, receptor inactivation (desensitization) induced by ACh that was observed in electrophysiological experiments with cells [179] was not observed in $^{22}Na^+$ ion efflux measurements made in *Electrophorus electricus* electroplax vesicles. One may wonder why these agonist-induced increases [172-173] in membrane permeability and receptor desensitization are so slow, since conformational changes of proteins, like channel opening and receptor desensitization can be expected to fall in the subsecond and sub-millisecond time region [165-166, 180-191, [192] and references cited there]. Matching binding and response processes, together with the unexpectedly slow process of receptor-mediated $^{22}Na^+$ ion exchange and increased maximal response *in vitro* may indicate, however, that the transmembrane $^{22}Na^+$ ion flux assay of bioelectrical response as quantified by Kasai and Changeux [172, 173] may be due to the interconversion between the desensitized and active states (*see* also paragraph *3.1.5.*) of a minor ACh receptor fraction present in the heterogeneous vesicle population.

Also slow, functional desensitization of neuronal nicotinic ACh receptor occurs when exposed to either stimulating or non-stimulating concentrations of nicotine in measurements on nicotine-stimulated $^{86}Rb^+$ ion efflux from a mouse brain nerve ending preparation (synaptosomes) [192]. Furthermore, [^{3}H]nicotine binding to midbrain particulate fractions displays a fast and a slow phase. Similar kinetic constants are derived from data on functional desensitization and receptor binding using a two-state model [192]. These and the above findings suggest that a slow sampling rate for investigation of agonist action on channel-forming membrane receptors may not provide true information neither on channel opening nor on fast desensitization. "This means that accurate concentration-response curves can be obtained only in cases where the agonist can be applied very rapidly, and makes binding experiments almost impossible" ([170], but see the application of rapid chemical kinetic method for measuring dissociation of radioligand from its binding site as described under paragraph *2.1.1.*, Fig. 4.).

3.1.3. Methodology allowing tracer ion translocation to be measured in the millisecond to second time region

The advantages of working with membrane vesicles, rather than with intact cells, have been addressed in a number of reviews [193-196]. The concentrations of solutions both inside and outside the vesicles can be controlled, internally by equilibrating the vesicles with defined buffer for an appropriate period of time, and externally by suspending and diluting the vesicles in defined buffers. To investigate receptor-controlled cation flux through a determined number of ACh receptors was possible since *i)* the concentration of receptor sites can be determined by radioactive iodine-labeled α-bungarotoxin, which binds specifically and irreversibly to the ACh receptor

[197], *ii)* α-bungarotoxin inhibits functioning receptor [197], and *iii)* vesicles isolated from *Electrophorus electricus* electroplax exhibited ACh receptor function as measured by the translocation of radioactive tracer cations in the presence of acetylcholine analogue, carbamoylcholine [172, 173].

The major factor that contributes to the low rates of ACh receptor-mediated efflux of radiotracer ions that were measured [172, 173, 192] is the presence of leaky vesicles that contain the bulk of the exchangeable tracer ions, but that do not respond to ACh [198]. As a result, the loss of radiotracer from leaky vesicles obscures the receptor-mediated flux of cations from a small population of vesicles that contain only the minor compartment of exchangeable tracer ion [189, 198]. The vesicle preparation of Kasai and Changeux [172] was found to contain three types of vesicle, each with a characteristic half time for efflux [198, 199]. Sensitivity to carbamoylcholine is exhibited by the vesicles that are rather impermeable, with a half time for efflux of 330 min and that contain between 10 % and 20 % of the tracer ion content of the whole preparation [199]. In general, the ACh receptor-mediated radiotracer cation flux process is very fast compared with the passive flux [182, 183]. Thus, the half time of the overall radiotracer cation efflux process from a heterogeneous vesicle population depends heavily on the half-time value of the passive efflux and on the relative amounts of radiotracer in the vesicle populations containing active and desensitized receptors. The effects of vesicle heterogeneity and other aspects of ion flux measurements have been treated theoretically by Bernhardt and Neumann [200, 201].

To measure the rapid receptor-mediated flux before the slower flux not controlled by the receptor became important a new approach has been developed. Most importantly, the approach applies the techniques of rapid mixing of solutions [22] and measures influx to avoid exposure of the vesicles to radiotracer ions before the measurements were made. This methodology is combined with the use of selectively fractionated functional ACh receptor-containing membrane vesicles [199, 202]. Techniques of rapid mixing of solutions have been used extensively with proteins to follow reactions occurring in the time range of milliseconds to seconds [203-205]. Quench flow techniques have been applied to enzyme systems [206-209]. Transmembrane ion transport has been demonstrated using quench flow technique in studies of ATP synthesis by submitochondrial particles [210] and calcium ion uptake by sarcoplasmic reticulum vesicles [211-213].

With a continuous quench flow system [22], the reaction time depends on the flow rate and the distance between an initiating and a quenching event. However, due to increased retention in the reaction tube, there is an upper limit to the reaction time (tube length). This limitation in reaction time imposed by the quantity of reactants available, is not present in the pulsed-mode quench flow systems ([22, 208].

The pulsed-mode quench flow technique [22] superiors the continuous one since *i)* the quantity of sample recovered is independent of the reaction time, *ii)* the initiation and quenching events are controlled independently of each other, and the reaction time is preset with an electronic timer. Tubing arrangement for one pulsed incubation with two displacements used to follow fast dissociation of specifically bound [^{3}H]girisopam in brain membrane suspension is shown in Fig. 3. The pulsed-mode arrangement of the quench flow instrument was used in measurements with brain membrane suspensions and radiotracer detection which allow ligand dissociation, transmembrane flux of neurotransmitters and receptor-mediated ion translocation to be followed in the millisecond to minutes time region (Department of Neurochemistry, Chemical Research Center, Budapest). The quench flow instrument is of the multiple mixer type constructed as described by Cash and Hess [22] with the only difference that the linear displacement of pneumatically driven pistons (stroke time) is recorded on the monitor of a computer functioning as the timer with the use of a software program designed for quench flow experiments.

With the pulsed-mode quench flow technique, however, there is a lower limit of reaction time equal to the stroke time required to displace the reactants from their reservoirs (thermostatted syringes, Fig. 3.). Besides the physical properties of the mixer and solution viscosity, the mixing time (the shortest possible reaction time) is determined by the diffusion coefficient of the reactants [214]. In practice, the time resolution of the pulsed mode quench flow instrument is given by the shortest possible reaction time that can be obtained precisely. This was 5 ms with neuromuscular [215] and 100 ms with brain membrane suspensions.

3.1.4. Analysis of receptor-mediated transmembrane tracer ion flux measurements

An exchange of isotope takes place while the concentration of the receptor-permeable ion inside and outside the vesicles (or cells) remain constant. The increase in internal specific activity of tracer ion, M gives the rate of influx

$$dM_I/dt = J(M_O - M_I) \tag{15}$$

where M_O and M_I are the specific activities of the tracer ion outside and inside the vesicles (or cells), respectively, and J is the first order rate constant for ion exchange. For receptor-mediated isotope exchange J is given by

$$J = \bar{I}\,[R']\,\bar{A} \tag{16}$$

where $[R']$ is the concentration of receptor per internal volume, $\bar{A}$ is the fraction of the receptor which is in the open channel state and $\bar{I}$ is the *specific* reaction rate [205] of receptor-mediated ion translocation of tracer, M through the receptor channel [185]. It is a second order rate constant that is characteristic of the receptor and is independent of the type of ligand used and of other properties of the vesicles (or cells) that also influence the rates of ion translocation. If $M_1 > M_O$, the sign of dM_1/dt is negative and efflux occurs. If the receptor is desensitized by the neurotransmitter, the value of J is progressively attenuated in a first order decay

$$J = J_i e^{-\alpha t} \tag{17}$$

where J_i is the initial value of the first order rate constant for the ion exchange, α is the first order rate constant for desensitization and t is the time of exposure to the neurotransmitter.

The value of $\bar{I}$ is expressed in liter per mole per second. Transport of substances across membranes is usually expressed in terms of flux coefficients, which requires the knowledge of the surface area of the vesicles (or cells). The internal volume of a population of vesicles (or cells) is obtained from the equilibrium concentration of tracer ions in the vesicles (or cells) (*see* paragraphs *3.1.6* and *3.1.8*), while determination of the surface area requires knowledge of the average diameter of the vesicles (or cells). This is why radiotracer flux data are expressed in terms of rate coefficients. Comparison of these rate coefficients and the single channel conductance are discussed in paragraphs *3.1.6* and *3.1.8*.

3.1.5. Molecular mechanism of ACh receptor-mediated radiotracer cation translocation through the membrane by rapid chemical kinetic methods

Minimum mechanism which relates ligand-binding and channel opening steps, and rates of interconversion between active (A) and desensitized (D) ACh receptor forms to rates of ACh receptor-mediated ion flux suggested on the basis of kinetic measurements of ACh receptor-mediated $^{86}Rb^+$ ion flux in membrane vesicles isolated from the electric organ of *Electrophorus electricus* [183-184] is given by the scheme

$$
\begin{array}{ccccc}
K_1 & & K_1 & & \varPhi \\
2L+A \rightleftharpoons & L+LA & \rightleftharpoons & L_2A \rightleftharpoons L_2\bar{A} & \updownarrow J_m \\
& k_{21} \updownarrow k_{12} & & k_{43} \updownarrow k_{34} & \\
& L+LD & \rightleftharpoons & L_2D & \\
& & K_2 & &
\end{array}
\tag{18}
$$

The rate equation is given by

$$M_{1t}/M_{1\infty}=1\text{-exp-}\{(J_A/\alpha(1\text{-e}^{-\alpha t})+J_D t\} \tag{19}$$

where J_A and J_D are the rate constants for ion exchange associated with the active and desensitized forms of the receptor, respectively. Ion flux through the open channel proceeds with a rate constant, J_m and, the desensitization rate constant, α is given by

$$\alpha=\{(k_{43}[L]+k_{21}2K_2)/([L]+2K_2)\}+\Phi\{(k_{34}[L]^2+k_{12}[L]2K_1)/([L]^2(1+\Phi)+[L]^2K_1\Phi+K_1^2\Phi)\}$$

$$\tag{20}$$

The minimal mechanism (18) and rate equations (Eq. (19) and Eq. (20) account for the four types of measurements made *i)* ion flux without prior receptor desensitization, *ii)* ion flux after receptor desensitization, *iii)* desensitization of the receptor, and *iv)* reactivation of desensitized receptor, over a 2000-fold range of carbamoylcholine concentration. According to the minimal mechanisms, the ACh receptor can exist in two states, active and desensitized, both of which bind activating ligands. The ligand binding and channel opening processes are fast, and appear as equilibria. The interconversions between the active and desensitized states are first order, with rates comparable with ion flux equilibration and may be measured by this process. Channel opening occurs as a structural rearrangement of receptor protein when two ligand molecules are bound to the active state of the receptor. This allows cation flux through the channel, characterised by the ACh channel-specific rate constant, $\bar{I}$. Channel opening with one bound ligand is negligible. In contrast, desensitization with one as well as with two bound ligands is significant. Thus the minimal mechanism accounts for the different ligand concentration dependencies of ion flux and receptor desensitization [189]. The cyclic features of the scheme (19) allow the reactivation rate of the receptor to be independent of the desensitization process, as was pointed out by Katz and Thesleff [179]. The channel opening process, characterised by the equilibrium constant, perturbs the equilibrium between A, LA and L_2A, thereby affects ligand binding to the active receptor (*see* also paragraph *3.1.1.*). The absence of free desensitized form of the receptor, D implies that its contribution is not significant, however the absence of D in the absence of bound ligand is not excluded.

The model described above differs from all other models previously proposed [216, 217], in that all the equilibrium constant and protein isomerization rate constants pertaining to the interconversion between active and desensitized forms can be evaluated (K_1=1.9 mM, K_2=21 μM, Φ=2.8, $J_m=\bar{I}$ [R']=37 s^{-1}, k_{21}=0.46 s^{-1}, k_{12}=4.6 s^{-1}, k_{43}=0.001 s^{-1}, k_{34}=11.2 s^{-1}), and account for the ion translocation process over a three order of magnitude wide range of ligand concentration used [189, 215].

3.1.6. Specific reaction rate of the ACh receptor-mediated cation translocation

Determination of the concentration of the receptor per liter internal vesicle volume, $[R']$ requires the knowledge of the internal volume of vesicles. Since the amplitude of the influx of Na^+, K^+, Li^+ and $^{86}Rb^+$ ion are the same [218], from the knowledge of the specific activity of the radiotracer in the external solution, the internal volume of the vesicles can be determined from the amplitude of the $^{86}Rb^+$ ion influx [219]. The internal volume of the vesicles that contain functional receptors was found to be about 2 µl/mg membrane vesicle protein [185]. On the basis of the concentration of α-bungarotoxin binding sites, stoichiometry of α-bungarotoxin and activating ligand binding sites and that the binding of two ligand molecules is required to initiate ion translocation, the receptor concentration is 2.4 pmol/mg membrane vesicle protein. Accordingly, the receptor concentration per liter internal volume, $[R']$ is 1.2 µM. Considering the value of $\bar{I}[R']=37$ s^{-1}, the specific reaction rate of the ACh receptor-mediated ion translocation, $\bar{I}$ is equal to 3×10^7 $M^{-1}s^{-1}$ [185, 187].

A comparison between chemical kinetic measurements of ion translocation rate and patch clamp recordings of single cells [220] was performed [221]. The single-channel conductance, γ was equal to 53 pS, which corresponds to a $\bar{I}$ value of 5×10^7 $M^{-1}s^{-1}$ [221].

3.1.7. Molecular mechanism of GABA receptor-mediated radiotracer anion translocation through the membrane by quench flow method

First reports on measurements of transmembrane $^{36}Cl^-$ ion flux with membrane vesicle preparations from brain [222-226] allowed to study channel opening and desensitization of GABA receptors in brain membrane preparations, which can be mixed rapidly with solutions of known and controlled composition. Quench flow techniques developed by Cash and Hess [22] for measuring ACh receptor-mediated transmembrane radiotracer cation flux, were successfully applied in a series of measurements of GABA receptor-mediated transmembrane $^{36}Cl^-$ flux with brain membrane vesicle preparations in 1987 [227-230] and thereafter. Measurements of Cash and Subbarao [227-230] enabled the whole progress of GABA mediated chloride exchange to be followed, and its progressive attenuation of rate due to desensitization to be studied. These measured rates over a wide range of GABA concentration enabled the microscopic model of GABA receptor channel-opening and desensitization to be made, which accounted for all the results and explained why the dependence on GABA concentration of chloride exchange and desensitization rates are different.

$$K_1 \qquad K_1 \qquad \Phi$$

$$2L + A \rightleftharpoons L + LA \rightleftharpoons L_2A \rightleftharpoons L_2\bar{A} \; \updownarrow J_m \tag{21}$$

$$K_2 \qquad K_2$$

$$2L + A \rightleftharpoons L + LA \rightleftharpoons L_2A \tag{22}$$
$$k_1 \downarrow \qquad \downarrow k_2$$
$$LD \qquad L_2D$$

An active receptor, A, binds two ligand (agonists) molecules with microscopic dissociation constants, K_1 giving a double liganded species with isomerizes to give an open channel conformation with a channel-opening equilibrium constant, Φ^{-1}. Ion flux through the open channel proceeds with a rate constant J_m. In parallel, A can also bind the ligand with microscopic dissociation constant, K_2. The singly- and doubly-liganded species, LA and L_2A thus formed, desensitize with rate constants, k_1 and k_2, respectively. Desensitized conformations, LD and L_2D, do not form open channels [230].

These studies showed two phases of chloride exchange characterized by two rates of desensitization. The same minimal mechanisms, as given by Eq. (21) and Eq. (22), apply to both receptors. If we take the case of two types of active receptors in the membrane (A desensitizing with a rate constant, α, and B, desensitizing with a rate constant, β), the exchange rate at time t is given by

$$dM_{1t}/dt = (J_{iA} e^{-\alpha t} + J_{iB} e^{-\beta t})(M_O - M_{1t}) \tag{23}$$

Since the volume outside is relatively large M_O is not significantly changed by the flux. $M_O = M_\infty$ is approximated to be constant and equal to the equilibrium value of M_1. Rearranging and integrating gives

$$\int_0^t dM_{1t}/(M_O - M_{1t}) = \int_0^t J_{iA} e^{-\alpha t} \, dt + \int_0^t J_{iB} e^{-\beta t} \, dt \tag{24}$$

$$\ln (M_O - M_{1t}) = - J_{iA} e^{-\alpha t}/\alpha - J_{iB} e^{-\beta t}/\beta + C \tag{25}$$

For influx, initially $t = 0$ and $M_1 = 0$ and

$$C = J_{iA}/\alpha + J_{iB}/\beta + \ln M_O \tag{26}$$

hence,

$$M_{It}/M_{I\infty}=1\text{-exp-}\{J_{iA}/\alpha(1\text{-e}^{-\alpha t}) + J_{iB}/\beta(1\text{-e}^{-\beta t})\} \tag{27}$$

For efflux, initially $t=0$, $M_O=0$ and $M_I=M_{Ii}$ hence,

$$M_{It}/M_{Ii}=\text{exp-}\{J_{iA}/\alpha(1\text{-e}^{-\alpha t}) + J_{iB}/\beta(1\text{-e}^{-\beta t})\} \tag{28}$$

Rate constants in Eq. (28) are given by the expressions

$$J_A=J_{mA}/\{1+\Phi_A(1+K_{1A}/L)^2\} \tag{29}$$

$$J_B=J_{mB}/\{1+\Phi_B(1+K_{1B}/L)^2\} \tag{30}$$

$$\alpha=\{k_{2A} + (2K_{2A}\,k_{1A})/L\}/(1+K_{2A}/L)^2 \tag{31}$$

$$\beta=\{k_{2B} + (2K_{2B}\,k_{1B})/L\}/(1+K_{2B}/L)^2 \tag{32}$$

Subscripts A and B refer to each of two receptors A and B.

Progress of $^{36}Cl^-$ ion exchange and receptor desensitization can independently be measured with the same preparation (Fig. 10.). The influx is initiated by mixing the brain membrane vesicle suspension in buffer (225 μl) with an equal volume of buffer solution containing $^{36}Cl^-$ ion (15 μCi/ml) and GABA. After the incubation time indicated, the GABA receptor-mediated exchange is terminated by mixing (quenching) with the same volume of buffer solution, containing the GABA receptor antagonist, bicuculline methiodide (3 mM). The mixture is immediately filtered [231], washed with iced buffer (2x10 ml), and the radioactivity retained on the filter disc is measured by liquid scintillation counting. The unspecific isotope exchange is measured in the same way in the absence of GABA and subtracted to give the specific, GABA receptor-mediated influx (Fig. 10. A). The first-order plot of the GABA-induced loss of receptor activity (Fig. 10. B), shows two phases indicating two desensitization processes of GABA receptor from rat brain [226, 229].

Using rapid kinetic techniques measuring tracer $^{36}Cl^-$ ion flux on a timescale of milliseconds to seconds, the separate function of individual GABA receptor channels in mammalian brain can be resolved on the basis of their different active lifetimes on exposure to GABA (Table 2). Being on the same membrane [232], the faster and the slower desensitizing GABA receptors can function together with complementary roles in neurotransmission [230]. It is tempting to speculate about their probable structural differences underlying distinguishable kinetic [229], [233], and possibly, pharmacological [234] behavior.

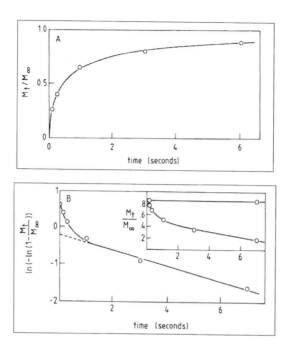

Fig. 10. Comparison of the progress of ^{36}Cl$^-$ ion exchange with the progress of receptor desensitization at 30 °C. A: Progress of GABA-specific ^{36}Cl$^-$ ion influx in the presence of 40 μM GABA, measured as described in the text. The line is a best fit computed from Eq. (27). B: First-order plot of the GABA-induced loss of receptor activity (desensitization). Inset: The fractional equlibration (M_t/M_∞) of ^{36}Cl$^-$ ion influx is measured in a short assay (320 ms) with saturating GABA concentration (1 mM) sequential to progressive times of preincubation with 40 μM GABA. The initial ^{36}Cl$^-$ ion influx, obtained with no GABA in the preincubation is not changed. The points are the means of triplicate determinations. The reproducibility is ±3 % (standard deviation) of the total GABA-induced ^{36}Cl$^-$ ion influx. After [230].

3.1.8. Specific reaction rate of the GABA receptor-mediated anion translocation

Similarity of the kinetics of GABA receptor-mediated ^{36}Cl$^-$ ion flux into cultured neurons from rat brain cerebellum [235] and the ^{36}Cl$^-$ ion influx measured with membrane vesicles from rat brain cortex has been revealed by computer simulation of data with cells using the rate expression derived for the vesicles [233]. This comparison was possible although the assay times used with cells (Fig. 11. lower panel) were longer than those used with membrane vesicles by comparing measurements at relatively low concentration of GABA (Fig. 11.). Lower EC_{50} values in isotope flux assays are no different with cells and membrane preparations and results from high values of receptor concentration per internal volume, leading to limitation of the influx by the internal volume, V_1, [232].

In accordance with studies of $^{86}Rb^+$ ion exchange mediated by ACh receptor from the electric organ of *Electrophorus electricus* [185], the rate

Table 2
Parameters characterizing kinetically distinguishable GABA receptors from rat brain cortex.

	Parameters	Faster desensitizing	Slower desensitizing
$^{36}Cl^-$ Ion flux	J_m (s^{-1})	12.1±2.1	2.55±0.45
	EC_{50}* (μM)	105±10	82±9
	K_1 (μM)	142±50	169±60
	Φ	0.27±0.15	0.14±0.11
	Activity ratio	4	1
Desensitization	k_2 (s^{-1})	α_{max}=21±2.2	β_{max}=1.35±0.18
	ln2/k_2=$t_{1/2}$	32 ms	533 ms
	k_1 (s^{-1})	1.0±0.2	0.10±0.03
	EC_{50}* (μM)	151±25	114±14
		53±7	K_2 (μM) 70±9

* half response GABA concentration. Data were taken from [230].

constant for $^{36}Cl^-$ ion exchange mediated by GABA receptor from rat brain cortex [232] increases with the receptor concentration per internal volume, $[R']$ as given by the Eq. (16). Thus, the above computer simulations (Fig. 11.) suggest that $[R']$ has similar value in cultured granule cells from rat cerebellum as in the membrane vesicles from rat cerebral cortex [233]. Moreover, similar rate parameters were obtained using a rapid flow system with a 10 ms time resolution that is developed for making measurements with single cells isolated from the cerebral cortex of embryonic mice [236]. The evaluated Φ constants allowed to calculate the conditional probability, that at a given GABA concentration the GABA receptor channel is open. It was also determined by single-channel current-recording technique, and was in good agreement with the conditional probability value obtained for the faster desensitizing GABA receptor (0.56 *vs.* 0.61, respectively with 100 μM GABA) [236].

As outlined above, the specific rate for anion translocation through the GABA channel, $\bar{I}$, can be derived from the knowledge of the molecular mechanism of receptor responses. Considering that $J_m=\bar{I}[R']$, the rate constant if all the channels were open, can be obtained from rate measurement (Table 2), independent determination of $[R']$ in measurements of internal volume accessed by the GABA and of binding to the receptor by specific ligands (Table 3) can provide a value for $\bar{I}$ [233]. The quantity of chloride ion in the vesicles or cells is proportional to the radiation count (*DPM*), hence

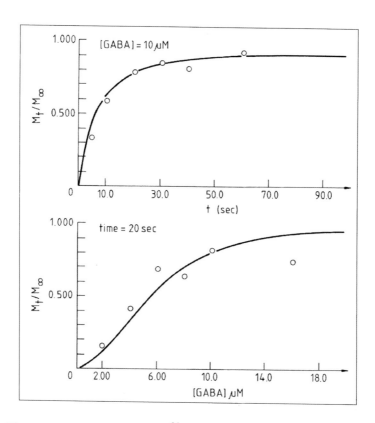

Fig. 11. The rates of GABA mediated $^{36}Cl^-$ exchange into cultured neurons from rat cerebellum have been compared with the rates of $^{36}Cl^-$ influx measured with membrane vesicles from rat cerebral cortex. The rates of chloride exchange with the cells, compared by using low GABA concentrations, were as fast as those seen with membrane vesicles. The time dependence as well as the GABA concentration dependence of the $^{36}Cl^-$ flux with the cells can be described by rate expressions derived for the vesicles using similar rate constants for ion-exchange as well as desensitization. After [233].

$$DPM_O/DPM_I = ([Cl^-]_O M_O V_O)/([Cl^-]_I M_I V_I) \tag{33}$$

where $[Cl^-]_O$ and $[Cl^-]_I$ is the ionic concentration and V_O and V_I the volume outside and inside, respectively. At equilibrium, $M_I = M_O$, hence

$$V_I = V_O (DPM_I [Cl^-]_O)/(DPM_O [Cl^-]_I) \tag{34}$$

Data summarised in Table 3 give similar results for [R'] supporting the conclusions from the ion-flux rate measurements.

Table 3
Comparison of GABA-specific volume ratio, V_I, the number of GABA receptor sites, B_{max} and receptor concentration per internal volume, $[R']$ in membrane vesicle suspension from rat brain cortex and cultured granule neurons from rat brain cerebellum $[Cl^-]_O=118$ mM and $[Cl^-]_I=5$ mM; Data were taken from [233] and [235].

	V_I (μl/mg protein)	B_{max} (pmol/mg protein)	$[R']$ (μM)
Membrane vesicles	13.7	0.96	0.07
Granule neurons	9.0	0.90	0.10

Thus, the specific reaction rate for GABA receptor-mediated $^{36}Cl^-$ ion translocation, $\bar{I} \sim 10^8$ $M^{-1}s^{-1}$ [233] can be derived, and is found $i)$ comparable with that for nicotinic ACh receptor-mediated $^{86}Rb^+$ ion translocation (3×10^7 $M^{-1}s^{-1}$, [185]) and, $ii)$ consistent with measurements of electrical conductance [236].

3.1.9. Assay time versus response time

The half-response concentration of functionally distinguishable GABA receptors for desensitization as well as for ion flux was in the range of 100 μM GABA (Table 2). This is, not to be confused with the EC_{50} of the assay, $i.e.$ GABA concentration for half-maximal $^{36}Cl^-$ influx or efflux in the membrane vesicles in the conditions and time of the assay, which is limited by factors other than channel opening. This was clear from the experimental $^{36}Cl^-$ flux data [229-230] and is discussed in more detail below [237].

Following the introduction of agonist-induced transmembrane $^{36}Cl^-$ ion flux assay as a measure of the open GABA receptor channel [238] about a hundred of studies using various brain membrane preparation and cultured neurons have been reported (for a review see [233], [238-242]). The majority of these studies used a single concentration of agonist (in the presence of absence of drugs to be tested) at a constant assay time. These types of measurements of GABA receptor 'activity' are suffered from incomplete tracer ion flux due to different factors, such as equilibration of the tracer, completed assay time, and receptor desensitization [237]. Desensitization of GABA receptor channel (for reviews see [41], [229], [233], [243-245] and references cited therein), known as the progressive attenuation of channel opening which occurs in the presence of GABA, other orthosteric agonists and positive allosteric modulators, such as the 1,4-benzodiazepine derivative, chlordiazepoxide ($Librium$) [239-241] limits $^{36}Cl^-$ ion influx. The nature of the problem is illustrated by the curve in Fig. 12.

The presence of GABA (endogenous or added) evokes a loss of $^{36}Cl^-$ ion influx even with very short assay times.

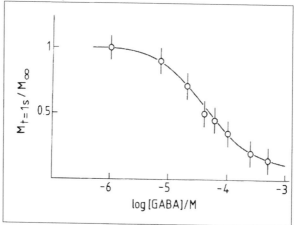

Fig. 12. Desensitization may be faster than the completion of the trans-channel ion exchange. The fractional equilibration (M_t/M_∞) of $^{36}Cl^-$ ion influx in rat cerebrocortical membrane vesicles is measured in a short assay (320 ms) with saturating GABA concentration (1 mM) sequential to *one-second pre-incubation* with increasing concentration of GABA. The points are the mean of three determinations. Bars indicate standard deviation. Data were taken from [229].

Computer simulation of the dependence of the half-maximal $^{36}Cl^-$ ion exchange (*apparent* potency [237] or affinity [170]), EC_{50} on assay time shows (Fig. 13.), that increasing assay time decreases the value of EC_{50} (Fig. 13. A), to a lower limit given by

$$M_{1t}/M_{1\infty}=1\text{-}exp\text{-}(J_{iA}/\alpha+J_{iB}/\beta) \tag{35}$$

where by the fractional ion exchange becomes independent of assay time (Fig. 13. B). These simulations have several implications. Firstly, assay time influences the measured dissociation constant, because the receptor becomes desensitized on exposure to GABA. Along the top-line (Fig. 13. A), the exchange reaction is stopped because desensitization is complete as given by the Eq. (36). This has the effect of decreasing the measured dissociation constant, depending on the GABA concentration, since response of a *population* of receptors is being measured. It is important to note, that if there were no desensitization, the EC_{50} would continue to decrease to approach zero [237].

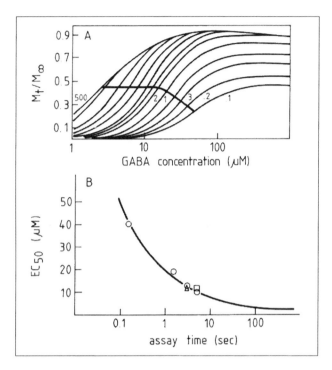

Fig. 13. Assay time changes the apparent potency of and the maximal response to GABA. A: Variation of EC_{50} ($M_{1t}/M_{1\infty}=0.5$) and maximal $^{36}Cl^-$ ion influx ($M_{1\infty}$) on GABA concentrations (10-1000 µM) at various assay times (0.1-500 s). The curves were computed using Eq. (27) and values of the rate constants for ion exchange and desensitization given in Table 2. B: EC_{50} decreases with increasing assay times. Experimental points (Δ, [226]; o, [230]; □, [246]) fits the line computed from the simulations shown in A. Adapted from [237].

This result may suggest a role for GABA receptor desensitization in keeping a fraction of receptor population for activation. Moreover, the evolutionary conservation of desensitization is a good hint that it *is* biologically important, not only in GABA receptors, but in other members of the ligand-gated receptor channel superfamily [11-13] with different ligand specificities. According to the Eq. (17), the shift of the response curve to decreasing concentration with increasing assay times to lower concentrations could be caused by an increase of [R']. This finding suggest, that a change in [R'] in various models of GABA receptor channel function which occur in animals *in vivo* due to drugs, experience (*e.g.* learning) or disease with no alteration in the properties of the receptor may cause a shift in EC_{50} in this type of assays. Alternatively, a large fraction of the receptor converted by GABA to a higher affinity state, which does not give significant open channel. The same applies for ligand binding experiments, involving measurements of binding to the receptor following its exposure to the ligand. The consequent ligand-induced changes in the receptor increase its affinity for the ligand. Also, changing slopes of simulation curves in Fig. 13. A, indicate two maximums whereby *i)* the receptor approaches saturation with GABA at short assay times (assay time-limited $^{36}Cl^-$ ion flux), and *ii)* tracer ion flux approaches equilibration at longer assay times (internal volume-limited $^{36}Cl^-$ ion flux) [237].

As addressed before (*see* paragraph *2.1.2.*) non-linearity of Hill plots often occurs, and the slope at half-saturation is non-integral. This is the case for the transmembrane ion flux assay for channel forming GABA receptors discussed here. The simulation of Hill plots, that is $\log\{M_{1t}/(M_{1t,max}-M_{1t})\}$ as a function of $\log[GABA]$ where $M_{1t,max}$ is the maximal influx obtained with assay time, t, showed that assay time alters the slope of the Hill plot (n_H) between 1 and 2.7, although two binding sites mediating GABA response were assumed [237].

3.1.10. Kinetic multiplicity of native receptor channels

Advances in the identification, synthesis and reconstitution of the subunits of the GABA receptor have demonstrated that a great variety of GABA-responsive oligomers composed of homologous subunit variants might exist having different subunit specificities (*see* for example [247]). The findings outlined in the previous paragraphs suggest, that in spite of the great variety of GABA receptor channels made possible by the large number of homologous subunits processed with cDNA technique (*see* for example [248]), the kinetic approach distinguished, on the basis of desensitization rates, two major types of native GABA receptor channels in the mammalian brain: the more abundant faster desensitizing and the slower desensitizing ones [229, 230], [236], [249].

Once multiplicity of native GABA receptor channel function is recognized, it is necessary to establish and investigate the comparative drug specificity of these GABA receptor channels with different lifetimes. Applying rapid kinetic technique to investigate the effect of pentobarbital on GABA-specific $^{36}Cl^-$ ion uptake by cortical membrane vesicles, it was observed that with anesthetic concentration of pentobarbital both, the faster and slower desensitizing GABA receptor channels displayed a higher rate of chloride exchange and of desensitization. However, higher than anesthetic concentration of pentobarbital (1 mM) completely inhibited channel opening of the slower desensitizing GABA receptor channel although this was subsequently reactivated by a conformational change other than channel opening and desensitization [234]. This was the first observation on differences in chemical mechanisms underlying drug action at GABA receptor channels with different active lifetimes. When ^{36}Cl$^-$ ion flux, as studied by conventional mixing and sampling techniques (assay time 7 s), into these membrane vesicles induced by chiral agonists, *i.e.* (+)[*S*]dihydromuscimol and (-)[*R*]dihydromuscimol were compared, a concentration dependent enantioselectivity was observed [250]. While (+)[*S*]dihydromuscimol showed complete response within 1.5 log units of its concentration, (-)[*R*]dihydromuscimol showed ~85 % response in 2 log units of its concentration. One explanation of this difference is that two recognition sites mediate the response mediated by GABA receptor channel [230] with the (+)[*S*]dihydromuscimol and only one recognition site on GABA receptor with the (-)[*R*]dihydromuscimol. An alternative explanation of this result is that (-)[*R*]dihydromuscimol, but not (+)[*S*]dihydromuscimol, discriminates between GABA receptor channels with different lifetimes, one of them having an affinity decreased by an order of magnitude. The pharmacological action of benzodiazepines [251-254] is due to their effects on GABA receptor channels [233], [255-258]. Benzodiazepine analogues enhance both channel opening [238] [259, 260] and desensitization [239], [261, 262]. These anxiolytics can give rise to dependency and withdrawal effects in humans and animals ([241]) and references cited therein). Rats were made tolerant to chlordiazepoxide for 15 days with an implanted osmotic pump. Dependence on GABA concentration of ^{82}Br$^-$ ion exchange in cerebrocortical membrane from tolerant *vs.* naive rats in the absence and presence of chlordiazepoxide was followed between 0.1-30 s. In the presence of chlordiazepoxide (150 μM), the rates of the faster and slower desensitizing GABA receptor-mediated initial halide exchange, J_A and J_B, as well as their desensitization rate, α and β, were increased ~3-fold with 10 μM GABA. Under the conditions, the progress of halide ion exchange was similar for naive and tolerant rat. With tolerant rat in the absence and presence of chlordiazepoxide, J_A, J_B and β reverted to the same values as with naive rat, but α remained enhanced in the absence of chlordiazepoxide [241]. The specific enhancement of desensitization of the faster desensitizing receptor with tolerant rats demonstrates separate control *i)* of desensitization and channel opening, and *ii)* of desensitization of the faster and slower desensitizing GABA receptor channels. If the explanation is an increased affinity for the first GABA bound [241], then the channel opening and desensitization must be mediated by different GABA binding sites: at least different structural domains and possibly different subunits [241]. It follows, that GABA receptor channels with short and long lifetimes may be structurally different. Some supporting evidence comes from the observation that two GABA receptors could be separated on the basis of their different charge densities [263] indicating possible difference in their glycosylation or phosphorylation states. In addition, two distinguishable putative intracellular domains were observed in a variety of recombinant GABA receptor channels [264].

Structural differences underlying kinetic (functional) differences of GABA receptor channels may be inferred from rate measurements of $^{36}Cl^-$ ion exchange induced by 40 μM GABA in cerebrocortical membrane vesicles in a series of buffered salt solutions made up of different D_2O-H_2O (0-70 % D_2O) using rapid kinetic techniques [265]. Incubation times below and above 100 ms were achieved with continuous and pulse-mode quench flow techniques [22], respectively. The initial rate constants for chloride ion exchange, J_A and J_B, as well as their desensitization rate, α and β, were derived by data processing according to the minimal mechanism of GABA receptor channel opening and desensitization using Eq. (23). Different solvent isotope effect was observed for J_A, J_B and α. Ion flux in the slower phase was *completely* inhibited by the presence of 20 % D_2O, whereas ion flux occurring in the faster phase was not affected. The shape of the curve relating decrease of fractional α to increasing deuterium mole fraction (Fig. 14.) indicates, that desensitization of GABA receptor channel with short lifetime might involve a specific hydrogen bonding step [266]. The extreme large and steep deuterium isotope effect observed with J_B is unprecedented, which could not be accounted for the effects of difference in mass number or hydrogen tunneling [267]. It can be due, however, to stronger deuterium bonding [268] that would hinder conformational isomerization rendering slower desensitizing channel to open. Substitution of deuterium for exchangeable hydrogenic sites in the faster desensitizing GABA receptor has a differential effect. These substantial differences in [D_2O] dependence indicate separate structural prerequisites *i)* for desensitization and channel opening, and *ii)* for channel opening of slower and faster desensitizing GABA receptor oligomers. It is often assumed, that selectivity and pore size are related in a way, that highly selective channels have the smallest pore [169]. It is proposed therefore, that the distinguishable, shorter and longer open lifetimes observed with GABA receptor channels may be due to different open channel diameters from distinct, tetrameric and pentameric, subunit combinations [233], [265]. The markedly different channel diameter estimates reported, ~6 Å [269, 270] and ~3 Å [271], might be conjectured.

It is known that transition metal cations, such as endogenous [272, 273] and exogenous Cu^{2+} [274] and Zn^{2+} [275] ions can inhibit GABA receptor binding and channel function non-competitively. Also, Zn^{2+} ion was proposed to affect a slowly desensitizing GABA receptor channel [276]. Heterooligomeric GABA receptor channels composed of $\alpha 1$ and $\beta 1$ subunits expressed in kidney cells and studied by whole-cell recording techniques were sensitive to Zn^{2+} ion, whereby those with $\gamma 2$ subunits were relatively insensitive [277, 278]. Additional fast kinetic flux measurements on the effects of Zn^{2+} ion on GABA-responsive transmembrane chloride ion flux may reveal differential sensitivity of faster and slower desensitizing GABA receptor channels to Zn^{2+} ion would be of particular interest.

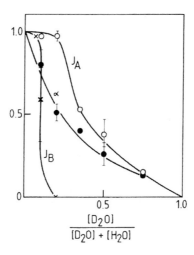

Fig. 14. Effects of D_2O on the initial rate constant for the GABA-responsive $^{36}Cl^-$ ion translocation through the faster (J_A: o) and slower (J_B: x) desensitizing GABA receptor channels and rate constant for desensitization of the faster desensitizing receptor, α (●). Error bars indicate standard deviation. Adapted from [265].

Extra-synaptic and synaptic benzodiazepine binding sites in cerebrocortical membranes being non-sensitive and sensitive to GABA, respectively [279], as well as the existence of δ subunit-containing extra-synaptic and γ2 subunit-containing synaptic GABA receptor oligomers in cerebellar granule neurons with slow and fast kinetic properties, respectively [280], may provide a clue to identify slower desensitizing GABA receptor channel as extra-synaptic and faster desensitizing GABA receptor channel as synaptic. However, it is demonstrated that the biphasic decrease of receptor activity could not be attributed to vesicle heterogeneity and must be due to desensitization processes with different rates [232]. In addition, different cerebrocortical membrane preparations with different amount of synaptic membrane regions do obey similar biphasic desensitization kinetics [249]. Conclusively these findings suggest, that GABA receptor channel oligomers with different desensitization kinetics distribute uniformly in the neuronal membrane. It is to note in this respect, that the electron microscopic image of structurally homologous ACh receptor channels in frozen tubular vesicles from the postsynaptic membrane of the electric ray *Torpedo marmorata* [165] shows the presence of two forms with thinner and wider cross-sections, with and without the cytosolic 43 kD protein attached, respectively. These forms are present in a ratio of approximately 2. The more abundant thinner form was identified as a pentameric structure composed of three different subunits [165]. The form showing wider cross-section may indicate *i)* the same pentameric structure embedded differently in the membrane, *ii)* the same subunit composition with the 43 kD protein attached differently (incorporated), and *iii)* a different oligomeric structure involving more than five subunits.

When glutamate-responsive transmembrane sodium ion flux was measured in rat brain hippocampal membrane vesicles with different rapid chemical kinetic methods, such as the quench flow technique with radiotracer ($^{22}Na^+$ ion) detection [281] and the stopped-flow technique with fluorescence detection [162] in the presence of the glutamate transport inhibitor, dihydrokainate, cation flux proceeded in two phases, also. The faster stopped flow technique, however, gave rise to faster channel opening and desensitization rate constants [162]. Rate constants for ACh [191] and GABA [282] receptor channel opening were also increased with increasing time-resolution of rapid chemical kinetic methods, but those of desensitization not. Dependence of measured rate constants on sampling rate suggests non-Markovian kinetics of channel opening [104] substantiated by structural information on a multitude of ACh receptor channel conformations imaged in the open state [283]. Differences in sampling rate dependence of measured desensitization rate constants may indicate, that desensitization of glutamate receptor channels proceeds from an open channel conformation [104]. In any case, kinetic multiplicity of GABA and glutamate receptor responses suggests a role for desensitization in the control of signaling *via* inhibitory and excitatory membrane receptor channels of different lifetimes.

3.1.11. Differences between electrophysiological (patch clamp) and quench flow measurements

Patch clamp measurements are of fluctuations in a system with receptor at equilibrium. Fluctuation are visible because of the low number of receptor molecules are present. *Quench flow* measurements follow relaxation of receptor populations to equilibrium after a concentration jump, from the initial non-equilibrium (equilibrium with no transmitter) to a final state. It should be understood, that the equilibration what taking place in the isotope exchange, is that of isotopic *specific activity*. This is not the same as net transport of chloride. Opening up the channels will only usually equilibrate the chloride concentration with the membrane potential, bringing it to a potential, which may be near to zero. But this occurs in a much shorter time than that of the quench flow measurements being made and involves an extremely small *net* transfer of ion, negligible compared with the radiotracer *exchange*. What quench flow measurements are following is the transmembrane isotope exchange, which, at any time *t* proceeds at a rate proportional to open channels.

In *patch clamp* measurements, the integration of several types of stochastic events is done, by following channels in the patch for a relatively long time. In *quench flow* measurements this is done by following a large number of receptors and analysing the results as a chemical reaction.

To model events at a synapse from *patch clamp* measurements, non-equilibrium events must be calculated from measurements of a receptor system at equilibrium. *Quench flow* results correspond more directly to the events in a synapse, such as large number of receptors, perturbation of equilibrium and subsequent relaxation.

Patch clamp measurements have normally been made with relatively low, single neurotransmitter concentrations. *Quench flow* measurements are made over the whole range of response, up to saturation with neurotransmitter. This is because in patch clamp studies activation occurs with micromolar neurotransmitter concentrations, whereas half-response of the receptor in quench flow measurements occurs about 100 μM. Incidentally, concentrations and dissociation constants are much suitable for synaptic transmission, for receptor occupancy. However, there are reasons for the patch clamp results to be low. Firstly, the receptor becomes desensitized on exposure to the neurotransmitter. That is, a large fraction of the receptor is converted by the neurotransmitter to a higher affinity state, which does not give open channel. This has the effect of decreasing the measured dissociation constant by a factor, which can be as high as 1000. Desensitization can be rapid, for example the half time for desensitizing of a GABA receptor is 30 ms. Most of patch clamp measurements have been made with some prior exposure to the neurotransmitter. Removal of most of the receptor activity by desensitization does not preclude measurement of single channels in a patch clamp: indeed it may even be required, in order to limit the number of active (capable of forming open channels) receptor complexes. Secondly, estimates of the response to the neurotransmitter, at saturating concentrations in many electrophysiological experiments have tended to be low, for reasons related to the measurement of electrical current, which do not apply to chemical techniques and isotope tracing. This gives the impression of saturation at a lower concentration than the true one.

Observations with *patch clamp* have been made of single channel fluctuating at equilibrium with time resolution of less than a millisecond. Events in shorter time than the resolution used may appear as flickers or are time averaged. The high resolution of patch clamp measurements results from the small number of molecules rather than the time resolution, which is not always shorter than 1 ms. Mixing techniques, for example *quench flow*, are limited to times of a few ms and above, depending on the volumes mixed. This is sufficient to follow channel opening and receptor desensitization processes. The isotope exchange at any time is averaged due to the number of receptors under study.

Patch clamp measurements of channel opening and closing events allow elucidation of forward and backward steps of the receptor present. These measurements might reflect processes of desensitization and ligand dissociation as well as of channel closing itself. In addition, conversions between different open or different closed states are sometimes modeled, to obtain a fit. Interpretation depends on the correct assignment of single channel events. In *quench flow* measurements, desensitization rates and initial isotope exchange rates (a measure of pre-equilibrium controlled open channel) are fit to the progress of isotope exchange. Alternatively, the decrease of the specific rate (desensitization) in a pre-incubation is fitted to the decrease of isotope exchange in single assay incubation due to changes in the pre-incubation. The time range of the processes studied can be selected by using the appropriate assay time. A much simpler model is fitted to more direct observations (the properties of a population are directly observed): the number of parameters to be fitted is less.

Both techniques are limited to the fitting of a minimal kinetic working model, which might not reveal all the states. Apart from the differences between the membranes studied, the comparison of patch clamp single channel data with quench flow data would depend on various assumptions made in the analysis. A better approach would be to ask whether a common model could account for the quench flow results as well as the single channel results, since some determined numbers are softer than others. In the two systems in which this has been reported, ACh receptor from *Electrophorus electricus* and GABA receptor from mammalian brain, it could.

While *patch clamp* measurements in combination with flow techniques have been applied with increasing success, there are still aspects, which can only be addressed with *quench flow* methods in combination with the experimental model. These are *i)* the unstirred surface layer, giving a diffusion limit to the receptor, is smaller with membrane vesicles than with cells, and *ii)* the possibility of identifying new, previously undetected receptor subtypes in cell regions, such as nerve terminals, which are different from those located on the soma (accessible to patch clamp).

Acknowledging enthusiastically that model systems with controlled molecular biology are of great importance, as are model systems of cultured cells and as are *patch clamp* experiments, the rationale for investigating native membrane in *quench flow* measurements is to investigate events and changes, which *do* occur in the brain.

3.2. Inhibition of response

Because GABA receptor channels are now recognized to be members of a superfamily of homologous protein complexes, including ACh, glycin and serotonin receptor channels [11-13], appreciation of their functional inhibition is relevant to channel forming neurotransmitter receptors in general. Conformationally restricted enantiomeric pairs of antagonist have the potential to discriminate between the two characterized GABA receptor channels with different active lifetimes. Moreover, by making measurements with varying concentrations of GABA and the antagonist and using the initial rates of the kinetically resolved receptor responses and the appropriate derived equations, various hypothetical models of antagonism, *e.g. competitive* vs. *non-competitive* models can be distinguished. The simplest of which (*competitive*) hypothesize that functional antagonism proceeds *via* binding of antagonist to recognition sites for GABA.

3.2.1. Competitive model

This is the model in which inhibitors first bind to recognition sites for GABA of the active receptor orthosterically, before the receptor channel opens. An active receptor, A, binds two ligand (agonist) molecules with microscopic dissociation constants, K_1, giving a double liganded species with isomerizes to give an open channel conformation with a channel-opening equilibrium constant, Φ^{-1}. Ion flux through the open channel proceeds with a rate constant J_m. In parallel, A can also bind two inhibitor (antagonist) molecules with microscopic dissociation constant, K_2. In addition, the single liganded receptor, LA, can also bind the inhibitor, with K_2. The single and double liganded species, IA, I_2A and ILA thus formed do not form open channels.

$$\quad\quad K_1 \quad\quad\quad K_1 \quad\quad \Phi$$

$$2L + A \leftrightarrows L + LA \leftrightarrows L_2 A \leftrightarrows L_2 \bar{A} \updownarrow J_m \tag{36}$$

$$\quad\quad K_2$$

$$2I + A \leftrightarrows I + IA \leftrightarrows I_2 A \tag{37}$$

$$\quad\quad K_2$$

$$I + LA \leftrightarrows ILA \tag{38}$$

In the presence of the inhibitor the ion flux is given by the expressions

$$J = J_m [R'] / \{1 + \Phi(1 + K_1/L)^2 + B\} \tag{39}$$

$$B = \Phi([I]^2 K_1^2 / [L]^2 K_2^2) + \Phi([I] K_1 / [L] K_2)(1 + K_1/[L]) \tag{40}$$

Diagnostic curves for planning measurements to determine reversible equilibrium dissociation constant for inhibition (K_2) of GABA receptor channel in the active state based on the competitive model, Eq. (36), Eq. (37), Eq. (38), and corresponding equations, Eq. (39), Eq. (40) are shown by Fig. 15. It is to note, that alternative models in which inhibitors affect the receptor only after the channel has opened, by blocking it ([284] reviewed by [285]) as well as the 'regulatory site' mechanism in which the inhibitor first binds to a regulatory site on the receptor before, and after, the channel opens followed by a conformational change to a form, that does not lead to open channel [191] have been proposed for inhibition of ACh receptor channel function.

3.2.2. Specificity and stability of the inhibitor, bicuculline

Mechanisms underlying functional inhibition of GABA receptor channels were investigated using the antagonist, a phthalide isoquinoline derivative plant alkaloid, bicuculline [286], (+)[1S,9R]BIC, where BIC stands for bicuculline. (+)[1S,9R]BIC has an *erythro* relative configuration, a structural requirement for antagonism of GABA receptor function [286-287]. An about two-fold difference was found in the concentration of (+)[1S,9R]BIC and that of the racemic mixture to displace [^{3}H]GABA from receptor [287] predicting enantioselectivity in inhibition of GABA-responsive ^{36}Cl$^-$ ion flux.

Specificity of (+)[1S,9R]BIC as an antagonist of GABA receptor binding and channel function has been tested in rat brain membrane [288]. (+)[1S,9R]BIC was about 70 times more potent than (-)[1R,9S]BIC as an inhibitor of [^{3}H]GABA binding and was about 100 times more potent than (+)[1S,9R]norBIC. In measurements of GABA-responsive ^{36}Cl$^-$ ion flux, (+)[1S,9R]norBIC was much less potent than (+)[1S,9R]BIC. The observed increase in binding and in inhibition of function caused by N-methyl substitut ion in (+)[1S,9R]norBIC was attributed to different conformations for (+)[1S,9R]norBIC and (+)[1S,9R]norBIC, as demonstrated by proton NMR nuclear Overhauser effect measurements. In measurements of inhibition of [^{14}C]GABA uptake, the relative values of the inhibition constants are 0.3~0.3<1<3.3 for (-)[1R,9S]norBIC, (+)[1S,9R]norBIC, (+)[1S,9R]BIC and (-)[1R,9S]BIC, respectively [289]. The GABA uptake inhibition constant was found to be higher than 100 μM for (+)[1S,9R]BIC.

For ^{36}Cl$^-$ ion flux measurements with (+)[1S,9R]BIC, stock solutions of (+)[1S,9R]BIC freshly prepared in pH 3.5 HCl and kept on ice in dark were diluted into the iced buffer pH 7.5 containing GABA and the tracer before the particular experimental sessions with (+)[1S,9R]BIC. Pilot experiments showed that the buffer capacity is high enough to jump pH 3.5 to working pH 7.5 in the quench flow experiment. The rate constant of (+)[1S,9R]BIC hydrolysis was determined using (+)[1S,9R]BIC solution (46 μM) prepared as before, but in the absence of GABA and the tracer. The hydrolysis of (+)[1S,9R]BIC was followed up to several days by monitoring changes at λ_{max}=292 nm and λ_{max}=326 nm, characteristic of phthalide ring opening in (+)[1S,9R]BIC. The molar extinction coefficient of (+)[1S,9R]BIC was found to be 6216±152 M^{-1}cm^{-1}. The rate constant, 3×10^{-3} hr^{-1} corresponds to less than 3 % decrease of (+)[1S,9R]BIC in 1.25 hr, which is the time for one experimental session.

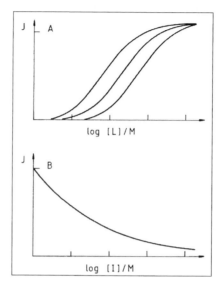

Fig. 15. Diagnostic plots for determination of reversible equilibrium dissociation constant for competitive inhibition (K_2) of GABA receptor channel in the active state based on the competitive model, Eq. (36), Eq. (37), Eq. (38), and corresponding equations, Eq. (39), Eq. (40). A: Varying [L] with constant [I], curves from left to the right represent [I]$_0$ ([I]=0) and different, but constant inhibitor concentrations, [I]$_1$ and [I]$_2$. B: Varying [I] with constant [L]. [L]=[GABA], [I]=(+)[1S,9R]BIC.

3.2.3. Pulsed flow $^{36}Cl^{-}$ ion flux techniques for measurements of GABA receptor channel-mediated response inhibition

As outlined before fast reaction techniques are necessary to study GABA receptor channel-mediated chloride ion flux because the rate constant for chloride translocation is high (~10^8 M^{-1}s^{-1}) and the desensitization of the receptor is rapid. The chloride flux is complete within a second on exposure to saturating concentration of GABA.

The quench flow machine used in measurements allows mixing of small volumes (>0.225 ml) covering reaction times from 5 ms to minutes with relatively low pressure (<60 psi) and at variable temperature. The apparatus allows applications and different combinations of rapid kinetic techniques such as for reaction times less than the stroke time continuous quench flow and pulsed quench flow for reaction times greater than the stroke times (flux techniques). Using more than one mixer, pre-incubation can be made with the pulsed or continuous flux techniques or a combination of the two (*see* also paragraph *3.1.3.*) depending on the incubation times. In this way the tracer ion flux with the neurotransmitter may be used as an assay for following receptor desensitization or inhibition during pre-incubation with ligands (agonists or antagonists) for various times (pre-incubation-flux techniques). GABA receptor channel-mediated chloride ion-exchange and receptor desensitization were measured in fresh suspensions of native neuronal plasma membrane vesicles directly prepared from the rat cerebral cortex using $^{36}Cl^-$ isotope tracer and the quenched flow technique in pulsed mode as described above. GABA-responsive specific chloride channels were opened on rapid mixing with GABA and closed on mixing with a quench solution containing the quaternary salt, (-)([1S,9R]BIC methiodide [290]. Typically, 0.225 ml aliquot of the membrane suspension (0.75 mg of protein/ml) mixed with an equal volume of buffer (containing 145 mM NaCl, 5 mM KCl, 1 mM $CaCl_2$, 1 mM $MgCl_2$, 10 mM glucose, 10 mM HEPES, pH 7.5) in the presence and absence of GABA (2 mM) with 15 μCi/ml of $^{36}Cl^-$ ion (>10 mCi/g chloride in NaCl solution, Amersham) gives 2000 and 4000 counts in 10 minutes (200 and 400 cpm), respectively, determined by allowing an incubation time of 6 s at 30°C (equilibration of GABA receptor channel-mediated $^{36}Cl^-$ ion uptake).

The membrane vesicle preparation takes up GABA [289], [291]. The molecular transport of GABA, an electrogenic, Na^+ and Cl^- ion dependent (symport) process (*see* section *3.3.*), might interfere with GABA receptor channel-mediated processes in GABA-responsive $^{36}Cl^-$ ion flux measurements. The possible contribution of GABA transport-coupled $^{36}Cl^-$ ion influx was determined in measurements of ^3H- and ^{14}C-labeled GABA uptake. The contribution of GABA transport-coupled $^{36}Cl^-$ ion influx was evaluated for 10 s with 40 μM GABA using the experimental first-order rate parameters which predict transmembrane GABA flux in the whole range of equilibration [291]. Under the conditions, the contribution of GABA transport-coupled $^{36}Cl^-$ ion flux to the GABA receptor channel-mediated $^{36}Cl^-$ ion flux, is negligible, *i.e.* 0.1 %.

In order to obtain data from different membrane vesicle preparations strictly comparable, the flux experiment with a single concentration of the inhibitor requires three experimental sessions with different conditions each, *i.e.* *baseline* (no GABA added), *no inhibitor* (GABA) and *inhibitor* (GABA with the inhibitor) ones. In addition, the $M_{I\infty}$ value for the particular membrane vesicle preparation is determined with saturating concentration of GABA at one incubation time (steady-state: 6-10 s).

The progress of $^{36}Cl^-$ ion influx, M_{It}, was followed in the absence and presence of varying concentration of (+)[1S,9R]BIC (1-75 μM) with 10 μM (Table 4) and 40 μM (Table 5). With each concentration of (+)[1S,9R]BIC and GABA, rate parameters, which predict inhibition of short-lived and long-lived

GABA receptor response were obtained by curve fitting using Eq. (27). Rate parameters obtained by fitting Eq. (28) to the progress of $^{36}Cl^-$ ion influx induced by 10 µM GABA in the absence and presence of varying concentration of (+)[1S,9R]BIC (Table 4) suggest that the change of α with the concentration of (+)[1S,9R]BIC where α is assumed to increase and J_A to stay relatively unchanged (fit 2), there is two concentration ranges of (+)[1S,9R]BIC (2-4 µM and 12-16 µM), in which seemingly local increase of α occur. The decrease of J_A with the concentration of (+)[1S,9R]BIC where α is assumed to stay constant (fit 1) showed similar concentration ranges of (+)[1S,9R]BIC, in which local recovery of J_A occurred. When J_A/α values are considered, local decreases of fast phase inhibition can also be observed in the above ranges of (+)[1S,9R]BIC concentration. In these partial recoveries from fast phase inhibition, the recovery

Table 4.
$M_{lt}/M_{l\infty}$ fraction of $^{36}Cl^-$ ion influx induced by 10 µM GABA as a function of time in the absence and presence of varying concentration of (+)[1S,9R]BIC ([I])

[I] (µM)	J_B (s^{-1})	J_B/β	fit1 J_A (s^{-1})	α (s^{-1})	J_A/α	fit2 J_A (s^{-1})	α (s^{-1})	J_A/α
0[a,b,c]								
1[a]	0.016	0.36	0.27	0.72	0.38	0.54	1.7	0.33
2[a]	0.008	0.18	0.35	0.72	0.49	0.50	1.2	0.42
4[a]	10^{-7}	0.00	0.45	0.73	0.62	0.53	0.9	0.60
8[b]	10^{-7}	0.00	0.15	0.71	0.21	0.45	2.6	0.18
9[c]	10^{-7}	0.00	0.14	0.72	0.19	0.53	4.5	0.12
12[b]	10^{-7}	0.00	0.19	0.70	0.27	0.44	1.9	0.23
16[b]	10^{-7}	0.00	0.28	0.74	0.38	0.47	1.3	0.37
27[c]	10^{-7}	0.00	0.07	0.72	0.10	[d]	[d]	[d]

The no inhibitor rate constants were, for (a) J_A=0.54 s^{-1}, J_B=0.049 s^{-1}, α=0.72 s^{-1}, β=0.045 s^{-1}, J_A/α=0.75, J_B/β=1.09; for (b) J_A=0.47 s^{-1}, J_B=0.052 s^{-1}, α=0.72 s^{-1}, β=0.045 s^{-1}, J_A/α=0.65, J_B/β=1.16; for (c) J_A=0.53 s^{-1}, J_B=0.039 s^{-1}, α=0.72 s^{-1}, β=0.045 s^{-1}, J_A/α=0.73, J_B/β=0.87; (d) the possibility of fit 2 is uncertain because of the large SD value obtained. J. Kardos and D.J. Cash, to be published.

from an enhanced desensitization (decreasing α) may be assumed to be the alternative of the fast phase flux recovery (increasing J_A).

Results from flux experiments performed on a timescale up to 20 s raised the possibility of the slow phase recovery. Experiments with incubation times extended up to 30 s suggested that the slow phase started to recover at about 8 s of $^{36}Cl^-$ ion influx induced by 10 µM GABA, after a short time of no slow phase. Comparisons of the differences between the measured and the calculated values of the ratio, M_{lt}, in the presence over the absence of (+)[1S,9R]BIC when

t=8 s, t=20 s and t=30 s suggest partial slow phase recoveries by (+)[1S,9R]BIC in the concentration ranges 2-4 μM and 12-16 μM. Rate parameters obtained by fitting Eq. (28) to the progress of $^{36}Cl^-$ ion influx induced by 40 μM GABA in the absence and presence of varying concentration of (+)[1S,9R]BIC suggest that the slow phase is inhibited in the presence of 3 μM (+)[1S,9R]BIC. The value of the ratio J_B/β indicates an about 70 % inhibition of the slow phase, which is in a reasonable agreement with K_i=0.89 μM [289] and IC$_{50}$=3 μM [288] values published for (+)[1S,9R]BIC in measurements of the displacement of [3H]GABA specifically bound to sites in synaptic membrane suspensions by (+)[1S,9R]BIC. The fast phase was also inhibited considerably by increasing concentration of (+)[1S,9R]BIC to 20 μM (Table 5). Further increase of the concentration of (+)[1S,9R]BIC to 75 μM resulted in nearly complete inhibition of the fast phase. The results can fit an inhibition of J_A with α unchanged (**fit 1**) or they can fit an enhanced α (**fit 2**). Apparently, the values of the ratio J_A/α were unaffected by the alternative fits. Thus decreasing values for J_A/α represent the increasing inhibition of the fast phase.

Results obtained with these measurements of GABA receptor channel response suggest that the inhibition by the antagonist (+)[1S,9R]BIC looks as if this will certainly not be a simple competitive inhibition including the following possibilities: *i*) inhibition of short-lived GABA receptor response can fit either inhibition of trans-channel ion flux (J_A) or enhancement of receptor desensitization (α) whereas J_A/α unaffected, *ii*) partial recovery of short-lived and long-lived GABA receptor channels may occur in two ranges of antagonist concentration, *i.e.* 2-4 μM and 12-16 μM. It may be conjectured, that variability of measurements on GABA-responsive $^{36}Cl^-$ ion flux in cultured cerebellar granule neurons was significantly reduced whereby response amplitude increased by pre-treating cells with 2 μM (+)[1S,9R]BIC [41], [292].

Table 5

$M_{1t}/M_{1\infty}$ fraction of $^{36}Cl^-$ ion influx induced by 40 μM GABA as a function of time in the absence and presence of varying concentration of (+)[1S,9R]BIC ([I]).

				fit1			fit2		
[I] (μM)	J_B (s^{-1})	β	J_B/β	J_A (s^{-1})	α (s^{-1})	J_A/α	J_A (s^{-1})	α (s^{-1})	J_A/α
0	0.29	0.30	0.97	2.6	3.3	0.79	2.6	3.3	0.79
20	10^{-7}	0.30	0.00	0.61	3.3	0.18	2.6	17	0.15
75	10^{-7}	0.30	0.00	0.20	3.3	0.06	1.3	20	0.07

J. Kardos and D.J. Cash, to be published.

These partial recoveries distinguished by the different ranges of antagonist concentration may be attributed to different mechanisms as to recovery of trans-channel flux and inhibition of desensitization. Results obtained so far, include the possibility, that enhanced receptor desensitization may be the mechanism underlying inhibition of GABA receptor response by the antagonist. To prove or reject this hypothesis is one of the most interesting challenges in this field.

3.3. Solute transport in excitable membranes

The free energy (ΔG_{diff}) released by the diffusion of a solute or required for its transport in the opposite direction is given by the Gibbs equation,

$$\Delta G_{diff} = RT \ln([L_2]/[L_1]) \tag{41}$$

where $[L_1]$ and $[L_2]$ stand for the concentrations of the solute (ligand) on opposite sides of the neuronal membrane. Since neuronal membranes are characterised by their selective ion permeabilities, an additional term (ΔG_{emf})

$$\Delta G_{emf} = ZFV \tag{42}$$

where V is the transmembrane potential difference in volts, is to add to the total of free energy changes as given by the sum of Eq. (42) and Eq. (43)

$$\Delta G_{total} = RT \ln([L_2]/[L_1]) + ZFV \tag{43}$$

Since ionic currents are carried primarily by Na^+ and K^+ ion in neurons, the energy input required for their transport against their concentration gradient can be approximated [12] by the application of Eq. (44) to just these species,

$$\Delta G_{total} = RT \ln ([K^+]_i [Na^+]_o)/([K^+]_o[Na^+]_i) \tag{44}$$

Propagating signaling functions of neurons can only be possible if the ionic gradients that drive them are maintained. Assuming typical values of 50 and 12 for concentration gradients $[K^+]_i/([K^+]_o$ and $[Na^+]_o/[Na^+]_i$, respectively, ΔG_{total} is about 4 kcal/mol of Na^+ ion exchanged for K^+ ion. The exothermic process, adenosine triphosphate (ATP) hydrolysis yields about 12 kcal/mol allowing the exchange of approximately 3 mol of cation for each mol of ATP hydrolyzed. Substantiating Eq. (45), this is the ratio, which has been measured in various tissue preparations [12]. As in most other animal cells, the principal transport process in neurons is mediated by the ATP-dependent Na^+/K^+-pump, which concurrently extrudes Na^+ ion and accumulates K^+ ion [293]. The complex integral membrane protein, Na^+/K^+-pump can be viewed as a molecular machine, which can work at the expense of ATP hydrolysis and is termed therefore Na^+/K^+-ATPase.

3.3.1. Family of Na^+/Cl^- neurotransmitter transporters (symporters)

As indicated (Fig. 8.), translocation of released neurotransmitters and other diffusible substances including hormones and trophic factors ([11-13], for reviews *see* [294], [295] and references therein) from the synapse into the presynaptic neuron or into glia or other neurons is achieved by integral membrane protein specialized for molecular transport, called transporters. Besides receptor desensitization, transporter-mediated clearance of neurotransmitters after depolarization-induced release from nerve terminal (*see* paragraph *3.2.2.*) represents a major mechanism for spatio-temporal regulation of chemical signaling. Most of neurotransmitter transporters have a similar topography of twelve transmembrane helices belonging to distinct subfamilies: *i)* GABA transporters, *ii)* monoamine transporters (norepinephrine, serotonin, dopamine), *iii)* amino acid transporters (glycine, proline), *iv)* 'orphan' (NTT4) transporters, and *v)* bacterial transporters. Studies on neurotransmitter uptake indicated the existence of neurotransmitter-specific uptake systems, such as vesicular transmitters that transport neurotransmitters into intracellular storage vesicles from the cytosol (Fig. 8.), Na^+ and Cl^- ion-dependent (Na^+/Cl^-) transporters embedded in neuronal and glial plasma membrane, that function in the uptake of neurotransmitters from the extracellular space, and the less-specific amino acid transporters that have a role in the control of extracellular level of neurotransmitters. The other family, characterized by ten transmembrane helices includes the Na^+, Cl^- and K^+ ion-dependent transporters for glutamate, the major excitatory neurotransmitter in brain.

The most important questions in the field include "resolving the mechanism of transport, the structure of the transporters, and the interaction of each transporter in complex neurological activities" [294]. A milestone in our knowledge on transporters was, when the first cDNAs encoding the GABA transporters were cloned and sequenced [296-297] with the promise predicting transport mechanisms from structure. Multiple GABA symporters do exist in the brain [294], [298-301], the precise function of each is not entirely clear. Among these GAT1 is the only neurotransmitter symporter that was cloned based on its amino acid sequence, therefore it is considered as the prototype of this symporter family.

Molecular mechanisms of plasma membrane neurotransmitter symporters, like GAT1 have been inferred primarily through radiotracer flux assays. These include measurements of specific uptake of ^3H- and/or ^{14}C-labeled neurotransmitters in brain suspensions (*see* for example [289], [291]) and cell model systems (*see* for example [300], [301]). Alternatively, assays exploring, that neurotransmitter transport is coupled energetically to the inward movement of Na^+ and Cl^- ion flux have been performed to understand the molecular mechanism and permeation pathway at the molecular level [302]. It is claimed, that the resolution of radiotracer ion flux techniques limits the opportunities for the molecular dissection of symporter mechanisms [303]. This may be true with exceptions, like the successful application of quench flow radiotracer techniques to follow progress of GABA [289] and glutamate [281] transport, but is not the point (for a more detailed discussion of the issue *see* paragraph *3.1.11.*). Instead, synchronous recording of neurotransmitter flux and ionic current using patch-clamp recording techniques in combination with micro-amperometry has been developed and used to assess norepinephrine transporter function and regulation [303]. Data obtained this way revealed voltage-dependent norepinephrine flux that correlates temporally with antidepressant-specific symporter currents in the same patch. Furthermore, the temporal resolution of this powerful method allowed the observation of unitary flux events in the millisecond-subsecond time range linked with burst of norepinephrine symporter openings. These findings provide evidence for a high-throughput, channel-mode norepinephrine symporter mechanism as opposed to the much slower, carrier-mode (alternating-access, [295], [303]) operation. By contrast, utilising the *Xenopus oocyte* expression system in combination with electrophysiology and radiotracer uptake, a channel-mode like function could also be resolved [304]. It is found to be the predominant one for a recently cloned insect epithelial membrane protein related to mammalian Na^+/Cl^- neurotransmitter symporters, that mediate thermodynamically uncoupled amino acid uptake as well (CAATCH1, [304-305]). Collectively the above results suggest, that the total activity of symporter function, that is abundance and lifetime, what compromises selection of the most appropriate technique(s) to be applied.

3.3.2. Release of neurotransmitters in excitable membranes

The neurotransmitter, that has been taken up by its specific plasma membrane symporter into the cell cytosol redistributes in a neurotransmitter-specific compartment, *i.e.* in 'synaptic' vesicles (Fig. 8., [11]). This secondary transport process is operated by the symporter embedded in the 'synaptic' vesicle membrane. When stimulus arrives at the presynaptic nerve ending, the specialized release mechanism ('exocytosis', [11-13]), a Ca^{2+} ion-dependent process, controls its *fast* egress from the synaptic vesicle into the synaptic cleft [306-307]. 'Exocytosis' is a mechanism in which the membrane of the synaptic vesicle is thought to fuse with the plasma membrane of the presynaptic neuron, thereby releasing a 'quanta' of the neurotransmitter stored. Depolarization-induced reversal of plasma membrane symporter function may also contribute to

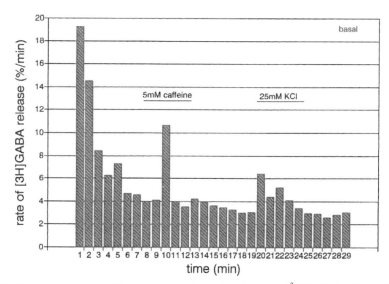

Fig. 16. Comparison of caffeine- and depolarization-induced [³H]GABA release from rat cerebrocortical membrane vesicles. After [314].

the amount of neurotransmitter released into the synaptic cleft [291], [308-312]. Real-time analysis showed that 'exocytosis' and symporter-operated release occurs in the subsecond range of time [307], [311], [314], however stimulus-secretion activity is so 'robust', that it can easily be followed by conventional sampling techniques, by analyzing the amount of radioactivity released from brain tissue after pre-incubation with labeled (³H, ¹⁴C, ¹²⁵I) neurotransmitters and/or neuromodulators within the minute to hours range of time under different conditions (see for example Fig. 16., and [313-320], just to mention a few).

ACKNOWLEDGEMENT

This chapter is dedicated to Dr. Ilona Kovács (1947-2003) for her invaluable contribution to the application and development of radiotracer methodology for studying binding and transport phenomena in biomembranes.

LIST OF ABBREVIATIONS
ACh   Acetylcholine
GABA   γ-Aminobutyric acid
(+)[1*S*,9*R*]BIC   Bicuculline

REFERENCES

[1] C. Miller and E. Racker, The Receptors. Vol. 1. General Principles and Procedures (Ed. R. D. O'Brian) Plenum Press, New York, 1979.

[2] R. Blumenthal and A. E. Shamoo, The Receptors. Vol. 1. General Principles and Procedures (Ed. R.D. O'Brian) Plenum Press, New York, 1979. p. 215.

[3] H.M. Berman, J. Westbrook, Z. Feng, G. Gilliland, T.N. Bhat, H. Weissig, I.N. Shindyalov and P.E. Bourne, *Nucleic Acids Res.*, 28 (2000) 235.

[4] P.A. Whittaker, Trends. Pharmacol. Sci., 24 (2003) 434.

[5] A.J. Clark, J. Physiol. (London), 61 (1926) 530.

[6] A.J. Clark, J. Physiol. (London), 61 (1926) 547.

[7] W.D.M. Paton and H.P. Rang, Proc. Roy. Soc. London Ser. B., 163 (1965) 1.

[8] K.-J.Chang, S. Jacobs and P. Cuatrecasas, Biochim. Biophys. Acta, 406 (1975) 294.

[9] S. Jacobs, K.-J.Chang and P. Cuatrecasas, Biochim. Biophys. Res. Commun., 66 (1975) 687.

[10] P. Cuatrecasas and M.D. Hollenberg, Adv. Prot. Chem., (1976) 251.

[11] D.G. Nicholls, Proteins, Transmitters and Synapses. Blackwell Science, Oxford 1994.

[12] Basic Neurochemistry. Fifth edition. (Eds. G.J. Siegel, B.W. Agranoff, R. W. Albers, P.B. Molinoff) Raven Press, New York, 1994.

[13] Basic Neurochemistry. Molecular, Cellular and Medical Aspects. Sixth edition. (Eds. G.J. Siegel, B.W. Agranoff, R. W. Albers, S.K. Fisher, M.D. Uhler) Lippincott-Raven Publishers, Philadelphia-New York, 1999.

[14] A. Goldstein, L. Aronow and S. M. Kalman, Principles of Drug Action: The Basis of Pharmacology, John Wiley and Sons, New York, 1974.

[15] G. Scathard, Ann. N. Y. Acad. Sci., 51 (1949) 660.

[16] I.M. Klotz, Science, 217 (1982) 1247.

[17] C. Salamon, E.J. Horváth, M.I.K. Fekete and P. Arányi, FEBS Lett., 308 (1992) 215.

[18] O.H. Lowry, N.J. Rosebrough, A.L. Farr and R.J. Randall, J. Biol. Chem., 193 (1951) 265.

[19] P.J. Munson and D. Rodbard, Anal. Biochem., 107 (1980) 220.

[20] P.R. Bevington, Data Reduction and Error Analysis for the Physical Sciences, McGraw-Hill, New York, 1969.

[21] M. Simonyi and I. Mayer, Edu. Chem., (1985) 52.

[22] D.J. Cash and G.P. Hess, Anal. Biochem., 112 (1981) 39.

[23] T. Akera and V-J.K. Cheng, Biochim. Biophys. Acta, 470 (1977) 412.

[24] J. Kardos, K. Maderspach and M. Simonyi, Neurochem. Int., 7 (1985) 737.

[25] A.V. Hill, J. Physiol., 40 (1910) iv.

[26] J. Monod, J. Wyman and J.-P. Changeux, J. Mol. Biol., 12 (1965) 88.

[27] A. Cornish-Bowden and D.E. Koshland JR., J. Mol. Biol., 95 (1975) 201.

[28] J-P. Changeux, J.C. Meunier and M. Hucket, Mol. Pharmacol., 7 (1971) 538.

[29] R. Miledi, P. Molinoff and L.T. Potter, Nature, 229 (1971) 554.

[30] C.B. Pert and S.H. Snyder, Science, 179 (1973) 1011.

[31] E.J. Simon, J.M. Hiller and I. Edelman, Proc. Natl. Acad. Sci. USA, 70 (1973) 1947.

[32] L. Terenius, Acta Pharmacol. Toxicol., 32 (1973) 317.

[33] Y. Cheng and W.H. Prusoff, Biochem. Pharmacol., 22 (1973) 3099.

[34] Handbook of Stereoisomers: Drugs in Psychopharmacology. (Ed. D.F. Smith) CRC Press, Boca Raton, Florida, 1984.

[35] R.P. Clausen, H. Bräuner-Osborne, J.R. Greenwood, M.B. Hermit, T.B. Stensbøl, B. Nielsen and P. Krogsgaard-Larsen, J. Med. Chem. 45 (2002) 4240.

[36] C.J. Grossmann, G.J. Kilpatrick and K.T. Bunce, Br. J. Pharmacol., 109 (1993) 618.
[37] L. Brehm, J.R. Greenwood, K.B. Hansen, B. Nielsen, J. Egebjerg, T.B. Stensbøl, H. Bräuner-Osborne, F.A. Sløk, T.T.A. Kronborg, P. Krogsgaard-Larsen, J. Med. Chem. 46 (2003) 1350.
[38] T. Hashimoto, S. Harusawa, L. Araki, O.P. Zuiderveld, M.J. Smit, T. Imaz, S. Takashima, Y. Yamamoto, Y. Sakamoto, T. Kurihara, R. Leurs, R.A. Bakker and A. Yamatodani, J. Med. Chem., 46 (2003) 3162.
[39] Johansen, T.N., Greenwood, J.R., Frydenvang, K., Madsen, U., Krogsgaard-Larsen, P.: Stereostructure-activity studies on agonists at the AMPA and kainate subtypes of ionotropic glutamate receptors. Chirality 15, 167-179 (2003).
[40] D. Belelli, A.L. Muntoni, S.D. Merrywest, L.J. Gentet, A. Casula, H. Callachan, P. Madau, D.K. Gemmell, N.M. Hamilton, J.J. Lambert, K.T. Sillar and J.A. Peters, Neuropharmacol., 45 (2003) 57.
[41] J. Kardos and A. Guidotti, Adv. Biochem. Psychopharmacol., 45 (1988) 161.
[42] A. DeLean, A.A. Hancock and R.J. Lefkowitz, Mol. Pharmacol., 21(1981) 5.
[43] B.T. Liang, L.H. Frame and P.B. Molinoff, Proc. Natl. Acad. Sci. USA, 82 (1985) 4521.
[44] P. McGonigle, K.A. Neve and P.B. Molinoff, Mol. Pharmacol., 30 (1986) 329.
[45] H. Kromann, F.A. Sløk, T.B. Stensbøl, H. Bräuner-Osborne, U. Madsen, and P. Krogsgaard-Larsen, J. Med. Chem. 45 (2002) 988.
[46] T.B. Stensbøl, P. Uhlmann, S. Morel, B. Eriksen, J. Felding, H. Kromann, M.B. Hermit, J.R. Greenwood, H. Bräuner-Osborne, U. Madsen, F. Junager, P. Krogsgaard-Larsen, M. Begtrup and P. Vedsø, J. Med. Chem. 45 (2002) 19.
[47] K.N. Klotz, S. Kachler, C. Lambertucci, S. Vittori, R. Volpini and G. Cristalli, Naunyn Schmiedebergs Arch. Pharmacol., 367 (2003) 629.
[48] K. Finlayson, T. Maemoto, S.P. Butcher, J. Sharkey, H.J. Olverman, Acta Pharmacol. Sin., 24 (2003) 729.
[49] R.C. Klein, S.E. Warder, Z. Galdzicki, F.J. Castellino and M. Prorok, Neuropharmacol 41 (2001) 801.
[50] P. Kahnberg, E. Lager, C. Rosenberg, J. Schougaard, L. Camet, O. Sterner, E. Østergaard Nielsen, M. Nielsen and T. Liljefors, J. Med. Chem., 45 (2002) 4188.
[51] G. Romeo, L. Materia, F. Manetti, A. Cagnotto, T. Mennini, F. Nicoletti, M. Botta, F. Russo, K.P. Minneman, J. Med. Chem., 46 (2003) 2877.
[52] A. Harms, E. Ulmer, K.A. Kovar, Arch. Pharm (Weinheim), 336 (2003) 155.
[53] J. M. Savola, M. Hill, M. Engstrom, H. Merivuori, S. Wurster, S.G. McGuire, S.H. Fox, A.R. Crossman and J.M. Brotchie, J. Mov. Disord. (2003) 872.
[54] I. Kovács, B. Lasztóczi, É. Szárics, L. Héja, Gy. Sági and J. Kardos, Neurochem. Int., 43 (2003) 101.
[55] P.W. Abel, D. Waugh and W.B. Jeffries, Methods. Mol. Biol., 73 (1997) 323.
[56] A. Phan-Chan-Du, C. Hemmerlin, D. Krikorian, M. Sakarellos-Daitsiotis, V. Tsikaris, C. Sakarellos, M. Marinou, A. Thureau, M.T. Cung and S. J. Tzartos, Biochemistry, 42 (2003) 7371.
[57] J.B. Zawilska, P. Niewiadosmki, J.Z. Nowak, J. Mol. Neurosci., 20 (2003) 153.
[58] I.A. Sharpe, L. Thomas, M. Loughnan, L. Motin, E. Palant, D.E. Croker, D. Alewood, S. Chen, R.M. Graham, P.F. Alewood, D.J. Adams and R. Lewis, J. Biol. Chem., 278 (2003) 34451.
[59] H. Lutjens, A. Zickgraf, H. Figler, J. Linden, R.A. Olsson and P.Scammells, J. Med. Chem., 46 (2003) 1870.
[60] M.P. Schaddelee, H.L. Voorwinden, E.W. van Tilburg, T.J. Pateman, A.P. Ijzerman, M. Danhof and A.G. de Boer, Eur. J. Pharmacol.,19 (2003) 13.

[61] G. Nyitrai, K.A. Kékesi, Zs. Emri, É. Szárics, G. Juhász G and J. Kardos, Eur. J. Pharmacol., 478 (2003) 111.

[62] J. Schmidt, J. Biol. Chem., 259 (1984) 1160.

[63] S. L. Linde, B. Hansen, O. Sonne, JJ. Holst and J. Gliemann, Diabetes, 30 (1981) 1.

[64] M. Berridge, V. Jiang and M. Welch, Radiat. Res., 82 (1980) 467.

[65] R. Deutzmann and G. Stocklin, Radiat. Res., 87 (1981) 24.

[66] A.E. Bolton and W.M. Hunter, Biochem. J., 133 (1973) 529.

[67] A.C. Doran, Y-P. Wan, A.S. Kopin and M. Beinborn, Biochem. Pharmacol., 65 (2003) 1515.

[68] N. Bosworth and P. Thomas, Nature, 341 (1989) 167.

[69] N. DeLapp, J.H. McKinzie, B.D. Sawyer, A. Vandergriff, J. Falcone, D. McClure and C.C. Felder, J. Pharmacol. Exp. Ther., 289 (1999) 946.

[70] A. Newman-Tancredi, D. Cussac, L. Marini and M.J. Millan, Mol. Pharmacol., 62 (2002) 590.

[71] S. Sen, V.-P. Jaakola, H. Heimo, P. Kivelä, M. Scheinin, K. Lundstrom and A. Goldman, Anal. Biochem., 307 (2002) 280.

[72] P. E. Brandish, L.A. Hill, W. Zheng and E.M. Skolnick, Anal. Biochem., 313 (2003) 311.

[73] J.J. Liu, D.S. Hartman and J.R. Bostwick, Anal. Biochem., 318 (2003) 91.

[74] R. Graves, R. Davies, G. Brophy, G. O'Beirne and N. Cook, Anal. Biochem., 248 (1997) 251.

[75] A. Cushing, M.J. Price-Jones, R. Graves, A.J. Harris, K.T. Hughes, D. Bleakman and D. Lodge, J. Neurol. Methods 90 (1999) 33.

[76] H. Bonge, S. Hallén, J. Fryklund, J.E. Sjöström , Anal. Biochem., 282 (2000) 94.

[77] R. Mallari, E. Swearingen, W. Liu, A. Ow, S.W. Young and S.-G. Huang, J. Biomol. Screening 8 (2003) 198.

[78] S.J. Culliford, P. McCauley, A.J. Sutherland, M. McCairn, J. Sutherland, J. Blackburn and R. Kozlowski, Biochem. Biophys. Res. Commun., 296 (2002) 857.

[79] J.C. McGrath, S. Arribas and C.J. Daly, Trends Pharmacol. Sci., 17 (1996) 393.

[80] N. Bandur and D.J. Triggle, Med. Res. Rev., 14 (1994) 591.

[81] A.P. Tairi, R. Hovius, H. Pick, H. Blasey, A. Barnard, A. Supranant, K. Lunström and H. Vogel, Biochemistry, 37 (1998) 15850.

[82] B.W. Madsen, C.L. Beglan and C.E. Spivak, J. Neurosci. Methods, 97 (2000) 123.

[83] A. Beaudet, D. Nouel, T. Stroh, F. Vandenbulcke; C. Dal-Farra; and J.P. Vincent, Braz. J. Med. Biol. Res., 31 (1998) 1479.

[84] S. Terrillon, L.L. Cheng, S. Stoev, B. Mouillac, C. Barberis, M. Manning and T. Durroux, J. Med. Chem., 45 (2002) 2579.

[85] P. Valloton, A.P. Tairi, T. Wohland, K. Friedrich-Benet, H. Pick, R. Hovius and H. Vogel, Biochemistry, 40 (2001) 12237.

[86] G. Turcatti, K. Nemeth, M.D. Edgerton, J. Knowles, H. Vogel and A. Chollet, Receptor Channels, 5 (1997) 201.

[87] G. Turcatti, S. Zoffmnn, J.A. Lowe III, S.E. Drozda, G. Chassaing, T.W. Schwartz and A. Chollet, J. Biol. Chem., 272 (1997) 21167.

[88] G. Turcatti, K. Nemeth, M.D. Edgerton, U. Meseth, F. Talabot, M. Peitsch, J. Knowles, H. Vogel and A. Chollet, J. Biol. Chem., 271 (1996) 19991.

[89] M. Macchia, F. Salvetti, S. Bertini, V. Di Bussolo, L. Gattuso, M. Gesi, M. Hamdan, K.N. Klotz, T. Laragione, A. Lucacchini, F. Minutolo, S. Nencetti, C. Papi, D. Tuscano and C. Martini, Bioorg. Med. Chem. Lett., 11 (2001) 3023.

[90] D. Hadrich, F. Berthold, E. Steckhan and H. Bonisch, J. Med. Chem., 42 (1999) 3103.

[91] I. Berque-Bestel, J.-L. Soulier, M. Giner, L. Rivail, M. Langlois and S. Sicsic, J. Med. Chem., 46 (2003) 2606.
[92] R.B. Raffa, Trends. Pharamacol. Sci., 6 (1985) 133.
[93] M.K.P. Lai, S.W.Y. Tsang, P.T. Francis, M.M. Esiri, J. Keene, T. Hope and C.P.L.-H. Chen, Brain Res., 974 (2003) 82.
[94] A. Rodriguez-Kern, M. Gegelashvili, A. Schousboe, J. Zhang, L. Sung and G. Gegelashvili, Neurochem. Int., 43 (2003) 363.
[95] C. Bonde, A. Sarup, A. Schousboe, G. Gegelashvili, J. Noraberg and J. Zimmer, Neurochem. Int., 43 (2003) 381.
[96] B.H. Harvey, C. Naciti, L. Brand and D.J. Stein, Brain Res., 983 (2003) 97.
[97] S.M. Vasconselos, D.S. Macedo, L.O. Lima, F.C. Sousa, M.M. Fonteles and G.S. Viana, Braz. J. Med. Biol. Res., 36 (2003) 503.
[98] S. Kumar, J.E. Kralic, T.K. O'Buckley, A.C. Grobin and A.L. Morrow, J. Neurochem., 86 (2003) 700.
[99] R.J. Tallarida and R.B. Raffa, Life. Sci., 51 (1992) PL-61.
[100] E.R.Lazarowski, R.C. Boucher and T.K. Harden, J. Biol. Chem., 275 (2000) 31061.
[101] U.S. Bhalla and R. Iyengar, Science 283 (1999) 381.
[102] P.S. Katz and S. Clemens, Trends Neurosci., 24 (2001) 18.
[103] H. Jeong, B. Tombor, R. Albert, Z.N. Oltvai and A-L. Barabási, Nature, 407 (2000) 651.
[104] J. Kardos and L. Nyikos, Trends Pharmacol. Sci., 22 (2001) 642.
[105] H.C. Cheng, J. Pharmacol. Toxicol. Meth., 46 (2002) 61.
[106] P. Cuatrecasas, Biochemistry, 12 (1973) 3558.
[107] W.E. Van Heyningen, Nature, 249 (1974) 415.
[108] R.J. Stockert, A.G. Morell and I.H. Scheinberg, Science, 186 (1974) 365.
[109] E.A. Barnard, The Receptors. Vol. 1. General Principles and Procedures (Ed. R.D. O'Brian) Plenum Press, New York, 1979. p. 247.
[110] C.W. Porter, T.H. Chiu, J. Wieckowski and E.A. Barnrad, Nature, 241 (1973) 3.
[111] C.W. Porter, E.A. Barnard and T.H. Chiu, J. Membr. Biol. 14 (1973) 383.
[112] C.W. Porter and Barnard, Exp. Neurol., 48 (1975) 542.
[113] A.S.W. Burgen, C.R. Hiley and J.M. Young, Br. J. Pharmacol., 51 (1974) 279.
[114] M.J. Kuhar and H.I. Yamamura, Nature, 253 (1975) 560.
[115] M.J. Kuhar and H.I. Yamamura, Brain Res., 110 (1976) 229.
[116] W.E. Stumpf, Introduction to Quantitative Cytochemistry, Vol. 2. (Eds. G.L. Wied and G.F. Bahr) Academic Press, New York, 1970. p. 507.
[117] C.B. Pert, M.J. Kuhar and S.H. Snyder, Life Sci., 16 (1975) 1849.
[118] R. Libelius, J. Neural Transm., 35 (1974) 137.
[119] C.B. Pert, M.J. Kuhar and S.H. Snyder, Proc. Natl. Acad. Sci. USA, 73 (1976) 3729.
[120] E.N. Cohen, K.L. Chow and L. Mathers, Anesthesiology, 37 (1972) 324.
[121] A.T. Bruinvels, J.M. Palacios and D. Hoyer, Naunyn-Schmiedeberg's Arch. Pharmacol., 347 (1993) 569.
[122] F. Biver, F. Lotstra, M. Monclus, S. Dethy, P. Damhaut, D. Wikler, A. Luxen and S. Goldman, Nucl. Med. Biol., 24 (1997) 357.
[123] H. Filthuth, J. Planar Chromatogr., 2 (1998) 198.
[124] H. Filthuth, Planar Chromatography in the Life Sciences (Ed. J.C. Touchstone) John Wiley, New York, 1990. p. 167.
[125] I. Klebovitch, Planar Chromatography (Ed. Sz. Nyiredy) Springer Scientific Publisher, Budapest, 2001. p. 293.

[126] I. Hazai and I. Klebovitch, Handbook of Thin-Layer Chromatography, Third Edition, Revised and Expanded (Eds.J. Sherma and B. Fried), Marcel Dekker, Inc., New York-Basel, 2003. p. 339.

[127] H. Lundquist, M. Lubberink and V. Tolmachev, Eur. J. Phys., 19 (1998) 537.

[128] D. Vera, W. Eckelman, Nuclear Med. Biol., 28 (2001) 475.

[129] D.W. McCarthy, R.E. Shefer, R.E. Klinkowstein, L.A. Bass, W.H. Margeneau, C.S. Cutler, C.J. Anderson and M.J. Welch, Nucl. Med. Biol., 24 (1997) 35.

[130] P.E. Kinahan, B.H. Hasegawa and T. Beyer, Seminars Nicl. Med., 33 (2003) 166.

[131] B.F. Hutton and M. Braun, Seminars Nucl. Med., 33 (2003) 180.

[132] C. Cohade and R.L. Wahl, Seminars Nucl. Med., 33 (2003) 228.

[133] S.B. Jensen, D.F. Smith, D. Bender, S. Jakobsen, D. Peters, E.O. Nielsen, G.M. Olsen, J. Scheel-Kruger, A. Wilson and P. Cumming, Synapse, 49 (2003) 170.

[134] E.S. Garnett, G. Firnau and C. Nahmias, Nature, 305 (1983) 137.

[135] H.N. Wagner, Jr., H.D. Burns, R.F. Dannals, D.F. Wong, B. Langstrom, T. Duelfer, J.J. Frost, H.T. Ravert, J.M. Links, S.B. Rosenbloom, S.E. Lukas, A.V. Kramer and M.J. Kuhar, Science, 221 (1983) 1264.

[136] W.C. Eckelman, R.C. Reba, W.J. Rzeszotarski, R.E. Gibson, T. Hill, B.L. Holman, T. Budinger, J.J. Conklin, R. Eng and M.P. Grissom, Science, 223 (1984) 291.

[137] C.D. Arnett, C.-Y. Shiue, A.P. Wolf, J.S. Fowler, J. Logan and M. Watanabe, J. Neurochem., 44 (1985) 835.

[138] S. Zijlstra, H. van der Worp, T. Wiegman, G.M. Visser, J. Korf and W. Vaalburg, Nucl. Med. Biol., 20 (1993) 7.

[139] L. Farde, E. Ehrin, L. Eriksson, T. Greitz, H. Hall, C.-G. Hedström, J.-E. Litton and G. Sedvall, Proc. Natl. Acad. Sci. USA, 82 (1985) 3863.

[140] R. Schloesser, P. Simkowitz, E.J. Bartlett, A. Wolkin, G.S. Smith, S.L.Dewey and J.D. Brodie, Clin. Neuropharmacol., 19 (1996) 371.

[141] M. Laruelle. A. Abi-Dargham, S. Simpson, L. Kegeles, R. Parsey, D.R. Hwang, Y. Zea-Ponce, I. Lombardo, R. Weiss, R. Van Heertum and J.J. Mann, Soc. Neurosci. Abstr., 24 (1998) 22.

[142] L. Farde, C. Halldin, S. Stone-Elander and G. Sedvall, Psychopharmacology, 92 (1987) 278.

[143] L. Farde, C. Halldin, L. Muller, T. Suhara, P. Karlsson and H. Hall, Synapse, 16 (1994) 93.

[144] S. Pappata, Y. Samson, C. Chavoix, C. Prenant, M. Maziere and J.C. Baron, J. Cereb. Blood Flow Metab., 8 (1988) 304.

[145] R.A. Koeppe, V.A. Holthoff, K.A. Frey, M.R. Kilburn and D.E. Kuhl, J. Cereb. Blood Flow Metab., 11 (1991) 735.

[146] B. Sadzot, H.S. Mayberg and J.J. Frost, Neurophysiol. Clin.,20 (1990) 323.

[147] B. Sadzot, H.S. Mayberg and J.J. Frost, Acta Psychiatr. Belg., 90 (1990) 9.

[148] M. Laruelle, J. Cereb. Blood Flow Metab., 20 (2000) 423.

[149] W.C. Eckelman, Nucl. Med. Biol., 25 (1998) 169.

[150] M. Carlsson and A, Carlsson, Schizophr. Bull., 16 (1990) 425.

[151] K.J. Friston, C.D. Fith, P.F. Liddle and R.S. Frackowiak, J. Cereb. Blood Flow Metab., 11 (1991) 690.

[152] J.R. Moeller, S.C. Strother, J.J. Sidtis and D.A. Rottenberg, J. Cereb. Blood Flow Metab., 7 (1987) 649.

[153] K.J. Friston, C.D. Fith, P.F. Liddle and R.S. Frackowiak, J. Cereb. Blood Flow Metab., 13 (1993) 5.

[154] A.V. Levy, F. Gomez-Mont, N.D. Volkow, J.F. Corona, J.D. Brodie and R. Cancro, J. Nucl. Med., 33 (1992) 287.
[155] S.H. Snyder, Science, 224 (1984) 22.
[156] G. Gaietta, T.J. Deernick, S.R. Adams, J. Bouwer, O. Tour, D.W. Laird, G.E. Sosinsky, R.Y. Tsien and M.H. Ellisman, Science, 296 (2002) 503.
[157] J. Kardos, I. Kovács, E. Simon-Trompler and F. Hajós, Biochem. Pharmacol., 41 (1991) 1141.
[158] E.J. Mody, B.D. Harris and P. Skolnick, Brain Res., 615 (1993) 101.
[159] J.J. Quinlan, S. Firestone and L.L. Firestone, Anesthesiology, 83 (1995) 611.
[160] B. Stensbøl, U. Madsen and P. Krogsgaard-Larsen, Curr. Pharmaceutical Design, 8 (2002) 857.
[161] D. Colquhoun, P. Jonas and B. Sakmann, J. Physiol., 458 (1992) 261.
[162] É. Szárics, G. Nyitrai, I. Kovács, J. Kardos, Neurochem. Int., 36 (2000) 83.
[163] B. Hahn, C.G.V. Sharples, S. Wonnacott, M. Shoaib and I.P. Stolerman, Neuropharmacol., 44 (2003) 1054.
[164] J. Wyman and D.W. Allen, J. Polymer Sci., VII (1951), 499.
[165] C. Toyoshima and N. Unwin, Nature, 336 (1988) 247.
[166] N. Unwin, C. Toyoshima and E. Kubalek, J. Cell. Biol., 107 (1988) 1123.
[167] C. Toyoshima and N. Unwin, J. Cell. Biol., 111 (1990) 2623.
[168] N. Unwin, J. Mol. Biol., 229 (1993) 1101.
[169] N. Unwin, Neuron, 3 (1989) 665.
[170] D. Colquhoun, Br. J. Pharmacol., 125 (1998) 924.
[171] J. del Castillo and B. Katz, Proc. Roy. Soc. Lond. B., 146 (1957) 369.
[172] M. Kasai and J.-P. Changeux, J. Membrane Biol., 6 (1971) 1.
[173] M. Kasai and J.-P. Changeux, J. Membrane Biol., 6 (1971) 24.
[174] M. Kasai and J.-P. Changeux, J. Membrane Biol., 6 (1971) 58.
[175] J. Cartaud, E.L. Benedetti, M. Kasai and J.-P. Changeux, J. Membrane Biol., 6 (1971) 81.
[176] B. Katz and R. Miledi, Nature, 226 (1970) 962.
[177] B. Katz and R. Miledi, J. Physiol., 224 (1972) 664.
[178] E. Neher and C.F. Stevens, Annu. Rev. Biophys. Bioeng. 6 (1977) 345.
[179] B. Katz and S. Thesleff, J. Physiol., 138 (1957) 63.
[180] G. Careri, P. Fasella and E. Gratton, CRC Crit. Rev. Biochem., (1975) 141.
[181] T. Heidman and J.-P. Changeux, Ann. Rev. Biochem., 47 (1978) 317.
[182] G.P. Hess, D.J. Cash and H. Aoshima, Nature, 282 (1979) 329.
[183] D.J. Cash and G.P. Hess, Proc. Natl. Acad. Sci. USA, 77 (1980) 842.
[184] H. Aoshima, D.J. Cash and G.P. Hess, Biochem. Biophys. Res. Commun., 92 (1980) 896.
[185] G.P. Hess, H. Aoshima, D.J. Cash and B. Lenchitz, Proc. Natl. Acad. Sci. USA, 78 (1981) 1361.
[186] H. Aoshima, D.J. Cash and G.P. Hess, Biochemistry, 20 (1981) 3467.
[187] G.P. Hess, E.B. Pasquale, J.W. Karpen, A.B. Sachs, K. Takeyashu and D.J. Cash, Biochem. Biophys. Res. Commun., 107 (1982) 1583.
[188] G.P. Hess, D.J. Cash and H. Aoshima, Ann. Rev. Biophys. Bioeng., 12 (1983) 443.
[189] D.J. Cash, H. Aoshima, E.B. Pasquale and G.P. Hess, Rev. Physiol. Biochem. Pharmacol., 102 (1985) 73.
[190] C. Franke, H. Parnas, G. Hovav and J. Dudel, Biophys. J., 64 (1993) 339.
[191] G.P. Hess, Arch. Physiol. Biochem., 104 (1996) 752.

[192] M.J. Marks, S.R. Grady, J.-M., Yang, P.M. Lippiello and A.C. Collins, J. Neurochem., 63 (1994) 2125.
[193] A.L. Lehninger, The Mitochondrion, Benjamin, New York, 1964.
[194] A.D. Bangham, M.M. Standish and J.C. Watkins, J. Mol. Biol., 13 (1965) 238.
[195] H.R. Kaback, Annu. Rev. Biochem., 39 (1970) 561.
[196] E. Racker (ed), Membranes of Mitochondria and Chloroplasts, Van Nostrand-Reinhold, New York, 1970.
[197] C.Y. Lee, Annu. Rev. Pharmacol., 12 (1972) 265.
[198] G.P. Hess, J.P. Andrews, G.E. Struve and S.E. Coombs, Proc. Natl. Acad. Sci. USA, 72 (1975) 4371.
[199] G.P. Hess and J.P. Andrews, Proc. Natl. Acad. Sci. USA, 74 (1977) 482.
[200] J. Bernhardt and E. Neumann, Proc. Natl. Acad. Sci. USA, 75 (1978) 3756.
[201] J. Bernhardt and E. Neumann, Neurochem. Int., 2 (1980) 243.
[202] A.B. Sachs, B. Lenchitz, R.L. Noble and G.P. Hess, Anal. Biochem., 124 (1982) 185.
[203] F.J.W. Roughton and B. Chance in S.L. Friess, E.S. Lewis and A. Weissberger (eds), Technique of Organic Chemistry, 2nd edn. Wiley, New York, 1963.
[204] B. Chance, R.H. Eisenhardt, Q.H. Gibson, K.K. Lonberg-Holm (eds), Rapid Mixing and Sampling Techniques in Biochemistry, Academic, New York, 1964.
[205] G.G. Hammes, Principles of Chemical Kinetics, Academic, New York, 1978.
[206] R.W. Lymn and E.W. Taylor, Biochemistry, 9 (1970) 2975.
[207] A. Martonosi, E. Lawinska and M. Oliver, Ann. N.Y. Acad. Sci., 227 (1974) 549.
[208] A.R. Fersht and R. Jakes, Biochemistry, 14 (1975) 3350.
[209] J.P. Froehlich and E.W. Taylor, J. Biol. Chem., 251 (1976) 2307.
[210] W.S. Thayer and P.C. Hinkle, J. Biol. Chem., 250 (1975) 8772.
[211] M. Kurzmack, S. Verjovski-Almeida and G. Inesi, Biochem. Biophys. Res. Commun., 78 (1977) 772.
[212] M. Sumida, T. Wang, F. Mandel, J.P. Froehlich and A. Schwartz, J. Biol. Chem., 253 (1978) 8772.
[213] S. Verjovski-Almeida and G. Inesi, J. Biol. Chem., 254 (1979) 18.
[214] H.L. Toor, in R.S. Brodkey (ed), Turbulence in Mixing Operations, Academic, New York, 1975.
[215] D.J. Cash, H. Aoshima and G.P. Hess, Biochem. Biophys. Res. Commun., 95 (1981) 1010.
[216] P.R. Adams, J. Membr. Biol., 58 (1981) 161.
[217] J.-P. Changeux, Harvey Lect., 75 (1981) 85.
[218] P.S. Kim and G.P. Hess, J. Membr. Biol., 58 (1981) 203.
[219] G.O. Ramsayer, G.H. Morrison, H. Aoshima and G.P. Hess, Anal. Biochem., 115 (1981) 34.
[220] E. Neher and B. Sakmann, Nature, 260 (1976) 779.
[221] G.P. Hess, H.-A. Kolb, P. Läuger, E. Schoffeniels and W. Schwarze, Proc. Natl. Acad. Sci. USA, 81 (1984) 5281.
[222] R.D. Schwartz, P. Skolnick, E.B. Hollingsworth and S.M. Paul, FEBS Lett., 175 (1984)193.
[223] G.M. Sánchez, M.C. Toledo and M.P. Gonzáles, Rev. Esp. Fisiol., 40 (1984) 375.
[224] K. Subbarao and D.J. Cash, Soc. Neurosci. Abstr., 11 (1985) 275.
[225] A.M. Allan, R.A. Harris, K. Subbarao and D.J. Cash, Fed. Proc., Fed. Am. Soc. Exp. Biol., 44 (1985) 1634.
[226] R.A. Harris and A.M. Allan, Science, 228 (1985) 1108.
[227] D.J. Cash and K. Subbarao, FEBS Lett., 217 (1987) 129.

[228] D.J. Cash and K. Subbarao, Life Sci., 41 (1987) 437.
[229] D.J. Cash and K. Subbarao, Biochemistry, 26 (1987) 7556.
[230] D.J. Cash and K. Subbarao, Biochemistry, 26 (1987) 7562.
[231] D.J. Cash, K. Subbarao, J.R. Bradbury and G.M. Meyes, J. Biochem. Biophys. Meth., 23 (1991) 151.
[232] D.J. Cash, R.M. Langer, K. Subbarao and J.R. Bradbury, Biophys. J., 54 (1988) 909.
[233] J. Kardos, Synapse, 13 (1993), 74.
[234] D.J. Cash and K. Subbarao, Biochemistry, 27 (1988) 4580.
[235] J. Kardos and K. Maderspach, Life Sci., 41 (1987) 265.
[236] N. Geetha and G.P. Hess, Biochemistry, 31 (1992) 5488.
[237] J. Kardos and D.J. Cash, J. Neurochem., 55 (1990) 1095.
[238] P. Serfőző and D.J. Cash, FEBS Lett., 310 (1992) 55.
[239] D.J. Cash and P. Serfőző, Eur. J. Biochem., 228 (1995) 311.
[240] D.J. Cash, P. Serfőző and K. Zim, J. Membr. Biol., 145 (1995) 257.
[241] D.J. Cash, P. Serfőző and A.M. Allan, J. Pharmacol. Exp. Ther., 283 (1997) 704.
[242] P. Lundgren, J. Strömberg, T. Bäckström and M. Wang, Brain Res., 982 (2003) 45.
[243] J. Kardos, Neurochem. Int., 34 (1999) 353.
[244] R.D. Schwartz, P.D. Suzdak and S.M. Paul, Mol. Pharmacol., 30 (1986) 419.
[245] D.A. Mathers, Synapse, 1 (1987) 96.
[246] M.D. Luu, A.K. Morrow, S.M. Paul and R.D. Schwartz, Life Sci., 41 (1987) 1277.
[247] A. Doble and I.L. Martin, Trends Pharmacol. Sci., 13 (1992) 76.
[248] S. Vicini, Neuropsychopharmacol., 4 (1991) 9.
[249] D.J. Cash and K. Subbarao, J. Membr. Biol., 111 (1989) 229.
[250] J. Kardos, I. Kovács, E. Simon-Trompler and F. Hajós, Biochem. Pharmacol., 41 (1991) 1141.
[251] W.E. Haefely, J.R. Martin and P. Schoch, Trends. Pharmacol. Sci., 11 (1990) 452.
[252] M.K. Ticku, Ann. Med., 22 (1990) 241.
[253] R.W. Olsen, M.H. Bureau, S. Endo, G. Smith, L. Deng, D.W. Sapp and A.J. Tobin, Adv. Exp. Med. Biol., 287 (1991) 355.
[254] B.E. Leonard, J. Psychiatr. Res., 27 Suppl. (1993) 193.
[255] A.J. Tobin, M. Khrestchatisky, A.J. Maclennan, M.Y. Chiang, N.J. Tillakaratne, W.T. Xu, M.B. Jackson, N. Brecha, C. Sternini and R.W. Olsen, Adv. Exp. Med. Chem., 287 (1991) 365.
[256] E.A. Barnard, A.N. Bateson, M.G. Darlison, T.A. Glencorse, R.J. Harvey, A.A. Hicks, A. Lasham, R. Shingai, P.N. Usherwood, E. Vreugdenhil and S.H. Zaman, Adv. Biochem. Psychopharmacol., 47 (1992) 17.
[257] T.M. DeLorey and R.W. Olsen, J. Biol. Chem., 267 (1992) 16747.
[258] R.L. McDonald and R.W. Olsen, Annu. Rev. Neurosci., 17 (1994) 569.
[259] D.W. Choi, D.H. Farb and G.D. Fischbach, Nature, 269 (1977) 342.
[260] R.L. McDonald and J.L. Barker, Nature, 271 (1978) 563.
[261] D. Mierlak and D.H. Farb, J. Neurosci., 8 (1988) 814.
[262] M. Farrant, T.T. Gibbs and D.H. Farb, Neurochem. Res., 15 (1990) 175.
[263] J.R. Moffet, A. Namboodiri and J.H. Neale, FEBS Lett., 247 (1989) 81.
[264] P. Montpied, E.I. Ginns, B.M. Martin, D. Stetler, A.-M. O'Carrol, S.J. Lolait, L.C. Mahan and S.M. Paul, FEBS Lett., 258 (1989) 94.
[265] J. Kardos, Neuroreport,3 (1992) 1124.
[266] R.P. Bell, The Proton in Chemistry, Cornell University Press, Ithaca, 1960.
[267] M. Simonyi and I. Mayer, JCS Chem. Commun., (1975) 695.
[268] G. Némethy and H.A. Sheraga, J. Phys. Chem., 41 (1964) 680.

[269] E.J. Moody, A.H. Lewin, B.R. deCosta, K.C. Rice and Ph. Skolnick, Eur. J. Pharmacol., 206 (1991) 113.

[270] J. Bormann, O.P. Hamill and B. Sakmann, J. Physiol., 385 (1987) 243.

[271] H.R. Parri,L. Holden-Dye and R.J. Walker, Exp. Brain Res., 76 (1991) 597.

[272] J. Nagy, J. Kardos, G. Maksay, M. Simonyi M., Neuropharmacol., 20 (1981) 529.

[273] J. Kardos, J. Samu, K. Ujszászi, J. Nagy, I. Kovács, J. Visy, G. Maksay and M. Simonyi, Neurosci. Lett., 52, (1984) 67.

[274] J. Kardos, I. Kovács, F. Hajós, M. Kálmán and M. Simonyi, Neurosci. Lett., 103 (1989) 139.

[275] J.J. Celentano,M. Gyenes, T.T. Gibbs and D.H. Farb, Mol. Pharamcol. 40 (1991), 766.

[276] P. Legendre and L. Westbrook, Mol. Pharmacol., (1991)

[277] T.G. Smart, S.J. Moss and R.L. Huganir, Br. J. Pharmacol., 103 (1991) 1837.

[278] A. Draguhn, T.A. Vendorn, M. Ewert, P.H. Seeburg and B. Sakmann, Neuron, 5 (1990) 781.

[279] J. Kardos, F. Hajós and M. Simonyi, Neurosci. Lett., 48 (1984) 355.

[280] Z. Nusser, W. Sieghart and P. Somogyi, J. Neurosci., 18 (1998) 1693.

[281] P. Serfőző and D.J. Cash, Eur. J. Biochem., 228 (1995) 498.

[282] R. Wieboldt, D. Ramesh, B.K. Carpenter and G.P. Hess, Biochemistry, 33 (1994) 1526.

[283] N. Unwin, Nature, 373 (1995) 37.

[284] D.C. Ogden and D. Colquhoun, Proc. Roy. Soc. London B, 225 (1985) 329.

[285] J.-L. Galzi, F. Revah, A. Bessis and J.-P. Changeux, Annu. Rev. Pharmacol., 31 (1991) 37.

[286] G.A.R. Johnston, GABA in Nervous System Function, E. Roberts, T.N. Chase and D.B. Tower (eds.), Raven, New York, 1976.

[287] J.Kardos, G. Blaskó, P. Kerekes, I. Kovács and M. Simonyi, Biochem. Pharmacol., 33 (1984) 3537.

[288] J. Kardos, T. Blandl, N.D. Luyen, G. Dörnyei, E. Gács-Baitz, M. Simonyi, D.J. Cash, G. Blaskó and Cs. Szántay, Eur. J. Med. Chem., 31, (1996) 761.

[289] J. Kardos, I. Kovács, T. Blandl, D.J. Cash, E. Simon-Trompler, N.D. Luyen, G. Dörnyei, M. Simonyi, G. Blaskó and Cs. Szántay, Eur. J. Pharmacol., 337 (1997) 83.

[290] M. Simonyi, G. Blaskó, J. Kardos and M. Kajtár, Chirality, 1 (1989) 178.

[291] J. Kardos, I. Kovács, T. Blandl, D.J. Cash, Neurosci. Lett., 182 (1994) 73.

[292] J. Kardos, Biochem. Pharmacol., 38 (1989) 2587.

[293] W.L. Stahl, Neurochem. Int., 8 (1986) 449.

[294] N. Nelson, J. Neurochem., 71 (1998) 1785

[295] D. Wipf, U. Ludewig, M. Tegeder, D. Rentsch, W. Koch and W.B. Fommer, Trends Biol. Sci., 27 (2002) 139.

[296] J. Guastella, N. Nelson, H. Nelson, L. Czyzyk, S. Keynan, M.C. Miedel, N. Davidson, H.A. Lester and B.I. Kanner, Science, 249 (1990), 1303.

[297] H. Nelson, S. Mandiyan and N. Nelson, FEBS Lett., 269 (1990) 181.

[298] L.A. Borden, Neurochem. Int., 29 (1996) 335.

[299] A. Schousboe and B. Kanner in Glutamate and GABA Receptors and Transporters: Structure, Function and Pharmacology, J. Egebjerg, A. Schousboe (Eds), Francis and Taylor, London, 2002.

[300] H.S. White, A. Sarup, T. Bolvig, A.S. Kristiensen, G. Petersen, N. Nelson, D.S. Pickering, O.M. Larsson, B. Frølund, P. Krogsgaard-Larsen and A. Schousboe, J. Pharmacol. Exp. Ther., 302 (2002) 636.

[301] A. Sarup, O. Miller-Larsson, T. Bolvig, B. Frølund, P. Krogsgaard-Larsen and A. Schousboe, Neurochem. Int., 43 (2003) 445.

[302] G. Rudnick, Neurotransmitter Transporters: Structure, Function and Regulation, Humana, Clifton, 1997.

[303] A. Galli, R.D. Blakely and L.J. DeFelice, Proc. Natl. Acad. Sci. USA, 95 (1998) 13260.

[304] M. Quick and B.R. Stevens, J. Biol. Chem., 276 (2001) 33413.

[305] D.H. Feldman, W.R. Harvey and B.R. Stevens, J. Biol. Chem., 275 (2000) 24518.

[306] W. Almers, Annu. Rev. Physiol., 52 (1990) 607.

[307] I. Kovács, É. Szárics, G. Nyitrai, T. Blandl and J. Kardos, Neurochem. Int., 33 (1998) 399.

[308] M.P. Blaustein and A.C. King, J. Membr. Biol., 30 (1976) 153.

[309] J.-P. Pin, T. Yasumoto and J. Bockaert, J. Neurochem., 50 (1988) 1227.

[310] T.J. Turner and S.M. Goldin, Biochemistry, 28 (1989) 586.

[311] D.G. Nicholls, J. Neurochem., 52 (1989) 331.

[312] S. Bernath, M.J. Zigmond, E.S. Nisenbaum, E.S. Vizi and Th.W. Berger, Brain Res., 632 (1993) 232.

[313] E.A. Milusheva, M. Barányi, T. Zelles, Á. Mike and E.S. Vizi, Eur. J. Neurosci., 6 (1994) 187.

[314] J. Kardos and T. Blandl, Neuroreport, 5 (1994) 1249.

[315] J. Kardos, L. Elster, I. Damgaard, P. Krogsgaard-Larsen and A. Schousboe, Neurosci. Res., 39 (1994) 646.

[316] E.A. Milusheva, M. Dóda, M. Barányi and E.S. Vizi, Neurochem. Int., 28 (1996) 501.

[317] Kovács I., Skuban N., Nyitrai G., Kardos J. Inhibition of [^{3}H]-D-aspartate release by deramciclane. European J. Pharmacology 381, 121-127 (1999)

[318] I. Kovács, N. Skuban, É. Szárics and J. Kardos J, Brain Res. Bull., 52 (2000) 39.

[319] P. Barabás, I. Kovács, R. Kovács, J. Pálhalmi, J. Kardos and A. Schousboe, J. Neurosci. Res., 67 (2002) 149.

[320] G. Nyitrai, K.A. Kékesi, Zs. Emri, É. Szárics, G. Juhász G and J. Kardos J. Eur. J. Pharmacol., 478, (2003) 111.

Radiotracer Studies of Interfaces
G. Horányi (editor)

Chapter 9

Instrumentation

A. Kolics

Blue29, Inc., Sunnyvale, CA 94089, USA

The use of radioactive tracers in interfacial processes dates back to the beginning of the twentieth century [1]. Since then enormous advance has been made on both methodological as well as instrumentation levels and so tracer techniques have grown into an indispensable tool for many research and industrial applications [2-3]. Compared to other analytical techniques, radioactive tracer methods accommodate experimental simplicity and provide straightforward data interpretation.

When the quantification of a tracer nuclide is performed ex situ, the measurement generally boils down to simple count rate measurement. There are well established experimental strategies and commercially available detection systems for the quantitative determination of most radioactive nuclides. In these situations, the task is to find the optimal detection condition, i.e. high signal to noise ratio with high sensitivity. The quantification becomes significantly more complicated when more than one radiotracer is used or there are other radioactive nuclides besides the intentionally labeled species. The proper choice of detectors as well as the signal processing units becomes a crucial aspect for successful measurement. The ultimate challenge is the quantification of labeled species *in situ.*

The *in situ* radiotracer techniques have experienced a rapid development since the early sixties and numerous reviews have been dedicated time to time to review the recent progress in the field [4-11]. While in most cases the focus is on the measurement of the concentration of labeled species at certain interfaces, the real question sometimes stays in the shadow. Specifically, what are we really measuring [12-15]? In order to answer this question the interaction of radiation with matter and the measuring system have to be analyzed in detail.

Consequently, this chapter will provide a brief overview on the detection systems, i.e. detectors and coupled electronics. Throughout this discussion, some basic principles of radiation interaction will be presented to highlight the governing factors and pitfalls in detector selection. Finally, measuring condition and data evaluation will be reviewed and further scrutinized in specific cases of *in situ* radiotracer methods.

1. DETECTORS AND SIGNAL PROCESSING ELECTRONICS

Detectors of three major types are used most frequently for measuring α, β- and $\gamma(X)$-radiation. These are gaseous counters, scintillation, and semiconductor detectors. While each detector category can be used for the measurement of α, β- or $\gamma(X)$-radiation their sensitivity, resolution, and especially their working principle differ significantly from each other.

1.1. Gas filled tubes

The most widely used gas-filled detectors are the proportional and Geiger counters. They are usually filled with a gas such as Ar or mixture of gases such as Ar and methane. This choice for filling gas is governed by the ratio of energy consumed for ionization per total energy loss. This ratio is the highest for noble gases. The counters are equipped with a thin window that can transmit low energy X- and γ-rays as well as β-radiation (Figure 1). Beryllium, mica, and mylar are the most frequently used materials for detector window. In the center of the tubes, there is a positively biased anode wire. A high voltage is maintained between the metal container (cathode) and the anode. Photons or electrons penetrate the window and pass into the gas inside where interactions with the gas atoms result in the creation of a number of ion pairs (electrons and partially ionized gas atoms).

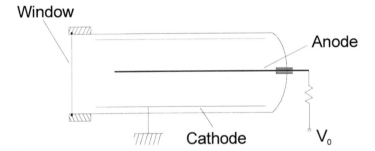

Fig. 1. Simplified view of an end-window gaseous counter.

At low anode voltage, only ion pairs directly produced by the ionizing radiation provide the signal at the anode. The gaseous counter operating in this voltage regime is called ionization chamber. While this setup provides energy selective measurement, due to the lack of multiplication, the output current or pulse height is very small. Therefore, this mode of operation is used for measuring high flux of strongly ionizing radiation.

At a certain applied voltage the electric field can accelerate the primary electrons to ionize the gas atoms or molecules in the chamber. This multiplication of electrons is called gas amplification and is referred to as Townsend avalanche. If the electrons and ions produced by ionizing radiation are multiplied in a manner that the total number of ion pairs is proportional to the number of primary ion pairs, the detector is called proportional counter. Consequently, proportional counters are energy selective devices. The multiplication in a proportional counter is generally in the range of 10^3 to 10^4.

Proportional counters are frequently used for X-ray measurements where moderate energy resolution is required. Their typical resolutions are about 16 to 20% full-width at half maximum (FWHM) for the 5.9 keV X-ray of ^{55}Fe [16]. For X-ray measurement, noble gases, such as xenon, krypton, neon and argon are commonly used. Xenon and krypton are selected for higher energy X-rays or to get higher efficiencies, while neon is selected when it is desired to detect low energy X-radiation in the presence of unwanted higher energy X-ray or γ-components. Sometimes gas mixtures, such as P-10 gas, which is a mixture of 90% argon and 10% methane, or noble gases, i.e. Xe or Ne with CO_2 are also applied. The organic compounds, called 'quenching' agents, stabilize the operation of the counter by minimizing the effect of secondary processes. Unlike noble gases, polyatomic organic molecules form radicals instead of producing further electron when hitting the cathode or absorb a photon. Due to their lower ionization energy compared to the main filing gas, they prevent ionized noble gases to reach the cathode. Gas pressures are generally around one atmosphere, while the typical dead time is around 1 μs. Before pulse height analysis, the signal from the proportional counter goes into a preamplifier

Upon further increasing the voltage, the ionizing radiation will not only generate electrons and ions but also excite atoms of the filling gas. These atoms later return to the ground state by emitting photons, which can further ionize gas atoms. Consequently, the electron multiplication becomes even greater, and the number of electrons collected will be independent of the initial ionization.

The Geiger counter produces a large voltage pulse that is easily counted without further amplification. Nevertheless, no energy spectrum analysis is possible since the output pulse height is independent of initial ionization. Geiger-Müller tubes are generally equipped with a thin mica window. The operating voltage, which is dependent upon the type of quenching agent are generally in the range of 500 to 1500V, is in a plateau region. This plateau is relatively flat over a range of bias voltage (typically 100-150 V wide).

The discharge produced by ionization must be quenched in order to return the detector into a neutral ionization state for the next pulse. This is achieved by using a fill gas that contains a small amount of halogen or organic molecules in addition to a noble gas. Furthermore, the voltage drop across a large resistor between the anode and bias supply will also serve to quench the discharge since the operating voltage will be reduced below the plateau. The Geiger counter is inactive for 50 to 200 μs after each pulse until the quenching is complete. Due to this long dead time, the counter is preferentially suited for low count rate applications.

1.2. Scintillation detectors

When ionizing radiation passes through matter it may excite the atoms and molecules of that medium. Those materials where the life-time of excited state is shorter than 10^{-3}s and the relaxation to ground state occurs via photon having an energy lower than the resonance absorption energy are called scintillators (or phosphors). Ionizing radiation such as α- and β-radiation excites the scintillator components directly, while γ-radiation generates electrons through photoelectric effect, Compton scattering or pair-production. These electrons will cause the scintillation.

The most important parameters of scintillators are density, average atomic number, size, purity, refraction index, fluorescence emission maximum, optical transmission, light yield, decay constant, linearity between particle energy and scintillation intensity, and resolution. Some of the main properties of most commonly used scintillators are compiled in Table 1 [17-18]. In general, low average atomic number scintillators are used for β-radiation detection, since both the backscattering of β-radiation as well as the generation of X-radiation (characteristic as well as bremsstrahlung) are strongly atomic number dependent processes. For instance the radiation energy loss of an electron can be described by the following equation [20]:

$$-\left(\frac{dE}{dx}\right) = \frac{e^4 Z^2 E \rho_A}{137(m_e c^2)^2} \tag{1}$$

where E is the energy of the electron, x is a distance traveled by the electron, ρ_A is the density of the medium, m_e the mass of electron, e is the charge of the electron, Z the atomic number of the medium, and c is the speed of light. The atomic number dependency of backscattered radiation has been described by several empirical formulas [21-24]. While the formulas differ significantly from each other, general consensus is that the backscattered intensity increases with the atomic number of the scattering material.

Obviously, high average atomic number medium causes significant scattering of electrons. Consequently, a large portion of the electrons impinging the scintillator will undergo backscattering, cf. they lose only a small portion of their kinetic energy before leaving the crystal. Therefore, the resulting energy spectrum will be distorted and exhibits an increased intensity in the low energy region. In the light of the above, it is not surprising that scintillators made of organic compounds or plastic are the most frequently used phosphors. In addition, plastic scintillators are easy to be manufactured into various sizes, they posses low light absorption to its emitted light, and short decay time. Due to this latter property, plastic scintillators can be advantageously used in applications where high radiation intensities are dealt with. Glass scintillators, while not ideal for β-radiation detection, are used due to their inertness to chemical environment as well as the ease of producing highly smooth and flat surfaces, which is especially critical at the thin gap technique [6-8]. Recently, $CaF_2(Eu)$ scintillator was introduced into radioactive labeling studies for β-radiation emitting nuclides [25]. This scintillator unites the advantageous properties of glass scintillators and plastic phosphors, i.e. high chemical stability with excellent light output (Table 1). Due to longer decay time of $CaF_2(Eu)$ crystals, these phosphors are used when lower radiation intensities are to be measured.

We should note that only very few nuclides emit pure β-or γ-radiation. The decay mode of most radioactive isotope includes β- or α-radiation with simultaneous emission of γ–photons to reach stable, ground state. This requires special detection setup. Specifically, the system has to be less sensitive to high energy radiation since most of the cases the radiation with high attenuation coefficient is measured. Discrimination can be made by energy selective measurement and/or with the application of thin scintillator to minimize the absorption of high energy components. For the measurement of β-radiation in presence of γ–radiation, the application of thin ($\leq$ 0.5 mm) plastic scintillators was proved to be useful [11, 15, 26-27]. When the tracer emits low energy X-radiation, the effect of high energy γ–photons can be minimized by employing thin NaI(Tl) phosphors with energy selective detection. In both cases the thickness of the scintillator is critical, since the high energy γ–radiation creates electrons of various energies. These electrons have a continuous energy spectrum from 0 keV to $E_{C,max}$ where $E_{C,max}$ is the maximum energy of electrons acquired in a Compton scattering process and can be expressed by the following formula:

$$E_{C,max} = \frac{E_\gamma^2}{E_\gamma + 0.255} \text{MeV} \tag{2}$$

These electrons create an increased background in the entire energy spectrum of β-particles since $E_{\beta,max} < E_{C,max}$ is valid for most nuclides. By reducing the thickness of the phosphor, smaller fraction of γ–radiation and induced secondary radiation is absorbed in the crystal. Such measurement strategies have already been introduced in the adsorption measurement X-and γ–radiation emitting nuclides, i.e. ^{57}Co [28], as well as ^{51}Cr [29-31].

The most widely used scintillator in X-ray or γ–radiation measurement is thallium doped NaI crystal. The high atomic number of iodine in NaI(Tl) gives good efficiency for γ–radiation detection. The best resolution is generally in the range of 7.5%-8.5% for the 662 keV γ–radiation of ^{137}Cs. While in low energy application, the energy resolution of this scintillator falls behind that of the proportional counters, the complete resolution of the 6.4 and 14.4 keV photons of ^{57}Co is still viable [28].

In an actual measuring system, the scintillator is coupled directly or through optical coupling to a photomultiplier tube (PMT). The emission maximum and refractivity index of the crystal are important factors in deciding the type of photomultiplier tube. Specifically, the emission maximum of the scintillator should be close to the absorption maximum of the photocathode in the photomultiplier tube. However, this strict selection guide can be significantly eased using certain wavelength shifters and optical coupling.

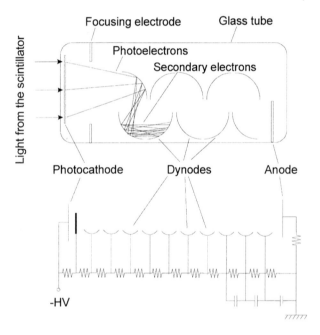

Fig. 2. Scheme of an end-window photomultiplier tube.

A PMT consists of a photocathode, a focusing electrode, a series of dynodes, and an anode in an evacuated glass enclosure (Figure 2). When a photon of sufficient energy strikes the photocathode, it ejects a photoelectron as a result of photoelectric effect. The photocathode material is usually a mixture of alkali metals, which make the PMT sensitive to photons throughout the visible and occasionally in the ultraviolet region. The photocathode is at a high negative voltage, typically -500 to -1500 volts. The photoelectrons are accelerated towards a series of additional electrodes since these dynodes are maintained at successively less negative voltage. The number of electrons is multiplied at each dynode creating a cascading effect. Due to this process, 10^5 to 10^8 electrons are created for each photoelectron that is ejected from the photocathode. The amplification depends on the number of dynodes and the accelerating voltage.

Table 1.
Some characteristic properties of scintillators [17-19]. *(NaI(Tl) = 100%. BGO = $Bi_4Ge_3O_{12}$; GOS = Gd_2O_2S; GSO = Gd_2SiO_5; LSO = Lu_2SiO_5, LuAP = $LuAlO_3$; YAP = $YAlO_3$)

Material	Density	Emission maximum	Integrated light output *	Decay time	Refraction index	Sensitivity to water
	g/cm^3	nm	%	µs		
Plastic	0.9-1.0	385-580	10-25	0.001-0.02	1.4-1.5	No
Anthracene	1.25	447	43	0.03	1.62	No
Liquid	0.9-1.0	385-430	15-35	0.002-0.004	1.4-1.5	No
$CaF_2(Eu)$	3.18	435	50	0.94	1.44	No
BaF_2	4.88	325/220	20/2	0.63/0.006	1.49	No
NaI(Tl)	3.67	415	100	0.23	1.85	Hygroscopic
CsI	4.51	315	6	0.005	1.80	Slightly hygr.
CsI(Na)	4.51	420	85	0.63	1.84	Slightly hygr.
CsI(Tl)	4.51	550	110	1.0	1.79	Slightly hygr.
$CsI(CO_3)$	4.51	405	60	2	1.84	Hygroscopic
BGO	7.13	480	11-20	0.6	2.15	No
$CdWO_4$	7.90	480	38	1/10	2.25	No
LiF(W)	2.64	430	4	40	1.4	No
LiI(Eu)	4.08	470	35	1.4	1.96	Hygroscopic
LuAP	8.34	360	30	0.018		No
LSO	7.40	420	63	0.04	1.82	No
GSO(Ce)	6.71	430	20-26	0.03-0.06	1.85	No
GOS	7.34	510	50	3		No
$PbSO_4$	6.20	350	10	0.1	1.88	No
CsF	4.64	390	4	0.005	1.48	Hygroscopic
CeF_3	6.16	300/340	11	0.005/0.02	1.68	No
YAP(Ce)	5.37	370	46	0.024		No
YAG(Ce)	4.57	550	15	70		No
Li-Glass (Ce)	2.42-2.66	395	13-30	0.05 – 0.07	1.06 – 1.57	No

As mentioned above, the PMT utilizes the photoelectric effect; therefore a PMT should have a small dark current that is produced by the thermal emission of electrons from its cathode. Consequently, the PMT has to be sealed off from ambient light. This is achieved by housing the scintillator and the photomultiplier tube in one light tight enclosure. In the thin gap technique the scintillator is part of the radiochemical cell and it is not covered with a light-proof layer. Consequently, the measuring cell is situated in light tight box.

The signal from the anode of the PMT goes into an amplifier and then into a signal discrimination unit. This signal can be further processed using a single or a multi-channel analyzer unit. When only one isotope is used in the measurement a single channel analyzer with proper differential discrimination is adequate.

1.3. Semiconductor detectors

The primary advantage of semiconductor detectors compared to other types of radiation counters lies in their superior energy resolution. This property can be especially important when several tracers emitting characteristic X-radiation or γ-photons. The limited application of semiconductor detectors in radiotracer studies of interfaces can be attributed to fact the vast majority of applications are in studies employing pure β-emitting nuclides with only one type of species being labeled. In most cases, spectroscopic analysis of β-radiation is not needed. Notwithstanding, there are many industrial problems such as corrosion and contamination issues in nuclear power plants that, in some cases, has to be addressed by interfacial studies in real-life solution environment. Such media generally contains several nuclides emitting γ-radiations with various energies. Another area, which may generate significant interest, is the application of semiconductor detectors with good spatial resolution capabilities. When the adsorbent surface is heterogeneous in composition, often it is more important to know the extent of accumulation locally than on the entire surface. This is usually the case at corrosion of alloys where certain metal components forms separate secondary phases in the bulk matrix. Nowadays, good charge coupled devices can reach a spatial resolution of couple of microns.

Semiconductor detectors have a P-I-N diode structure in which the intrinsic (I) region is created by depletion of charge carriers when a reverse bias is applied across the diode. When photons interact within the depletion region, charge carriers (holes and electrons) are freed and are swept to their respective collecting electrode by the electric field. The resultant charge is integrated by a charge sensitive preamplifier and converted to a voltage pulse with amplitude proportional to the original photon energy.

The most common X-ray and gamma-ray detectors use lithium-drifted silicon Si(Li) or lithium-drifted germanium Ge(Li) (Figure 3). In these detectors,

Li is incorporated into the semiconductor lattice by annealing the semiconductor with Li at a high temperature (~500°C). A voltage of approximately 1000 V is applied across the semiconductor material with two electrodes, and the electron cascade produced by a photoelectron is detected as an electrical pulse at the anode.

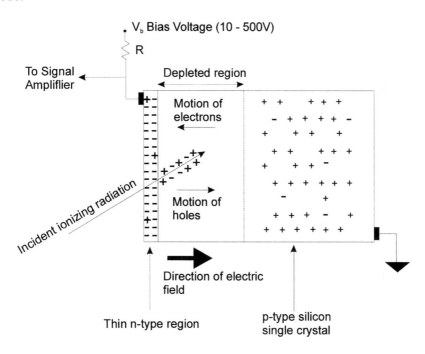

Fig. 3. Working principle of a semiconductor detector [32].

In addition to being more robust than gas-filled or scintillator detectors, these semiconductor detectors also provide a much higher resolution. Their only disadvantage is the need for cooling, usually with liquid nitrogen, to decrease the dark noise of the detector and current-to-voltage preamplifier.

The high energy resolution of semiconductor detectors is due to the small amount of energy required to produce a charge carrier and the consequent large "output signal" relative to other detector types for the same incident photon energy. At 3 eV/e-h pair the number of charge carriers produced in Ge is about one and two orders of magnitude higher than in gas and scintillation detectors. The resultant energy resolution in keV (FWHM) vs. energy for various detector types is illustrated in Table 2.

At low energies, detector efficiency is a function of cross-sectional area and window thickness while at high energies total active detector volume more or less determines counting efficiency. Detectors having thin contacts, e.g.

Si(Li), Low-Energy Ge and Reverse Electrode Ge detectors, are usually equipped with a Be cryostat window to take full advantage of their intrinsic energy response.

Table 2.
Energy Resolution (keV at FWHM) of different detectors [16]

Detector	Energy of radiation/keV		
	5.9	122	1332
Proportional Counter	1.2	—	—
X-ray NaI(Tl)	3.0	12.0	—
3" x 3" NaI(Tl)	—	12.0	60
Si(Li)	0.16	—	—
Planar Ge	0.18	0.5	—
Coaxial Ge	—	0.8	1.8

The most common medium for detector cooling is liquid nitrogen, however, recent advances in electrical cooling systems have made electrically refrigerated cryostats a viable alternative for many detector applications. In liquid nitrogen (LN_2) cooled detectors, the detector element (and in some cases preamplifier components), are housed in a clean vacuum chamber, which is attached to or inserted in a LN_2 Dewar. The detector is in thermal contact with the liquid nitrogen, which cools it to around −200 °C. At these temperatures, reverse leakage currents are very low, i.e. in the order of 10^{-9} to 10^{-12} amperes.

1.4. Signal processing electronics

The small current obtained at the anode of the PMT cause a voltage drop across the bias resistor generating a voltage pulse. This pulse is generally processed by a preamplifier. Two characteristic parameter of the output pulse from a preamplifier are the rise time and the decay time. The first is related to the collection time of the charge (maximum a few µs), while the second is representative of the RC time constant of the preamplifier. The decay time is generally around 50µs. In ideal case the preamplifier is located close to the detector in order to reduce the capacitance of the leads, which adversely affects the rise time. In addition, the preamplifier provides an impedance match between the detector and the amplifier. The amplifier is used to shape the pulse as well as further amplify it. The shaping is needed to shorten the long delay time of the preamplifier with preserving the important detector signal in the pulse rise time. This signal processing is achieved by differentiating the pulse from the preamplifier. Either the first or the second derivative of the signal is used, giving a unipolar or a bipolar shaped signal. The pulse from the amplifier is directed into single channel analyzer (SCA) unit with counter/timer module or into a multichannel analyzer (MCA).

An SCA is equipped with a lower level (baseline) and upper level discriminator. This unit will produce a signal if the input pulse falls within the preset discriminator levels. For β-radiation measurement it is generally better to set the discriminator window wide to obtain good statistic. When characteristic X-radiation is to be monitored the two discriminator levels have to be set to closely surround the X-ray peak to increase the signal to noise ratio. In the case of complex radiation fields such as with simultaneous emission of β- and γ-radiation or X-ray and γ-radiation, the discriminator levels have to be adjusted to improve the signal to noise ratio without significantly losing sensitivity. The pulses from the SCA go into a counter, which records the number of logic pulses on an individual basis. If the counter is used in combination with a timer (either built-in, or external), the number of pulses per unit of time is recorded.

The multichannel analyzer (MCA) is made up of an analog-to-digital converter (ADC), control logic, memory, and display. It collects pulses in all voltage ranges simultaneously and displays this information in real time. If the input pulse falls within the selected voltage range then it is transferred to the ADC. To each pulse, the ADC assigns a memory location, which is proportional to the pulse height of the signal. The memory content is increased by one digit for each pulse. Since the pulse height is proportional to the energy of the radiation the display reads the memory contents vs. memory locations, which is equivalent to number of pulses vs. energy.

High count rate application requires a pile up rejector together with a unit correcting for the live time. Live time correction is needed since the pulse conversion at each units of circuitry requires certain time during which period the radiation arriving to the detector is not counted. At high count rates this dead time becomes significant portion of the uncorrected data collection time. Similarly, pulse pileup occurs when a new pulse from the preamplifier reaches the amplifier before it finished processing the previous pulse. In order to circumvent this situation a Pile Up Rejection/Live Time Correction unit is used. This circuitry prevents the ADC from processing any composite pulses and turn off the live time clock until the system is ready to process the next pulse.

2. RADIATION MEASUREMENT AND DATA EVALUATION IN *IN SITU* RADIOTRACER STUDIES OF SOLID/LIQUID INTERFACES

2.1. Measuring setup

The two most widespread used radioactive labeling techniques are the "foil" and the "thin gap" methods. The major difference in the two techniques lies in the measuring cells or more particularly as to how the adsorbent is arranged with respect to the detector (see below).

2.1.1. The "foil" method

A simplified view of a measuring cell developed by Horányi et al. [33] is demonstrated in Figure 1. In this method a thin metallized plastic foil forms the bottom of the radiochemical cell below which a radiation detector is placed.

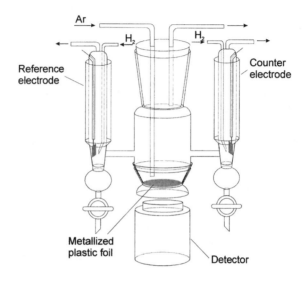

Fig. 4. Schematic view of the measuring cell used in "foil" technique [33].

The metal film, which is in contact with electrolytic solution containing the labeled species of interest, forms the working electrode of the cell. The foil and the metal film are thin enough to allow β-radiation from solution and from the adsorbate – if present on the electrode – to reach the detector. As seen from Figure 4, the scintillator is never exposed to the labeled solution as it is separated from the solution by the supporting foil of the cell as well as a light-proof metallized foil on the scintillator itself. While this system cannot be used for measuring very low energy β-radiation, its operation is very simple since it does not require dark chamber and lacks the need for significant manual intervention during measurement.

An alternative to the design in Figure 4 is a system where the bottom of the cell is a scintillator covered by a thin metal film [34, 35]. The metal layer serves as the working electrode in the measurements. Due to the poor adhesion of platinum or gold to glass firstly thin chromium layer is deposited onto the scintillator.

Another version of the 'foil' method is when the foil is mounted on the face of the detector, which is lowered by couple of mm into the solution [31, 36].

There are two major limitations using thin-foil electrodes, namely, the restriction of the radioactive labeling work to polycrystalline electrodes and the need for electrode roughening, especially for weakly adsorbing species at low coverage. Such rough surfaces are needed to increase the ratio of the signal (cps) from the electrode surface to that from the solution. Other adsorbents such as polymer films (e.g. Ref. 37), powdered metals/alloys (e.g. Ref. 38) sprinkled on the vacuum deposited noble metal layer or oxides (e.g. Ref. 39) dispersed on the bottom of the cell have also been applied.

When the labeled species emits high energy β-radiation, there is a possibility to use thin metal electrodes [40-43] instead of metallized plastic films. Namely, first Marcus et al. [40, 41] applied successfully very thin (< 30 μm) polycrystalline and monocrystalline nickel electrodes in the course of chloride adsorption studies on nickel surfaces. With this alteration the adsorption measurement can be performed on "real-life" alloys with proper crystallographic and grain structures. Another advantage is that the surface accumulation can be followed even at high current densities contrary to the vacuum deposited metal layer, which would peel off from the support under such condition. Such measuring principle was used for the study of phosphate accumulation on stainless steel and Al 1100 alloy surfaces [42, 43]. Despite the similarity between this measuring arrangement and the original "thin-foil" method, the Γ determination is somewhat different. This difference is the result of the different absorption behavior of radiation coming from the solution and the adsorbed phase [42] (see below).

2.1.2. The "thin-gap" method

In the section we discuss the working principle of the "thin gap" technique, though most of statements are valid for the "electrode lowering" method that was first introduced by Kazarinov in the sixties [44]. The key features of the 'thin-gap' scintillation technique are detailed in Figure 5. With the electrode in the "up" position the scintillator is sensitive to the β-radiation emitted only from radiolabeled species present in the bulk solution, since the adsorbed layer is too far away to be detected. By pressing the electrode down against the scintillator, the detector responds mostly to radioactive molecules adsorbed onto the electrode surface (Figure 5). The electrode also shields the detector from the radiation from the bulk solution. Only a thin layer of solution that remains in the gap between the electrode and the bottom of the cell will be measured together with the adsorbed layer. Therefore, the solution's counting rate from the labeled compound is significantly reduced, making it easier to detect the signal resulting from molecules adsorbed on the electrode. The thin-gap radiation component of the signal can be determined separately at potentials

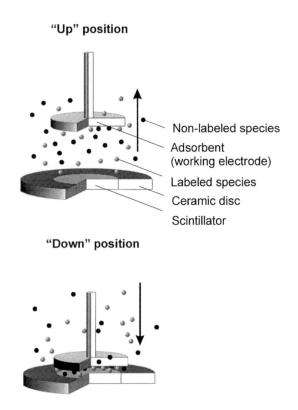

"Up" position

Non-labeled species

Adsorbent
(working electrode)

Labeled species

Ceramic disc

Scintillator

"Down" position

Fig. 5. Working principle of the thin-gap technique.

where adsorption does not occur, and subtracted from the overall counting rate measured at the adsorption potential, leaving just the adsorbate signal. The degree of background reduction by the electrode movement is described by the *"squeezing efficiency"* (S) in the absence of adsorption is [7-8]:

$$S = \frac{(I_{sol} - I_{gap})}{I_{sol}} \tag{3}$$

where I_{sol} and I_{gap} are the count rates measured in the up (solution) and down (squeezed) electrode position, respectively. In addition to maximize the signal-to-background ratio, the key to a successful experiment is to obtain reproducible counting rates from the radioactive solution in the "squeezed" gap. For this reason the I_{gap} is measured under an experimental condition where the adsorption of labeled isotope is essentially undetectable. In order to achieve gap thickness of 1.5 to to 2 µm, the electrode and scintillator has to be optically flat and smooth [7-8].

A radioelectrochemical cell working on the above principle and developed by Wieckowski is given Figure 6 [45]. Ten Kortenaar et al. [46], used recently an identical setup while a modified version of this cell was applied by Varga et al. [15].

An improved version of the "electrode lowering" technique was introduced by Poskus and Agafonovas [47]. In their setup, an inflated gaseous counter is used for radiation measurement. The counter window is a flexible and chemically inert plastic foil. The novelty of their system is the use of an excess pressure in the counter, which provides an excellent squeeze-efficiency.

3. RADIOACTIVE ISOTOPES IN *IN SITU* STUDIES OF INTERFACIAL PROCESSES

Isotopes applicable for *in situ* radiotracer studies are compiled in Table 3 [48]. While there are many more nuclides having desired type of radiation and appropriate radiation energy one must note that there are many more requirements that has to be fulfilled before an isotope can be considered acceptable as a tracer. Most important of such properties are half-life, the existence and decay-scheme of daughter nuclides, and available specific activity (Bq/g). Obviously, a nuclide with short half-life is impractical not only because its short period of availability but due to the necessity to compensate for decay-related intensity decrease during a radiotracer measurement. Isotopes with very long half-life have very low specific activity therefore they are inappropriate for labeling. Those radioactive isotopes, which decay to radioactive nuclides emitting radiation overlapping with that of the parent nuclides' are also neglected from consideration. This complication is most severe when the daughter nuclide has a shorter half-life than its parent isotope.

Even though the first successful application of "foil" method was the accumulation measurement of polonium on gold electrode [49] since then nuclides with alpha-decay have not been applied in *in situ* interfacial studies. The primary reasons for the lack of such studies are the stringent safety regulations for handling nuclides with alpha decay mode and the limited area of interest for these nuclides. In addition, the very limited range of α-particles may cause essential nuclear detection and instrumentation problems! Consequently, α-emitter nuclides are not included in Table 3.

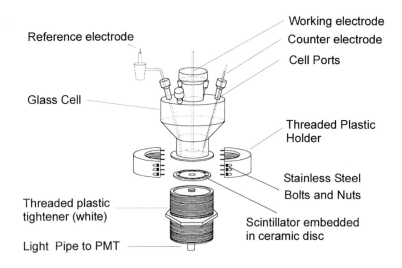

Fig. 6. Schematic view of the radioelectrochemical cell developed by Wieckowski [45].

4.   DIRECT MEASUREMENT OF INTERFACIAL PROCESSES IN SOLID/LIQUID SYSTEMS

In direct measurements, the radiation emitted from the nuclide provides the analytical signal. Due to the direct measurement, this class is represented by high signal-to-noise ratio. Consequently, this methodology finds its major application in *in situ* measurement of adsorption and absorption processes. In most cases only one type of ions or molecules is labeled. This setup requires a simple detection system since no energy selective detection is needed. Consequently, even GM counters can be selected for radiation intensity measurement. Due to the mechanical stability, easy handling, and versatility of solid state phosphors, nowadays most of the *in situ* radiotracer studies are performed with scintillators. Depending on the system arrangement, plastic or inorganic scintillators are used the most.

4.1. Measurement of labeled species emitting β⁻-radiation
Pure β-emitting nuclides.

Even though there are only a dozen or so pure β-radiation emitting nuclides (Table 3), notwithstanding they represent a very important group of isotopes. This category includes isotopes of carbon (^{14}C), phosphorous (^{32}P, ^{33}P), sulfur (^{35}S), and chlorine (^{36}Cl), which are the building blocks of organic or some of the most important, basic inorganic compounds. In addition the safety regulations are less stringent for the application of pure β-radiation emitters than any other radionuclides listed in the following sections. Consequently, simple chemical laboratories with minimal modification including special storage for radioactive isotopes and equipment for routine radiation monitoring are satisfactory for radiotracer work.

Table 3.
Radioactive nuclides applicable in "foil" and "thin-gap" techniques [48].

Pure β-emitting isotopes ($\eta_\gamma < 1\%$)	β(e),γ-emitting isotopes	Low energy X-ray emitting isotopes ($\eta_\beta < 1\%$)
^{3}H, ^{14}C, ^{32}P, ^{33}P, ^{36}Cl, ^{45}Ca, ^{63}Ni, ^{79}Se, ^{85}Kr, ^{89}Sr, ^{90}Y, ^{91}Y, ^{99}Tc, ^{107}Pd, ^{113m}Cd, ^{147}Pm, ^{169}Er, ^{185}W, ^{204}Tl	^{47}Ca, ^{46}Sc, ^{47}Sc, ^{59}Fe, ^{60}Co, ^{67}Cu, ^{199}Au, ^{203}Hg, ^{74}As, ^{75}Se, ^{86}Rb, ^{94}Nb, ^{95}Nb, ^{98}Tc, ^{110m}Ag, ^{111}Ag, ^{117m}Sn, ^{119m}Sn, ^{123}Sn, ^{122}Sb, ^{124}Sb, ^{126}Sb, ^{125m}Te, ^{127m}Te, ^{126}I, ^{133m}Xe, ^{134}Cs, ^{141}Ce, ^{144}Ce, ^{143}Pr, ^{149}Pm, ^{152}Eu, ^{154}Eu, ^{155}Eu, ^{160}Tb, ^{161}Tb, ^{170}Tm, ^{177}Lu, ^{181}Hf, ^{182}Ta, ^{186}Re, ^{191}Os, ^{192}Ir, ^{198}Au, ^{199}Au, ^{203}Hg, ^{210}Pb, ^{210}Bi	^{49}V, ^{51}Cr, ^{52}Mn, ^{53}Mn, ^{54}Mn, ^{55}Fe, ^{56}Co, ^{57}Co, ^{58}Co, ^{59}Ni, ^{65}Zn, ^{67}Ga, ^{71}Ge, ^{73}As, ^{121m}Te, ^{123m}Te, ^{124}I, ^{125}I, ^{131}Cs, ^{133}Ba, ^{153}Gd, ^{193}Pt, ^{195}Au, ^{206}Bi, ^{207}Bi

The surface concentration calculation is now well documented in the literature for these isotopes [5-11]. The calculation is the simplest for the "foil" method, when thin film adsorbents are used. The total radiation intensity builds up from the two major components:

$$I_{Tot} = I_{sol} + I_{ads} \tag{4}$$

where I_{sol} and I_{ads} are the intensity of radiation coming from the solution phase and adsorbed layer, respectively. The solution component can be expressed by Eq. 5.

$$I_{sol} = \alpha q I_0 C \int_0^K \exp(-\mu_{sol}\rho_{sol}x)\,dx \tag{5}$$

where α is a proportionality factor, q is the surface area (cm^2), I_0 is the molar activity of the labeled species (Bq mol^{-1}), ρ_{sol} is the density of the solution phase (g cm^{-3}), C is the chemical concentration of the labeled species (mol cm^{-3}), K is the thickness of the solution layer, and μ_{sol} is the mass absorption coefficient of the β-radiation (cm^2 g^{-1}) for the solution phase. After integration we can make a simplification since $\mu_{sol}\rho_{sol}K \gg 1$, therefore $1 \gg \exp(-\mu_{sol}\rho_{sol}K)$. Consequently, Eq. 4 can be rewritten into:

$$I_{sol} = \alpha q I_0 \frac{C}{\rho_{sol}\mu_{sol}}$$
(6)

The intensity coming from the adsorbed phase can be expressed as follows:

$$I_{ads} = \alpha q I_0 \gamma \Gamma$$
(7)

where γ is the roughness factor and Γ (mol cm^{-2}) is the surface concentration of the labeled species on the adsorbent surface. Combining the terms into Eq.4, we obtain:

$$I_{Tot} = \alpha q I_0 \left(\frac{C}{\rho_{sol}\mu_{sol}} + \gamma\Gamma \right)$$
(8)

or

$$\Gamma = \frac{I_{ads}}{I_{sol}} \frac{C}{\gamma\mu_{sol}\rho_{sol}}$$
(9)

The general protocol for the determination of Γ is the measurement of the natural background, then the solution background under a condition where the surface excess concentration is zero (or below the detection limit). With the help of μ_{sol}, ρ_{sol}, C, and I_{sol}, the product of $\alpha q I_0$ can be assessed. The radiation intensity increase during the measurement is only due to adsorption, if $q\gamma\Gamma <<$ CqK. Therefore, Γ can be obtained from combining Eqs. 4 and 8. Equation 6 also indicates that lower adsorbate concentration and high absorption coefficient provide lower detection limit. Considering that the mass absorption coefficient is between 5 and 350 cm^2 g^{-1} for the majority of application, the working concentration range is generally below 10^{-2} mol dm^{-3} for low energy radiation and 10^{-3} mol dm^{-3} for higher energy β-particles. The sensitivity of the measurement (S) that is defined as

$$S = \frac{\Delta I_{ads}}{\Delta\Gamma}$$
(10)

depends on the molar activity of the tracer, the surface area of substrate or the detector (whichever is smaller), detection geometry, the energy of the radiation, and the roughness of the electrode. The limit of quantitative determination (L_Q) [50] can be given as

$$L_Q = 50\left(1 + \sqrt{1 + \frac{I'_{sol}}{12.5}}\right) \tag{11}$$

where I'_{sol} is the uncorrected solution background. Considering a situation where $C = 10^{-4}$ mol dm^{-3}, $\gamma = 4$, $I'_{sol} = 5000$ cpm (from which 500 cpm corresponds to system background), $\rho_{sol} = 1.0$ g cm^{-3}, and $\mu = 250$ cm^{-1}, the limit of quantitative determination equals to 1051 cpm, which translates to 2.3×10^{-11} mol cm^2. The quantitative detection limit for surface concentration (Γ_Q) can be lowered by improving the detection geometry, increasing q and γ. It is well established in the literature that the absorption coefficient is inversely proportional to the radiation energy (Table 4). It can be seen from Eqs. 6 and 9 that the quantitative detection limit improves with decreasing radiation energy.

Table 4.
Empirical equations for the surface concentration calculation (cm^2 g^{-1}) for β-radiation.

Equation		Reference
$\mu = \dfrac{35Z}{M_A E_{\beta,max}^{1.14}}$	if $Z < 13$	20
$\mu = \dfrac{7.7Z^{0.31}}{E_{\beta,max}^{1.14}}$	if $Z \geq 13$	20
$\mu = \dfrac{8Z^{0.38}}{E_{\beta,max}^{(1.57-[Z/160])}}$		51
$\mu = \dfrac{0.0061Z^{0.366}}{E_{\beta,max}^{1.23}}$		52
$\mu = \dfrac{15.2Z^{1.333}}{M_A E_{\beta,max}^{1.485}}$		53
$\mu = \dfrac{15Z^{1.333}}{M_A E_{\beta,max}^{1.43}}$		54

At very low radiation energies (<100 keV) the contribution bremsstrahlung and characteristic X-ray generated in the adsorbent should also be considered [15, 55]. In extreme cases the primary radiation may be completely absorbed in the layers before reaching the detector, and only secondary radiation or other particles emitted during the radioactive decay are measured. Such a situation is presented earlier when authors described the measurement ^{125}I on thin platinum electrode using a glass scintillator [56, 57]. They claimed that the measurement was based on the measurement of 3.2 and 3.5 keV monoenergetic electrons. Considering that stopping power of 4 keV electrons in platinum is 14 MeV cm^2 g^{-1} [58], one may easily calculate that the electrons could not reach the scintillator through a 0.5 μm platinum film. It is more likely that the authors measured the low energy characteristic X-ray emitted by the tracer nuclide.

The calculation is somewhat more complicated in the case of adsorbent having thickness comparable to the range of radiation. Such situation was analyzed in Ref. 42, where phosphate accumulation was measured on stainless steel disks. Due to the significant scattering of β-radiation in the aqueous solution, the radiation spectrum of the solution phase is measurably shifted to lower energies. This results in a lower average energy, which translates to higher radiation absorption. Using stainless steel as radiation absorbent, the mass absorption coefficient of ^{32}P present in the adsorbed layer and distributed homogeneously in the solution phase were found to be 8 and 11 cm^2 g^{-1}, respectively [42]. In this case the surface concentration calculation is expressed using the following equations:

$$I_{sol} = \alpha q I_0 \exp(-\mu_{d,sol}\rho_d d) \frac{C}{\mu_{sol}\rho_{sol}} \tag{12}$$

$$I_{ads} = \alpha q I_0 \exp(-\mu_{d,ads}\rho_d d)\gamma\Gamma \tag{13}$$

$$\Gamma = \frac{C I_{ads}}{I_{sol}\mu_{sol}\gamma\exp(\Delta\mu_d\rho_d d)} \tag{14}$$

where ρ_d is the density of the adsorbent (g cm^{-3}), $\mu_{d,sol}$ and $\mu_{d,ads}$ are the mass absorption coefficients of the radiation coming from the solution and adsorbed phases, respectively. The term $\Delta\mu_d$ is obtained by subtracting $\mu_{d,ads}$ from $\mu_{d,sol}$. Suffix d indicates that the absorption coefficient is calculated for the adsorbent layer having a thickness of d.

The original equation has to be modified when powdered or porous adsorbents are used. In these cases one has to consider the absorption of radiation from the solution in the dispersed system as well as the self-absorption of solution and adsorbate radiations in the powdered layer:

$$I_{sol} = \alpha q I_0 C \exp(-\mu_{disp}\rho_{disp}k) \int_0^K \exp(-\mu_{sol}\rho_{sol}x)dx \tag{15}$$

after integration:

$$I_{sol} = \alpha q I_0 \exp(-\mu_{disp}\rho_{disp}k) \frac{C}{\mu_{sol}\rho_{sol}} \qquad (16)$$

$$I_{ads} = \frac{\alpha q I_0}{\mu_{disp}\rho_{disp}}\left(\varepsilon C + (1-\varepsilon)a_s\rho_m\Gamma\right)\left(1-\exp(-\mu_{disp}\rho_{disp}k)\right) \qquad (17)$$

where μ_{disp} is the mass absorption coefficient for the dispersed system (cm^2 g^{-1}), ε is the volume ratio of liquid in the porous layer, a_s is the specific surface area of the adsorbent (m^2 g^{-1}), ρ_m is the density of the adsorbent (g cm^{-3}), and k is the thickness of the porous layer (cm). When a membrane or coatings are applied the total intensity may consist of three components:

$$I_{Tot} = I_{sol} + I_{ads} + I_c \qquad (18)$$

The detailed mathematical expression of each terms are as follows:

$$I_{sol} = \alpha q I_0 C \exp(-\mu_c\rho_c d) \int_0^K \exp(-\mu_{sol}\rho_{sol}x)dx \qquad (19)$$

after integration:

$$I_{sol} = \alpha q I_0 \exp(-\mu_c\rho_c d) \frac{C}{\mu_{sol}\rho_{sol}} \qquad (20)$$

where μ_c is the mass absorption coefficient for the coating (cm^2 g^{-1}) and d is the thickness of the coating (cm). The intensity originating from the labeled species incorporated in the coating:

$$I_c = \alpha q I_0 C_c \int_0^d \exp(-\mu_c\rho_c x)dx \qquad (21)$$

after integration:

$$I_c = \alpha q I_0 C_c[1 - \exp(-\mu_c\rho_c d)] \qquad (22)$$

where C_c is the concentration of the labeled species in the coating. Finally, the relationship between the adsorption intensity and the surface concentration:

$$I_{ads} = \alpha q I_0 \exp(-\mu_c \rho_c d) \gamma \Gamma \tag{23}$$

In the case of the "thin gap" technique the calculation has only been established for the situation when the labeled species adsorbs on the smooth adsorbate. Evidently, this is the scenario where its advantage over the "foil" method can be utilized the best. When the electrode is lifted off from the scintillator the radiation intensity (I_{up}) is practically equal to the saturation count rate of the solution since the radiation from the electrode surface is completely absorbed in the solution layer between the electrode and scintillator:

$$I_{up} = I_{sol} \tag{24}$$

$$I_{sol} = \alpha q I_0 \frac{C}{\mu_{sol} \rho_{sol}} \tag{25}$$

In the "squeeze" position the total radiation intensity (I_{down}) is the sum of the intensities coming from the thin solution gap and the species adsorbed on the electrode surface:

$$I_{down} = I_{gap} + I_{ads} \tag{26}$$

The detailed expression of each term is given in Eqs. 27 and 28:

$$I_{ads} = \alpha q I_0 \gamma \Gamma \left(\exp(-\mu_{sol} \rho_{sol} x) + (f_b - 1) \exp(-\mu'_{sol} \rho_{sol} x) \right) \tag{27}$$

$$I_{gap} = \alpha q I_0 \frac{C}{\mu_{sol} \rho_{sol}} \left(1 - \exp(-\mu_{sol} \rho_{sol} x) \right) \left(1 + (f_b - 1) \exp(-\mu'_{sol} \rho_{sol} x) \right) \tag{28}$$

where f_b is the saturation backscattering factor, μ'_{sol} is the mass absorption coefficient of the backscattered radiation ($cm^2 \ g^{-1}$), and x is the thickness of the gap (cm). In several cases, the simplification of $\mu'_{sol} = \mu_{sol}$ can be made. Finally, the surface concentration can be calculated using Eqs.25 and 27:

$$\Gamma = \frac{I_{ads} C}{I_{sol} \gamma f_b \mu_{sol} \rho_{sol} \exp(-\mu_{sol} \rho_{sol} x)} \tag{29}$$

Lately, this equation, or more specifically the meaning of f_b, was slightly modified to take into account the light reflection of the electrode [59]. Namely, higher light output is obtained when the electrode is in "squeeze" position since the highly polished electrode reflects back the light generated in the scintillator. In the "lifted" position, the light reflection effect is significantly smaller. It was demonstrated that the correction factor is dependent on the electrode material, since different metals have different light reflection coefficient in the emission range of the scintillator (Figure 7). Canceling out this effect by employing a nontransparent film on the scintillator, the real f_b values can be determined (Figure 8). It has to be noted that the reflection effect can be eliminated if Γ is calculated using the equation given for I_{gap} and I_{ads} instead of I_{sol} and I_{ads}.

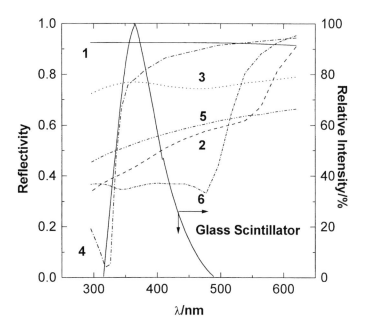

Fig. 7. The reflectivity –wavelength plot for several metals (1: aluminum, 2: copper, 3: rhodium, 4: silver, 5: platinum, 6: gold.) and the emission spectrum of a glass scintillator [59].

Practically, the same terms determine the sensitivity of "thin gap" method as described at the "foil" technique. In the case of L_Q the situation is somewhat different. Specifically, the background of I_{ads} is not I_{sol} but I_{gap}. Equation 29 is not adequate to assess the effect of experimental parameters on Γ_Q. For instance, the radiation energy has much less impact on the lowest detectable amount than in the case of the "foil" method. In fact Eq. 29 gives the impression that Γ_Q is inversely proportional to μ. Analyzing Eqs. 27 and 28 and considering that $\mu_{sol} = \mu_{sol}$ as well as taking the gap thickness equal to 1.5 μm one realizes that the expressions for I_{ads} and I_{gap} can be rewritten, since $\exp(-\mu_{sol}\rho_{sol}x)$ is practically equal to $1-\mu_{sol}\rho_{sol}x$:

$$\Gamma = \frac{I_{ads}C}{I_{gap}\gamma}\frac{x(f_b - f_b\mu_{sol}\rho_{sol}x + \mu_{sol}\rho_{sol}x)}{f_b(1-\mu_{sol}\rho_{sol}x)} \qquad (30)$$

Denoting the second part of the equation with φ and plotting this value against μ_{sol} we obtain a curve given in Figure 9. It is clear from this figure that the surface concentration depends only slightly on the mass absorption coefficient, i.e. the lowest and highest φ values differ by 3% from each other. Noteworthy is the fact that φ is almost identical to the gap thickness (since $1 \gg \mu_{sol}\rho_{sol}x$). Therefore, Eq. 30 can be simplified to:

$$\Gamma = \frac{I_{ads}}{I_{gap}}\frac{Cx}{\gamma} \qquad (31)$$

The slight dependence of Γ_Q on the energy of radiation is in fact one of the major advantage of this technique over the "foil" method.

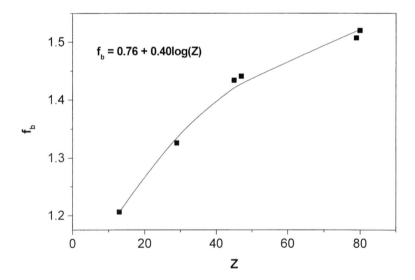

Fig. 8. Backscattering factor (f_b) measured in the "thin gap" setup using different electrodes [59].

One must keep in mind that the above calculations are based on several simplifications and assumptions as to the detection geometry and radiation scattering in the solution phase. Therefore, it is important to perform system calibrations as proposed in Refs. 12, 25, 26 to verify the validity of equations, i.e. check the accuracy of the values obtained from equations detailed above. Fortunately, in most cases the accuracy of the theoretically derived values are very good, i.e. less than 10-15% difference from the values obtained by calibration [25]. However, the necessity of calibration procedure becomes unavoidable when the radiotracer experiment is carried out in mixed radiation field.

The simplest way to calibrate the system, in the case of the foil method, is to form a thin of radioactive layer on a foil by using known amount of labeled solution. After layer is formed cover the film with another thin plastic foil assemble the cell and place inactive solution with composition similar to that used in the real experiment. Measure the count rate of the sample. This short procedure would be in fact enough for fast calibration since this establish the

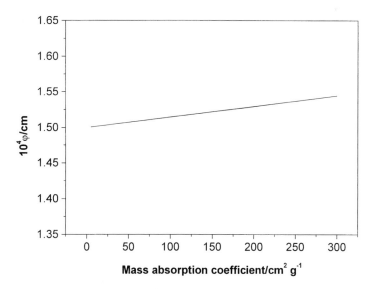

Fig. 9. Dependence of φ on the mass absorption coefficient.

relationship between I_{ads} and Γ. Since the intensity change during experiment is due to surface accumulation only, the intensity increase equals to I_{ads}. With a more thorough calibration one can get the linear absorption coefficient for the system that can be used in Eq. 8. In this case the calibration continues with the placement of a new, inactive foil onto the bottom of the cell and put radioactive solution into the cell using known amount of labeled species. The count rate measured in this setup will be representative for I_{sol} if no adsorption occurs on the plastic foil. The adsorption effect can be eliminated if we further increase the unlabeled species in the solution, i.e. decreasing the molar(or specific) activity. In this case correction has to be made since I_0 will be different for I_{ads} and I_{sol}. As a final comment on the calibration procedure, we note that the formation of thin radioactive film is critical in order to avoid self-absorption.

Calibration procedures for the "thin-gap" technique were discussed in [25]. In this procedure increasing amount of radioactive liquid is applied between the electrode and the scintillator to get the count rate vs. activity curve [25].

Nuclides emitting simultaneous β- and γ-radiation.
Most important representatives of nuclides falling into this category are given in (Table 3). The major measuring challenge originates from the complexity of radiation field. Specifically, γ-radiation creates electrons through Compton scattering and photoelectric effect. The contribution of secondary electrons is especially high at the radiation coming from the solution phase (Figure 10). While β-radiation cannot be detected beyond 1 mm of solution layer, γ-radiation with high energy can easily penetrate through even 100x thicker layer. Photons, which create electrons in the vicinity of the adsorbent surface, will be counted.

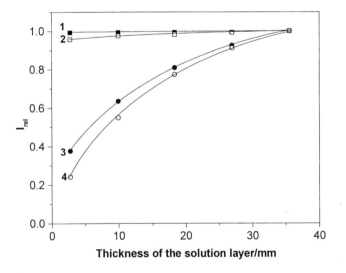

Fig. 10. Relative intensity of radiation from the solution phase in the case of ^{99}Tc (1, 2) and ^{60}Co (3, 4) isotopes using 0.5 mm thick (1,3) or 5 mm thick plastic scintillators.

Consequently, thicker solution layer at a given radioactive concentration will give higher background. This effect should be seriously considered at the "thin foil" method. Unfortunately, these secondary electrons have continuous energy spectra and most cases their energy range exceeds that of the primary β-radiation. We have to note that similarly to the solution phase, the adsorbent and especially the scintillator itself are media for secondary electron generation (Figure 11). Therefore, some simple strategies to counter-balance the adverse effect of γ-radiation are (i) minimizing solution layer in the measuring cell, (ii) application of thinner scintillator, and (iii) the use of energy selective measurement. Even under optimized detection condition, the calculation of surface concentration values is better based on experimentally setup empirical formulas rather than theoretically deduced equations (as given in the case of pure β-emitting nuclides). A detailed analysis of radiation condition is given in Ref. 26 in the case of ^{60}Co. This nuclide emits a low energy β-radiation with $E_{\beta,\max} = 312$ keV together with high energy γ-radiation ($E_{\gamma 1} = 1.17$ MeV, $E_{\gamma 2} = 1.33$ MeV).

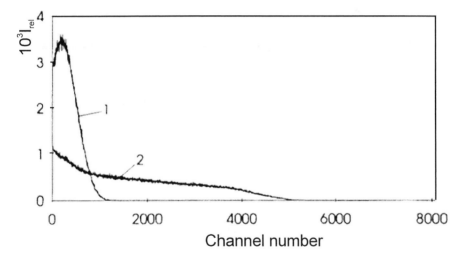

Fig. 11. Energy spectra of radiation from the solution phase measured with 5 mm thick plastic scintillator. (1: ^{99}Tc, 2: ^{60}Co)

In the case of "thin-gap" technique, the contribution of secondary radiation is even more complex and the extent of this effect is a strong function of adsorbent thickness, atomic number and density of adsorbent, geometry of cell, volume of labeled solution in the cell, as well as the energy of the radiation.

Consequently, the evaluation of surface concentration from equations presented above is not recommended. Due to the contribution of Compton electrons generated in the scintillator as well as in the thick electrode, the application of gaseous counters [47] is especially advantageous since they have low sensitivity for high energy γ-radiation. As a result of the low density and atomic number of the filling gas, secondary electron generation will be significantly smaller than in a solid state detector.

4.2. Measurement of labeled species emitting γ- or X-radiation

The vast majority of radioactive nuclides emit γ- or X-radiation simultaneously with other type of ionizing particles. This group is divided into two groups according to their energy in order to emphasize the difference in the measuring condition.

Measurement of labeled species emitting low energy γ- or X-radiation.
There are only few nuclides, which emit low energy X-radiation (Table 3). These nuclides decay mostly through an electron capture process. A majority of the isotopes also emit β-particles and high energy γ-radiation. Since the attenuation of X-and γ-radiation can also be described by the exponential function that is well known for β-radiation [20], the application of low energy X-ray emitting nuclides looked conceivable. Varga et al. [60] introduced this methodology. They measured silver accumulation on a stainless steel by utilizing the low energy X-ray emitted by the different silver isotopes. The mass attenuation of X-radiation at different energies in three different media is shown in Figure 12 [61]. As seen the upper energy limit is in the range of 10 keV, which corresponds to a mass attenuation coefficient of 5 cm^2 g^{-1} in water [61]. It should be noted however that the applicability of a given isotope, unlike in the case of β-labeling, very strongly depends on the adsorbent material. For instance, the mass attenuation coefficient of a 3 keV X-ray in water is ~200 cm^2 g^{-1} while in gold it is around 2000 cm^2 g^{-1}. Such an increase or even higher happens when the emitted X-ray energy is only slightly higher than the energy difference of electrons at different shells in the adsorbent material. As in the case of β-radiation detection, the sensitivity of measurement increases with the decrease of radiation energy. Indirectly, but Figure 12 also indicates that the maximum applicable adsorbent thickness depends even more strongly on the atomic number of the adsorbent than in the case of β-radiation (compare with Table 4).

While the disturbing high energy γ-radiation can be successfully filtered out, the effect of β-radiation is more difficult to deal with. This problem mainly results from the continuous energy spectrum of β-radiation. In addition, many scintillators used for X-ray detection has high average atomic number inducing significant backscattering effect on β-particles. In this process, electrons loose only a fraction of their initial energy in the scintillator, hence increasing the intensity of low energy portion of the spectrum. Since the intensity of both components, i.e. both β-and X-radiations, increase with adsorption, the surface concentration calculation has to be performed using empirical relationship instead of theoretical formulas.

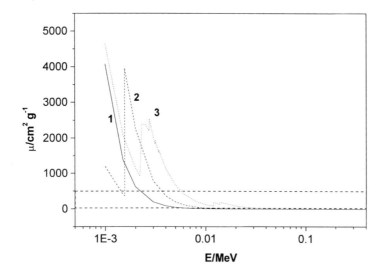

Fig. 12. Mass attenuation coefficient of X-radiation in various media. (1: water, 2: aluminum, 3: gold) [61].

In a recent paper the modeling of radiation condition that arose during the measurement of ^{65}Zn was described [62]. The established model showed very good correlation with the experimental data.

Measurement of labeled species emitting high energy γ-or X-radiation
The majority of radionuclides listed in the second and third columns in Table 3 also emit high energy γ-or X-radiation. This may seem to an appealing alternative to measurement of β-or X-radiation in mixed radiation field. Nevertheless, radiotracer techniques based on the direct measurement of high energy X-radiation requires special shielding to lower the solution background since the radiation attenuation coefficient of high energy γ-or X-radiation is very low. Consequently, the 'foil' method that relies on background intensity reduction through self-absorption cannot be used for these categories of

isotopes. (Note that most nuclides emit other ionizing radiation, which can be measured in this detection setup.) The "electrode lowering" and "thin gap" techniques can be considered useful when the high energy γ-or X-radiation of the radiotracer is to be measured. Up till now, the detection condition has not been analyzed in detail. However, it is likely that instead of gaseous counters that possess low sensitivity for high energy γ-radiation, solid sate detectors such as scintillation or semiconductor detectors are preferred. The electrode or adsorbent will serve as a shield to minimize the intensity from the solution. However, one must realize that at a γ-radiation energy of 1 MeV, the mass attenuation coefficient even in Pt is only 0.066 $cm^2 g^{-1}$. Under this condition the intensity reduction by a 1 cm thick electrode is just 75%, indicating that the background will be significantly higher than in the case of pure β-radiation.

5. INDIRECT MEASUREMENT OF INTERFACIAL PROCESSES IN SOLID/LIQUID SYSTEMS

In the '90s several new "unconventional" measuring alternatives were introduced in the in situ radioactive labeling studies. One of such methodology was the application of β-backscattering in the *in situ* measurement of metal deposition processes [63, 64]. Since this technique does not fall strictly into the radiotracer methods we do not discuss this approach here (see Chapter 5 for more detail).

Another interesting approach to follow surface accumulation processes is based on the measurement of bremsstrahlung or characteristic X-ray generated by β-emitting nuclides in the adsorbent [55, 65]. This conversion of radiation is especially helpful when the labeled isotope emits very low energy β-radiation or the thickness of the adsorbent is comparable or exceeds the range of β-particles. These possibilities are based on the fact, that at energies higher than couple of keV the attenuation of X-radiation in matter can be orders of magnitude lower than that of the original β-radiation. For instance electrodeposition of ^{99}Tc can be followed with submicrogram/cm^2 detection limit when the adsorbent layer would cause 30 orders of magnitude decrease in β-radiation intensity [55]. This technique is well suited for the measurement of multilayer deposition, though on rough substrates monolayer or submonolayer coverages can also be assessed [55, 65].

6. MEASUREMENT OF ABSORPTION AND INCORPORATION PROCESSES

The *in situ* measurement of absorption or incorporation processes is based on the observation that the energy spectrum of β-radiation changes with the extent of absorption. For instance in the "thin foil" setup the radiation spectrum of the solution phase, differs slightly from the adsorbed phase and even more from the spectrum given by the absorbed nuclides. The radiation coming from the solution phase is shifted to lower energies. Such effect is even more relevant when high energy β-emitting nuclides are used.

During the measurement of the incorporation process the radiation intensity and spectrum of the solution phase remain constant. When the adsorption attains a steady state value the changes in radiation spectrum will only be due to the incorporation process.

Such detection strategy can be advantageously used when the thickness of the absorbing layer is close to the range of the radiation, or the extent of adsorption is low compared to the absorption [67]. Examples of such application can be found in Refs. 68, 69, where the incorporation of chloride ions into corrosion protective coating or phosphate ion migration into ion-exchange membrane were described.

7. *IN SITU* MEASUREMENT OF ADSORPTION AT THE SOLID/GAS AND FLUID/FLUID INTERFACES

Adsorption measurement at the liquid/gas and solid/gas interfaces is mostly performed with detection setup similar to that of the "foil" technique. At the "surface count" method, for instance, the detector is positioned above the liquid surface facing the liquid/gas interface. In some cases the radiation intensity of both the adsorbed as well the gas phase are measurement simultaneously. This is especially important in situations when the radiation intensity of the gas phase, i.e. the "solution" background, changes during adsorption. In other words, the amount of labeled species adsorbed at the solid/gas interface is comparable to amount of the same component in the gas phase.

A measuring arrangement used in positron emission tomography was also used to determine the exact position of adsorbate on the solid phase (see Chapter 3 for further details.) These studies are restricted to isotopes with β^+ decay mode since the photons generated in the annihilation of positrons are measured. The experiments are performed in a coincidence mode in order to clearly identify the position of radiation source.

LIST OF ABBREVIATIONS

ADC	Analog-digital converter
FWHM	Full width at half maximum
MCA	Multi-channel analyzer
PMT	Photomultiplier tube
SCA	Single-channel analyzer

REFERENCES

[1]   G. Hevesy, F. Paneth, Z. Anorg. Chem., 82 (1913) 223.
[2]   K. H. Lieser, Nuclear and Radiochemistry : Fundamentals and Applications, Wiley-VCH, Weinheim, Germany (2001).
[3]   W. D. Ehmann, D. E. Vance, Radiochemistry and Nuclear Methods of Analysis, Wiley Interscience, New York, NY (1993).
[4]   G. Horányi, Electrochim. Acta, 25 (1980) 43.
[5]   V.E. Kazarinov and V.N. Andreev, in Tracer Methods in Electrochemical Studies, Comprehensive Treatise of Electrochemistry Vol. 9, edited by E. Yeager, J. O'M. Bockris and B.E. Conway (Plenum Press. New York. 1984), pp. 393-443.
[6]   P. Zelenay and A. Wieckowski. in Electrochemical Interfaces: Modern Techniques for In-Situ Surface Characterization, edited by T. Abruna (VCH. New York. 1991), pp.
[7]   E.K. Krauskopf and A. Wieckowski, in Radiochemical Methods to Measure Adsorption at Smooth Polycrystalline and Single Crystal Surfaces, Frontiers of Electrochemistry edited by P.N. Ross and J. Lipkowski (VCH. New York. 1992), pp. 119-169.
[8]   M. Gamboa-Aldeco, K. Franaszczuk and A. Wieckowski. Radiotracer Study of Electrode Surfaces, The Handbook of Surface Imaging and Visualization edited by A.T. Hubbard (CRC. Boca Raton. 1995), pp. 635-646.
[9]   G. Horányi, Rev. Anal. Chem., **XIV** (1995) 1.
[10]   G. Horányi: "Radiotracer Studies of Adsorption/Sorption Phenomena at Electrode Surfaces" in Interfacial Electrochemistry. Ed. by Andrzej Wieckowski. Marcel Dekker, Inc., New York, NY (1999) pp. 477-491.
[11]   K. Varga, G. Hirschberg, P. Baradlai, M. Nagy: Surface and Colloid Science (Ed. by E. Matijevic). Plenum Press, New York, Vol. 16, 2001. pp.341-393
[12]   A. Kolics, G. Horányi, J. Electroanal. Chem., 372 (1994) 261.
[13]   G. Horányi, A. Kolics, Elektrokhimya, 31 (1995) 105.
[14]   D. Poskus, J. Electroanal. Chem., 442 (1998) 5.
[15]   G. Hirschberg, Z. Németh, and K. Varga, J. Electroanal. Chem., 456 (1998) 171.
[16]   Canberra Inc.. Technical brochure.
[17]   RMD Inc. Technical brochure.
[18]   Maketech International. Technical brochure.
[19]   Amcryst Technical brochure.
[20]   A. Vértes, I. Kiss, Nuclear Chemistry. Akadémiai Kiadó. Budapest (1987).
[21]   P. Lerch: Anal. Chem., 27 (1955) 921.
[22]   R. H. Müller: Anal. Chem., 29 (1957) 969.
[23]   T. Lengyel: Acta Chim. Hung., 21 (1959) 51.
[24]   Industrial Application of Radioisotopes, G. Földiák (Ed.), Akadémiai Kiadó, Budapest (1986).
[25]   A. Kolics, A. Wieckowski: J. Phys. Chem. B., 105 (2001) 2588
[26]   A. Kolics, E. Maleczki and G. Horányi, J. Radioanal. Nucl. Chem. Articles, 170 (1993) 443.
[27]   A. Kolics and K. Varga: Electrochim. Acta, 40 (1995) 1835.
[28]   A. Kolics and G. Horányi, J. Appl. Radiat. Isot., 47 (1996) 551.
[29]   L. Gáncs, R. Buják, L. Reichstetter, Z. Németh: Magy. Kém . Foly., 108 (2002) 334.
[30]   K. Varga, I. Szalóki, L. Gáncs, R. Marczona: J. Electroanal. Chem., 524-525 (2002) 168.

[31] L. Gáncs, A. S. Besing, R. Buják, A. Kolics, Z. Németh, A. Wieckowski: Electrochem. Solid-State Let., 5 (2002) B16.
[32] N. Tsoulfanidis: Measurement and Detection of Radiation. Taylor & Francis; 2nd edition (1995).
[33] G. Horányi, J. Solt, and F. Nagy, J. Electroanal. Chem., 31 (1971) 87.
[34] A. Wieckowski, J. Electrochem. Soc., 122 (1975) 252.
[35] V. Jovancicevic, J. O'M. Bockris, J. L. Zelenay, T. Mizuno, J. Electrochem. Soc., 133 (1986) 2219.
[36] H. Wroblowa, M. Green, Electrochim. Acta, 8 (1963) 679.
[37] G. Inzelt and G. Horányi, J. Electroanal. Chem., 200 (1986) 405.
[38] K. Varga, E. Malecki and G. Horányi. Electrochim. Acta, 33 (1988) 25.
[39] G. Horányi and P. Joó, J. Colloid Interface Sci., 227 (2000) 206.
[40] J.-M. Herbelin, N. Barbouth and P. Marcus, J. Electrochem. Soc., 137 (1990) 3410.
[41] P. Marcus and J.M. Herbelin, Corrosion Sci., 34 (1993) 1123.
[42] A. Kolics and G. Horányi, Electrochim. Acta, 41 (1996) 791.
[43] A. Kolics, P. Waszczuk, L. Gáncs, Z. Németh, and A. Wieckowski: Electrochem. Solid-State Let., 3 (2000) 369.
[44] V. E. Kazarinov, Elektrokhimija, 2 (1966) 1170.
[45] A. E. Thomas, Y.-E. Sung, M. Gamboa-Aldeco, K. Franaszczuk, and A. Wieckowski, J. Electrochem. Soc., 142 (1995) 476.
[46] M. V. ten Kortenaar, Z. I. Kolar, J. J. M. de Goeij, and G. Frens, Langmuir, 18 (2002) 10279.
[47] D. Poskus and G. Agafonovas. J. Electroanal. Chem., 393 (1995) 105.
[48] D. C. Kocher, Radioactive Decay Data Tables. Report DOE/TIC 11O26. Technical Information Center V. S. Department of Energy, Washington, D.C. 1981.
[49] F. Joliot, J. Chym. Phys., 27 (1930) 119.
[50] L.A. Currie, Anal. Chem., 40 (1968) 586.
[51] T. Baltakmens, Nucl. Inst. Methods, 82 (1970) 264.
[52] A. Ranganathaiah, R. Gowda, B. Sunjeevaiah, Indian. J. Phys., 52A (1978) 120.
[53] H. W. Thümmel, Isotopenpraxis, 12 (1976) 240.
[54] L. Herforth, H. Koch, Praktikum der Radioaktivitat und Radiochemie. Johann Ambrosius Barth. German edition. Verlag der Wissenschaften Leipzig, 1992.
[55] A. Kolics and G. Horányi: J. Electroanal. Chem., 376 (1994) 167.
[56] J. O'M. Bockris, Maria Gamboa-Aldeco, M. Szklarczyk, J. Electroanal. Chem., 339 (1992) 355.
[57] J. O'M. Bockris, Maria Gamboa-Aldeco: J. Electroanal. Chem., 372 (1994) 272.
[58] M. J. Berger, J.S. Coursey, and M.A. Zucker, ESTAR, PSTAR, and ASTAR: Computer Programs for Calculating Stopping-Power and Range Tables for Electrons, Protons, and Helium Ions (version 1.2.2). [Online] Available: http://physics.nist.gov/Star. National Institute of Standards and Technology, Gaithersburg, MD (2000).
[59] A. Wieckowski and A. Kolics: J. Electroanal. Chem., 464 (1999) 118.
[60] K. Varga, E. Maleczki, E. Házi and G. Horányi, Electrochim. Acta, 35 (1990) 817.
[61] J. H. Hubbell, S. M. Seltzer, Tables of X-Ray Mass Attenuation Coefficients and Mass Energy-Absorption Coefficients (version 1.03). [Online] Available: http://physics.nist.gov/xaamdi. National Institute of Standards and Technology, Gaithersburg, MD (1997).
[62] I. Szalóki, K. Varga and R. Van Grieken, Spectrochimica Acta Part B: Atomic Spectroscopy 55 (2000) 1029.
[63] G. Horányi, J. Electroanal. Chem., 370 (1994) 67.

[64]   A. Kolics and G. Horányi, J. Electroanal. Chem., 374 (1994) 101.
[65]   A. Kolics and G. Horányi: Electrochim. Acta, 40 (1995) 2465.
[66]   G. Aniansson, J. Phys. Chem., 55 (1951) 1286.
[67]   L. Méray and Z. Németh: Appl. Radiat. Isot., 49 (1998) 13.
[68]   Z. Németh, L. Erdei, and A. Kolics, J. Radioanal. Nucl. Chem. Letters, 199 (1995) 265.
[69]   Z. Németh, L. Erdei and A. Kolics, Corrosion Sci., 37 (1995) 1163.

Radiotracer Studies of Interfaces
G. Horányi (editor)

Chapter 10.1

The role of interfacial phenomena in the contamination and decontamination of nuclear reactors

K. Varga

Department of Radiochemistry, University of Veszprém, H-8201 Veszprém, P.O. Box: 158, Hungary

1. INTRODUCTION

The adequate interpretation of radioactive contamination and decontamination issues emerged in nuclear industry requires a thorough and wide-ranging explanation of the interfacial phenomena in various (mainly construction material/solution) heterogeneous systems. As demonstrated in e.g. Ref. [1], a wide variety of experimental techniques is available for the study of sorption (deposition) phenomena and for the characterization of the structure and state of the surfaces and interfaces. On reading the previous chapters of this book, it is probably obvious that the application of radioactive-tracer technique elaborated by G. Hevesy in 1913 provides the most direct tool, and consequently, has played a key role in the understanding of contamination-decontamination problems of nuclear reactors for many decades.

1.1. Sources and significance of the radioactive contamination at nuclear reactors

Since the time when the first nuclear reactor was constructed and became critical in Chicago on December 2, 1942, many hundreds of nuclear power plants have been built throughout the world, mostly for electricity generation. Further uses have been to propel ships (dominantly naval vessels), to produce radioisotopes, and to a limited extent, to supply heat. Numerous additional reactors have been designed and built solely for weapon production and for education or research. We are here directly concerned only with the use of reactors for electricity generation.

At the end of 2002, there were 441 nuclear power reactors operating in 30 countries, representing a total capacity of 359 GW(e), 16% of global electricity generation and 7% of global primary energy use [2]. Of 33 nuclear reactors currently under construction world wide 20 are located in Asia. Seventeen countries depend on nuclear power for at least a quarter of their electricity. France and Lithuania get around three quarters of their power from nuclear energy, while Belgium, Bulgaria, Hungary, Japan, Slovakia, South Korea, Sweden, Switzerland, Slovenia and Ukraine obtain one third or more. The greatest growth in nuclear electricity production in 2002 was in Japan. An important demand at nuclear power plants in large countries like the United States and the Russian Federation is the license extensions of up to 20 years each beyond the original 30 years period.

Considering the design of a controlled nuclear chain reacting system, it is worth noting here that a variety of nuclear reactors is in use in the world today. Fig. 1 shows a "family tree" of the nuclear rectors in operation at the end of 2002 as well as those reported to be under construction as tabulated in [2-5]. Owing to the advantages and drawbacks of different concepts, more than a dozen different types of nuclear reactors have been developed and tested. However, only a few types are in common use. Over 77% of the currently operating nuclear power plants are of light water reactor (LWR) type, i.e. pressurized water reactor (PWR) or boiling water reactor (BWR). In this chapter we therefore discuss the main sources and the significance of the radioactive contamination of LWR type nuclear power reactors.

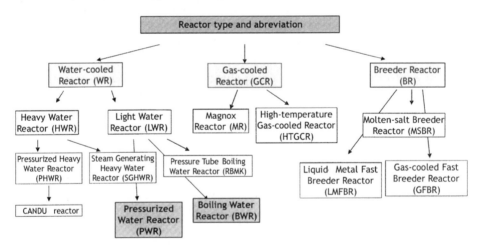

Fig. 1. Classification of nuclear reactor types in common use

In an "ideal" nuclear reactor all fission products and actinides formed are contained in fuel elements. In practical reactors listed in Fig.1 there are however independent processes through which radionuclides leave the reactor vessel, causing *radioactive contamination* problems. It is important to realize that the primary carrier of undesired radioactivity in all case is the coolant. In the cooling circuit of LWR type reactors there are at least three main sources of the radioactive contamination [4-6]:

(i) Fission products and actinides released from faulty fuel elements;

(ii) Activation of feed water and its impurities;

(iii) Activation of corrosion products formed from construction materials.

Under normal operating conditions, (when there is no fission product release due to fuel cladding failure), the majority of radioactive contamination in the primary loop is caused by various radioactive corrosion products (see Ref. [5-15] and references therein). The most important corrosion product radionuclides in the primary coolant of PWR and BWR type reactors are ^{60}Co, ^{58}Co, ^{51}Cr, ^{54}Mn, ^{59}Fe (as well as ^{110m}Ag in some Soviet-made VVER-type reactor). The two cobalt isotopes are known to be the predominant contaminants, but the ^{60}Co bears special importance owing to its relatively long half-life (5.27 years) and high-energy γ-photons (1.17 and 1.33 MeV). These two factors work towards ^{60}Co causing up to 80% of the radiation exposure to the reactor operating personnel. Careful treatment of the contamination phenomena at this specific case requires the comprehensive investigation of solid (construction material)/liquid (coolant) interfaces. It is to be noted that in spite of the stringent efforts made to control all radioactivity, small releases of radionuclides produced in (i) and (ii) processes are unavoidable, too. Such releases, gaseous as well as liquid, may cause further radioactive contamination at the following interfaces: gas/liquid, solid/liquid and solid/solid.

As a consequence of the above radioactive contamination problems, *decontamination procedures* have been elaborated in order to reduce the radiation field (dose rate) prior to in-situ inspection, maintenance or repair of the full cooling system or its components. *As used in this chapter, decontamination is defined as decrease or removal of unwanted radionuclides from surfaces.* The main criterion for decontamination is the amount of residual radioactivity (dose rate) on the surface; however, qualification of a given technology is strongly associated with the following technical issues: corrosion and recontamination of treated surfaces, waste management and cost-benefit analysis [6-8].

On summarizing the significance and current status of decontamination methods, it should finally be highlighted that not only the normal operation but also the decommissioning of nuclear power plants require an extensive (and commercial) application of rather sophisticated decontamination techniques. The increasingly significant global impact of the decommissioning on decontamination methodology is supported by the fact that:

(i) Nuclear facilities which have been retired from operation and are either awaiting or undergoing decommissioning are as follows: 115 power and research reactors, 5 reprocessing facilities, 14 fuel fabrication plants, and 60 mines [16].

(ii) Of the world's 441 operating nuclear power reactors, 345 have been in operation for 15 and more years, while 128 have been in operation for more than 25 years [2-3].

1.2. General problems of radioactive contamination-decontamination studies

The corrosion-corrosion prevention and radioactive contamination-decontamination problems perpetuate the interest in the study of the fundamental aspects of contamination and corrosion phenomena of constructional materials used in nuclear power stations. As described earlier, cooling circuits of LWR type nuclear reactors may become contaminated with radioactive isotopes during their normal operation. The most significant contributors to the radioactivity of surfaces are radionuclides from corrosion products (^{58}Co, ^{60}Co, ^{54}Mn, ^{51}Cr, ^{59}Fe, as well as ^{110m}Ag). It is of further interest to note that several aggressive anions such as Cl$^-$ and SO$_4^{2-}$ accumulated from the secondary feed water of PWRs (e.g. so-called hide-out at steam generators) may cause significant corrosion failure of various surfaces of special importance. Therefore, the knowledge of the fundamental aspects of these processes may contribute to the development of more efficient corrosion protection and decontamination procedures [1, 6-15, 17].

It is well known that corrosion and contamination processes in the primary cooling circuit of nuclear reactors are essentially interrelated: the contaminant isotopes are mostly corrosion products activated in the reactor core, and the contamination takes place on surfaces which were modified by the corrosion. Also, the two countermeasures (decontamination and corrosion-prevention) are connected to each other by similar ways: the usual and effective decontamination methods cause a very limited extent of corrosion, while a successful corrosion-prevention method can retard the contamination, too. Establishing the inevitable links between corrosion and contamination phenomena should lead to a new approach in handling the above problems. A schematic presentation of this approach is shown in Fig. 2, in which it is essential to emphasize the importance of the electrochemical aspects of the contamination and decontamination processes.

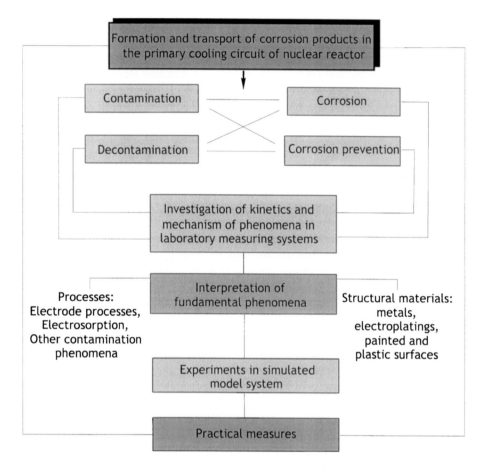

Fig. 2. Comprehensive view of the study of contamination-decontamination, corrosion-corrosion prevention problems that occur in nuclear power plants

An inspection of relevant literature data reveals that the significance of electrochemical aspects has not been taken into account in the evaluation of standardized investigation methods of contamination processes and decontamination procedures for a long time. Recently, industrial practice has proven that a number of contamination-decontamination issues cannot be satisfactorily interpreted, if the adequate explanation of important electrochemical (corrosion/corrosion prevention) effects is disregarded. In order to provide a deeper insight into the above problem, Fig. 3 shows an industrial example for the inevitable links among corrosion - corrosion prevention and contamination – decontamination phenomena.

As illustrated in Fig. 3, replacement of the feed water distributing systems of numerous steam generators (SGs) in four units of a VVER type nuclear power plant was considered to be unavoidable as a consequence of the detrimental corrosion-erosion processes during an operation period of 15 years or more. Owing to the significant radioactivity of corrosion product radionuclides accumulated on primary surfaces of stainless steel heat exchanger tubes, dose rate reduction was required prior to the maintenance work. In order to reduce the radiation field extensive chemical decontamination of SGs at three units was performed by making use of a version of the AP-CITROX technology. However, the chemical decontamination procedure did exert an undesired transformation on the stable constituents of surface layer, leading to the formation of a "hybrid" structure of amorphous and crystalline phases. The formation of above mobile oxide-layer increased significantly the amount of corrosion products in the primary circuit, resulting in magnetite deposition on fuel assemblies. As deposits blocked the cooling channels, the flow rate of water coolant through the reactor core decreased. Consequently, the power capacity of three nuclear reactor units had to be reduced, and full core fuel replacement became necessary. This example reveals that the lack of some fundamental information on the kinetics and mechanism of contamination-decontamination and corrosion-corrosion prevention phenomena can even cause serious outage of nuclear power reactors.

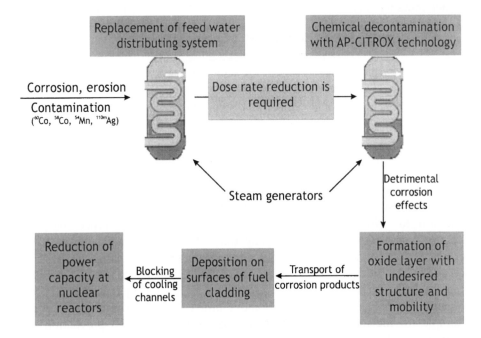

Fig. 3.   An industrial example for the inevitable links among corrosion – corrosion prevention and contamination – decontamination phenomena

In response to the need for the better understanding of the kinetics and mechanism of processes mentioned above, the first step is the application of laboratory measuring systems (see Fig. 2.). The next step required is the investigation of the phenomena in simulated model systems. The methods used on laboratory scale should be able to combine advantages of the experimental techniques of radiochemistry, electrochemistry and corrosion science to explain the relevant problems. Therefore, one has to choose investigation methods from the potentially great number of possibilities that can answer most of the posing questions. Among these methods, simultaneous use (coupling) of in-situ radiotracer and electrochemical techniques is considered to be one of the most powerful tools.

The aim of the present chapter is to give a brief overview on the methodology and applicability of radiotracer methods for the investigation of contamination-decontamination phenomena at solid/liquid heterogeneous systems. In addition, we shall highlight – through selected experimental results – significant aspects of the contamination processes of main corrosion product radionuclides. An important objective of this paper is to present a classification of decontamination techniques, and to demonstrate the applicability of chemical decontamination procedures in nuclear power plants.

2. AN OVERVIEW ON THE EXPERIMENTAL TECHNIQUES USED IN CONTAMINATION-DECONTAMINATION STUDIES

The contamination-decontamination phenomena occurring at the solid/liquid interfaces have been extensively studied for decades (see e.g. [1, 5-15, 18-21] and references cited therein). It would be difficult to offer a comprehensive survey of this field owing to the fact that research teams with significant past and scientific background in the study of contamination-decontamination processes have been published their results rarely, or even have not put forward any information. Further difficulty is the treatment of the diverse literature material according to identical considerations. With the exception of a few papers dealing with all aspects of contamination-decontamination phenomena [6-11, 18], most of the publications report new findings on industrial contamination-decontamination problems or decontamination methods. It should also be noted that in the qualification of contamination-decontamination behavior of surfaces of industrial importance, standardized investigation methods have been playing a predominant role. Figure 4 is compiled from the available literature data and gives an overview on the methods used for the study of radioactive contamination-decontamination processes in solid/liquid heterogeneous systems.

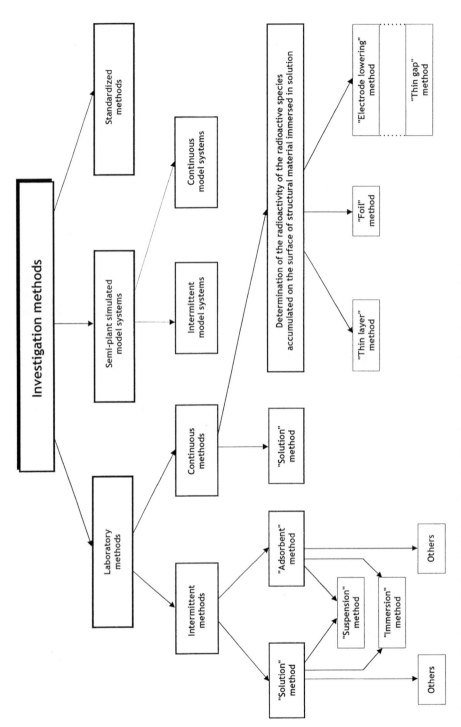

Fig. 4. Investigation methods used for the study of radioactive contamination – decontamination processes.

It is obvious from this figure that three main groups of experimental techniques can be distinguished. Specifically:
(i) laboratory methods;
(ii) semi-plant simulated model systems;
(iii) standardized methods.
For the purpose of this review a brief survey on the methodology of the above groups will be presented below.

2.1. Laboratory methods

In fact, the laboratory investigation of radioactive contamination phenomena is a radioactive-tracer study of the accumulation processes of various radionuclides on construction materials of industrial importance. The requirements regarding to the adsorbents are beyond demands of the radiotracer electrosorption researches, as (i) the knowledge of contamination-decontamination behavior of various coatings and plastic materials besides metals are also of considerable importance; (ii) it is full of meaning at the preparation of the adsorbents that their surface properties do not alter from the original behaviors of structural materials studied.

Monographs [1, 18], review articles [14, 20-21] and original papers (see e.g. [15, 22-41]) published on this topic cover experimental techniques, important subfields, and research trends. A comparison of the published laboratory techniques (see Fig. 4.) with those outlined in Chapters 4 and 9 reveals that most of the methods used in electrochemical studies are directly applicable to the investigation of contamination-decontamination processes at solid/liquid interfaces. With exception of some data measured in the past 15 years [14-15, 37-41], experimental results have been obtained by using the so-called "intermittent" methods. As seen in Fig. 4, this group can be divided into two subgroups; namely, the "solution" and the "adsorbent" methods. The basic principle of the classification is that the ex-situ measurement is carried out by determining the radioactivity either on the adsorbent surface ("adsorbent" method) or in aliquot volumes of the solution containing the contaminants ("solution" method). It should be noted that there is a continuous version of the "solution" method at which the radioactivity of the solution is detected continuously throughout the contamination-decontamination processes. Since the fundamental methodology as well as the general advantages and shortcomings of the "intermittent" experimental techniques have been analyzed in Chapters 4 and 9, this section deals in detail with radioactive labeling methods used for ex-situ studying contamination processes, exclusively.

As shown in Fig. 4, the "immersion" [22-23] and "suspension" [24-33] methods form the main part of the "intermittent" techniques. In case of the "immersion" methods the compact sample is kept in contact with the solution containing the radioactive contaminants for a given period of time, then the immersed specimen is taken off and its radioactivity (intensity, counting rate or dose rate) is determined. It is of interest to highlight in this case that the radioactivity of not only the adsorbent, but that of the solution is often detected.

Radiotracer techniques that are based on the application of powdered samples suspended in the solution represent an important subgroup in the study of contamination-decontamination phenomena. They are called "suspension" methods. Determination of the radioactivity of the radionuclides accumulated onto the powdered adsorbent is performed after the separation of the two phases mostly by measuring the radioactive concentration of the contaminants in the solution. A change in the radioactivity of the solution yields information about the decreasing solution concentration of investigated radionuclides and, consequently, the amount of species accumulated at the surface. This technique, widely used in analytical and colloid chemistry, was first applied for studies of cobalt sorption (^{60}Co contamination) by Kurbatov and coworkers in 1951 [24]. The application of the "suspension" methods in the field of contamination-decontamination is based on the hypothesis that oxide layers formed on metal surfaces of industrial importance play a predominant role in the accumulation of radioisotopes, and the effect of bulk metal on contamination as well as the depth distribution (migration) can be neglected [26]. I view of these assumptions, the uptake of various radionuclides (mostly ^{60}Co, ^{137}Cs) is studied on metal-oxides removed from the surface of construction material or synthesized artificially. Besides radiotracer method other techniques such as spectrometry, ellipsometry are also used for determination of the amount of the contaminants. In this area, the pioneering work of research group headed by Matijevic (see e.g. [25-26]) should be mentioned.

In addition to ex-situ investigation methods listed above, there were other attempts to provide plausible explanations on the mechanism of various contamination-decontamination phenomena. Such methods, which cannot be classified into the above mentioned subgroups (consequently, called "others" in Fig. 4.) include procedures elaborated for investigation of contamination processes on e.g. steel wool [34], polystyrene [35] or polyethylene glasses [36].

Finally, the in-situ techniques constitute an important subdivision of radiolabeling methods, which enable the continuous study of the radioactivity of radionuclides on the surface of structural material immersed in a solution (see Fig. 4). All these methods utilize the "thin layer" principle put forward by Aniansson [42], and are considered to be the most advanced technical solutions for the investigation of contamination-decontamination phenomena in solid/liquid heterogeneous systems from both a radiochemical and an electrochemical standpoint. The very basis of the "thin layer" principle is that application of radioisotopes emitting β-particles or γ- (X-ray) photons of energy not higher than 20 keV allows one to measure a significant radioactivity (intensity) surplus on the electrode surface above a small solution background. Methodologically this means that, owing to the absorption and self-absorption of β- and low energy ($E_\gamma \leq 20$ keV) γ-particles, the detector can "see" radiation coming from a thin solution layer only, making the intensity of the solution background small. When radionuclides accumulate on the surface in the course of a sorption process, they arrive into the "visual field" of the detector. Therefore, intensity excess can be measured which is proportional to the surface excess concentration of contaminants studied.

As shown in Fig. 4, in-situ radiotracer methods can be divided into three subgroups according to the technical realization of the "thin layer" principle. The methodology, special designs, instrumentation and applicability of the in-situ methods have been reviewed periodically [1, 43-52], and in addition, described in previous chapters (Chapters 5 and 9). Therefore, it should be realized only that the progress in the methodology and technical solutions of the so-called "foil" and "thin-gap" ("electrode lowering") in-situ methods has reached recently a certain level, where the investigation of contamination and corrosion processes on construction materials of industrial importance becomes possible [14-15, 37-41]. In the next section we shall demonstrate their applicability for the qualitatively new approach to contamination processes caused by ^{60}Co and ^{110m}Ag radionuclides.

2.2. Semi-plant simulated model systems

Based on the model test law, simulated model systems have been developed in order to obtain fundamental findings about various radioactive contamination-decontamination phenomena under industrial conditions (see e.g. [6-8, 18, 53-66]).

As regards the methodological aspects and measuring techniques used for detection, two main groups can be distinguished; namely, the "continuous" and "intermittent" methods (see Fig. 4).

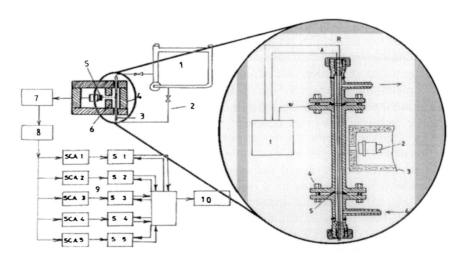

Fig. 5. Continuous semi-plant simulated model system with on-line measuring technique [53]: (1) recirculation loop, (2) by-pass for measurement, (3) controlled probe section, (4) lead shielding, (5) Ge(Li) detector, (6) lead collimator, (7) preamplifier, (8) amplifier, (9) pulse height analyzer, (10) computer. In the insert, the controlled probe section for electrochemical measurements is shown: (1) potentiostat, (2) Ge(Li) detector, (3) lead shielding, (4) fixation, (5) sealing and insulation (6) flow of radioactive coolant, (R) reference electrode, (A) auxiliary electrode, (W) working electrode.

The fundamental idea of the "continuous" methods is to construct a controlled specimen section, which may either be made artificially or part of a recirculation loop detached from the cooling circuit of a nuclear reactor. In this system, information about the contamination-decontamination processes is gained under industrial circumstances (temperature, pressure, flow rate, pH and composition of radioactive coolant etc.) by on-line measuring technique. Such very expensive setup was built only in a few research centers (see e.g. [6, 53-57]), and can be considered to be a semi-plant version of the in-situ "thin-layer" laboratory method. The schematic presentation of the semi-plant instrument, widely used for investigation of the contamination of corrosion product radionuclides (mainly ^{60}Co) by Lister et al. [53], is shown in Fig. 5. The special controlled probe section illustrated in the insert in Fig. 5, provides a simple and rapid way to carry out complex radiochemical-electrochemical experiments.

It is worth mentioning the studies performed in continuous model systems by Japanese researchers [56-57]. From methodological standpoint these systems belong under the continuous version of the "solution" method.

In case of the so-called "intermittent" semi-plant studies the measuring procedures and/or the sampling are carried out with the interruption (step-by step) of the contamination and/or decontamination processes, and several important parameters (such as electrode potential, corrosion state of the surface) cannot be controlled after removing the structural material from the radioactive solution. This method is considered to be the semi-plant version of the "immersion" technique performed in autoclaves. Despite its limitations, this semi-plant technique has been frequently used by numerous research groups, all over the world (see Ref. [58-66] and references cited therein). An advanced design of the "intermittent" semi-plant systems is illustrated schematically in Fig. 6 [63].

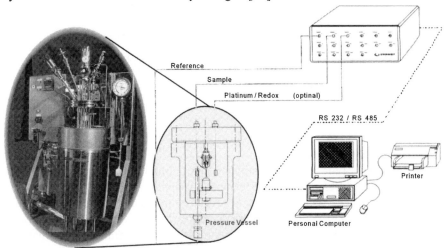

Fig. 6. An advanced design of the "intermittent" semi-plant model systems utilized for contamination and electrochemical experiments [63]

Finally, there are "intermittent" semi-plant model systems in which the sampling and/or the measurement of various parameters (e.g. surface radioactivity) are intermittent although the investigation procedure is performed in a continuous system [62, 64-66]. This version of the "intermittent" techniques attempts to unify the advantages of both the "adsorbent" and the continuous "solution" methods, and yields acceptable results primarily in studies of the efficiency and secondary corrosion effects of decontamination procedures. Scheme of an "intermittent" pilot-plant circulation system elaborated recently for the decontamination of heat exchanger tubes of SGs [66] is shown in Fig. 7.

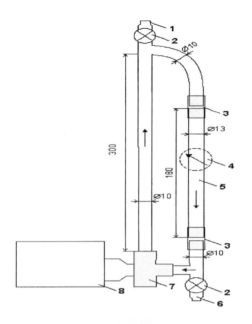

Fig. 7. Scheme of an "intermittent" pilot-
plant circulation system used for
the decontamination of heat
exchanger tubes of SGs [66]:
1 – deaerator stud,
2 – bullet valve,
3 – silicon tube,
4 – flow rate meter,
5 – stainless steel specimen,
6 – filling up/draining stud,
7 – centrifugal pump,
8 – pump Drive.

2.3. Standardized methods

The simplest way to give a quantitative approximation on the contamination-decontamination features of structural materials (paintings, coverings, metals etc.) under well-controlled conditions is a surface treatment with standardized procedures. In countries having considerable nuclear capacity, standards have been elaborated and utilized for the characterization of various surfaces with regard to their contamination-decontamination behaviors. The main characteristics of six standardized procedures are summarized in Table 1[18]. As obvious from Table 1, these standards assure normative conditions for performing both the contamination and decontamination processes, and provide definition of the terms used for quantification of surfaces to be studied. It should be noted that all standardized methods are basically classified as laboratory-scale technique (a version of the "adsorbent" methods). They give only empirical information about contamination-decontamination phenomena, and in addition, none of them pay attention to most circumstances of industrial importance (e.g. electrochemical parameters, considerable aspects of solution and surface chemistry).

Table 1.

A compilation of standardized methods [18]

  (a) Characteristic features of the contamination procedures

Country (Reg.No)	Radionuclides used as contaminants	Sample size (mm)	Conditions of procedure
USA (ANSI NS 12-1974)	mixture of fission products (^{137}Cs, ^{106}Ru, ^{144}Ce) and ^{95}Zr dissolved in 8 mol dm^{-3} HNO$_3$. pH of solution used for contamination: 4.0	65x125x3	- solution volume: 0.2 cm^3 - time: no data available - temperature: 25 °C - surface area: 1 cm^2
UK (BS4247)	mixture of ^{134}Cs and ^{60}Co. radiochemical concentration: 185 MBq dm^{-3}, solution pH= 5±0.2	140x50x13	- solution volume: 0.8 cm^3 - time: 24-28 h - temperature: 21±3 °C - surface area: 0.78 cm^2
Germany (DIN 25415)	mixture of ^{137}Cs and ^{60}Co in solution containing 10^{-5} mol dm^{-3} carriers and 10^{-4} mol dm^{-3} HNO$_3$ (pH = 4.0±0.2)	70x150x2	*Adsorption procedure* - solution volume:~0.6 cm^3 - time: 2 h - temperature: room temperature - surface area: no data available
France	mixture of fission products (^{90}Sr/^{90}Y, ^{137}Cs, ^{239}Pu) in 0.1 mol dm^{-3} HNO$_3$. radiochemical concentration: 15 MBq dm^{-3}	100x100x1	- solution volume: 1 cm^3 - time: 24 h - temperature: 23±2 °C - surface area: 15 cm^2
Japan	^{35}S, ^{32}P, ^{131}I or ^{140}Ba and other fission products in HNO$_3$ solution	no data available	- solution volume: 0.1 cm^3 - time: 1 h - temperature: 20 °C - surface area: no data available
Hungary (MSZ-05 22.7662-83)	mixture of ^{131}Cs and ^{60}Co in solution containing 10^{-5} mol dm^{-3} carriers and 10^{-4} mol dm^{-3} HNO$_3$ (pH = 4.0 ±0.2)	40x40x2	*Adsorption procedure* - solution volume:~0.6 cm^3 - time: 2 h - temperature: room temperature - surface area: ~12 cm^2 *Drying procedure* - solution volume: 0,1 cm^3 - time: 1 h - temperature: 40 °C - surface area: ~0.78 cm^2

(b) Characteristic features of the decontamination procedures

Country (Reg. No)	Detected radiation	Decontamination steps	Solutions used for decontamination	Terms* used for quantification
USA (ANSI NS 12-1974)	γ	3 consecutive steps (time:10 min each)	1. water at 25 °C 2. mixture of acids at 25 °C 3. mixture of acids at 80 °C (composition of mixture: 0.4 mol dm^{-3} (COOH)$_2$, 0.05 mol dm^{-3} NaF, 0.03 mol dm^{-3} H$_2$O$_2$).	DF
UK (BS4247)	γ	3 consecutive steps (time: 10 min each)	1. solution type A at pH =9 (detergent, Na$_3$PO$_4$, EDTA) 2. solution type B at pH=5 (detergent, EDTA-Na$_2$) 3. water at 25 °C	DF
Germany (DIN 25415)	γ	3 consecutive steps (time:2.5 min each)	1. water at 25 °C 2. detergent solution at 25 °C 3. 1 mol dm^{-3} HCl or 1 mol dm^{-3} HNO$_3$ at 25 °C	DF
France	γ	4 consecutive steps (2 steps water jet; 2 steps brushing)	1. water jet at 10-25 °C 2. water jet at 10-25 °C 3. brushing 4. brushing	Dx
Japan	β, γ	2 consecutive steps	1. water at 25 °C 2. HNO$_3$ + H$_3$PO$_4$ in case of ^{32}PO$_4^{3-}$ H$_2$SO$_4$ in case of ^{35}SO$_4^{2-}$ HJ in case of ^{131}I$^-$ HNO$_3$ in case of ^{140}BaCl$_2$	no data available
Hungary (MSZ-05 22.7662-83)	γ	3 consecutive steps (time:2.5 min each)	1. water at 25 °C 2. detergent solution at 25 °C 3. 1 mol dm^{-3} HCl or 1 mol dm^{-3} HNO$_3$ at 25 °C	DF

* *Definition of terms used for quantification is as follows:*
Decontamination factor (DF)

$$DF = \frac{\text{surface activity (or intensity or dose rate) measured after contamination}}{\text{surface activity (or intensity or dose rate) measured after decontamination}}$$

Residual surface activity in percentage (Dx)

$$Dx(\%) = \frac{1}{DF} \cdot 100$$

3. KINETICS AND MECHANISM OF CONTAMINATION PROCESSES OF MAIN CORROSION PRODUCT RADIONUCLIDES (SELECTED RESULTS)

As outlined in Section 1, formation, presence and deposition of corrosion product radionuclides (such as ^{60}Co, ^{58}Co, ^{51}Cr, ^{54}Mn, ^{59}Fe and/or ^{110m}Ag) in the cooling circuit of LWR type nuclear reactors throw many obstacles in the way of normal operation. The two cobalt isotopes are known to be the predominant contaminants, but the ^{110m}Ag radionuclide (originating from the welding material of some Soviet VVER-type PWRs) may also contribute significantly to the radiation exposure of the operating personnel. Moreover, the many γ-photons of different energy emitted by ^{110m}Ag hinder the continuous detection of other solution contaminants, leading to troubles in the monitoring of e.g. fission product release (if any) [14-15]. Therefore, it can easily be understood that knowledge of the fundamental aspects of ^{60}Co, ^{58}Co, and ^{110m}Ag deposition is needed, which may facilitate the elaboration of more efficient surface prevention and/or decontamination procedures.

In this section, recent findings obtained in our laboratory on two selected areas of contamination phenomena are reviewed in order to demonstrate the potentialities of the combined application of in-situ radiotracer and electrochemical methods.

3.1. ^{60}Co contamination on austenitic stainless steel

The accumulation of Co containing species on oxide layers formed on various steel surfaces is rather complex and may be influenced by numerous parameters (such as pH, composition and concentration of solution, electrode potential, structure and chemical composition of surface, technological parameters like temperature and pressure). Thus, it is no wonder that the formation and accumulation of cobalt on different steel surfaces and under various conditions have been the scope of many studies so far (see e.g.[1, 6-13, 15, 18, 21] and references cited therein). In spite of the great effort put out by both Western and Eastern research groups many issues of cobalt contamination are still open to debate. In order to contribute to the better understanding of the fundamental aspects of cobalt contamination, now we give a brief overview of some findings obtained by in-situ radiotracer studies of cobalt sorption on austenitic stainless steel.

The in-situ radiotracer experiments, aimed to study the kinetics and mechanism of cobalt contamination on steel surfaces, were performed by using a version of the in-situ "electrode-lowering" radiotracer method, following the protocol described in our previous papers [15]. All the experiments were carried out at room temperature in a model solution of the primary cooling circuit of

VVER-type nuclear reactors. Disk-shaped electrodes (diameter: 10.0 mm; thickness: 2-3 mm) were cut from austenitic stainless steel type 08X18H10T (GOST 5632-61). The stainless steel samples were polished to optical quality with emery paper and diamond paste (down to 0.25 μm). The ^{60}Co isotope having a molar activity of 3.3×10^{12} Bq·mol^{-1} was used. Throughout the experiments the cobalt concentration was set to 1×10^{-6} mol·dm^{-3} in the solution phase. The surface excess values (Γ) of sorbed cobalt species were calculated from the intensity of the β-radiation emitted by ^{60}Co ($E_{\beta max}=318$ keV) by making use of the relationships in [15].

In the insert in Fig. 8 the time dependence of cobalt sorption can be seen at austenitic stainless steel surface. The sorption exhibits saturation feature after about 20 hours, although the amount of cobalt accumulated on the surface is extremely low. The maximum surface excess measured was about 2.5×10^{-12} mol cm^{-2} which is approximately 0.5% of a monolayer coverage calculated assuming close packing of cobalt hydrolysis products ($\Gamma_{ML}= 4.3 \times 10^{-10}$ mol cm^{-2}) [67]. In spite of the low level of contamination the parallel measurements reveal good reproducibility as may be seen in the insert to Fig. 8. The 20% difference in the Γ values may most likely be attributed to the different roughness factors of the given steel samples.

Potential dependence results of cobalt contamination on austenitic stainless steel are shown in Fig. 8 (the surface excess values depicted in this figure were measured after 30 minutes contamination period). In the potential range of -0.10 V - 1.10 V (the passive region of this steel sample) only slight changes in the surface excess can be observed (curve 1). On the negative-going branch one may see minor increase in the Γ values. It may be the result of the reduction of oxygen traces to OH$^-$, which promotes the formation of cobalt hydrolysis products possessing much higher sorption ability than the Co^{2+} ions [15, 39-40]. In these laboratory-scale investigations the inert gas (high-purity argon) bubbling through the solution removes vast majority of the dissolved oxygen from the solution, while in a PWR-type nuclear reactor the reducing water chemistry (addition of NH$_3$ and/or N$_2$H$_4$) may help to overcome this problem.

Interesting potential dependence results were obtained upon polarizing the austenitic steel surface above 1.10 V. The corresponding curves are 1' and 1" in Fig. 8. During a period of 30 minutes the surface excess values increased by one and half orders of magnitude and kept on growing upon cathodic polarization, too. This conspicuous behavior was investigated in separate voltradiometric experiments in the course of the continuous cyclic polarization of the sample in the range of 1.10-1.40 V. The resulting curves are depicted in Fig. 9. It is clear from the curves that (i) the cobalt layer is growing continuously on the surface; (ii) the bigger the cobalt amount on the surface, the bigger the cobalt intake during the cathodic-going polarization.

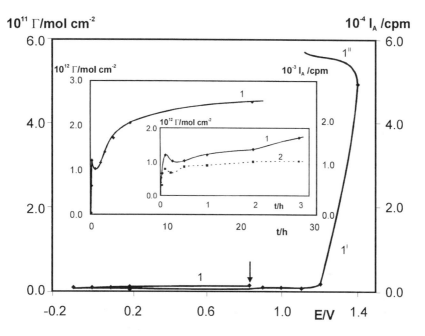

Fig. 8. Potential dependence of ^{60}Co sorption on austenitic stainless steel in model solution containing 1×10^{-6} mol dm^{-3} Co(II). Arrows indicate the open-circuit potential values.
Curve 1: polarization into negative direction starting from open-circuit potential (E=0.83 V). Curve 1': after negative polarization potential was set to E=0.83 V and polarization into positive direction. Curve 1": finishing polarization after reaching E=1.40 V.
Time dependence of ^{60}Co sorption on austenitic stainless steel in the model solution at open-circuit potential is shown in the insert. Open-circuit potential values are 0.84 (1) and 0.83 (2). Curves 1 and 2 were obtained under identical experimental conditions, but on different samples.

Similar findings were reported by Kolics and Varga on polycrystalline gold surface in a borate buffer solution [40]. Therefore, the potential dependent deposition of cobalt above E=1.10 V are almost identical on a noble metal and a stainless steel surface (which is in transpassive state), giving a strong indication of the unique sorption feature of the cobalt species rather than the surfaces studied. Moreover, it should be noted that such phenomenon has only been experienced in solutions containing borate species, however, the exact role of borate ions (and complexes) in this process is not clear at the moment [39-40].

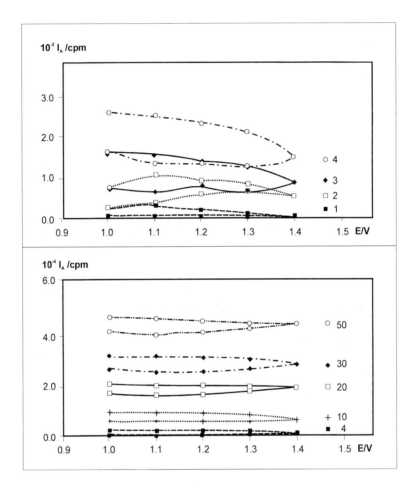

Fig. 9. Accumulation of ^{60}Co on austenitic stainless steel from model solution containing 1×10^{-6} mol dm^{-3} Co(II) upon cyclic polarization in the range 1.0-1.4 V vs. RHE. Arabic numbers beside the curves correspond to number of cycles.
Scan rate: 10 mVs^{-1}

Considering the potential diagram of Co-containing species at pH=7.7 (see Fig. 10) the possible mechanism of the high cobalt deposition found above 1.10 V is as follows [15, 40]:

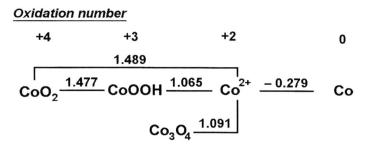

Fig. 10. Potential diagram of Co-containing species at pH = 7.7

(a) The oxidation of previously deposited Co(II) species to Co(IV) may proceeds at anodic potentials according to the following equations:

(1) Formation of Co(III)-oxide:

$$Co^{2+} + 2\ H_2O \quad \Leftrightarrow \quad CoOOH + 3\ H^+ + e^- \tag{1}$$
$$3\ Co^{2+} + 4\ H_2O \quad \Leftrightarrow \quad Co_3O_4 + 8\ H^+ + 2\ e^- \tag{2}$$
$$3\ Co(OH)_2 \quad \Leftrightarrow \quad Co_3O_4 + 2\ H^+ + 2\ H_2O + 2\ e^- \tag{3}$$
$$Co_3O_4 + 2\ H_2O \quad \Leftrightarrow \quad 3\ CoOOH + H^+ + e^- \tag{4}$$

(2) Formation of Co(IV)-oxide:

$$Co^{2+} + 2\ H_2O \quad \Leftrightarrow \quad CoO_2 + 4H^+ + 2\ e^- \tag{5}$$
$$CoOOH \quad \Leftrightarrow \quad CoO_2 + H^+ + e^- \tag{6}$$

(b) The reduction of Co(IV) in the course of the cathodic-going polarization is coupled with the build up of Co^{2+} ions from the solution phase in the range of 1.00 V – 1.60 V (embedding of Co(II) into the surface Co-oxide layer):

$$2CoO_2 + Co^{2+} + 2\ e^- \quad \Leftrightarrow \quad Co_3O_4 \tag{7}$$

The above selected results of the laboratory-scale studies have underlined the importance of the strict control of operating parameters in the water-cooled nuclear reactors. It has become obvious that: (i) The passive behavior of the austenitic stainless steel at open-circuit conditions, the slightly alkaline pH and the reducing water chemistry can be considered to be optimal to minimize the ^{60}Co contamination. (ii) The highly potential dependent deposition of various Co-oxides at E > 1.10 V offers a unique possibility to elaborate a novel electrochemical method for the decrease or removal of cobalt traces from borate-containing coolants contaminated with ^{60}Co and/or ^{58}Co radionuclides. The column-like adsorption system elaborated for the semi-plant decontamination of primary side coolants provides very promising preliminary results [15], however, the optimization of the experimental conditions requires further investigations.

3.2. Accumulation of ^{110m}Ag on various steel surfaces

The contamination caused by ^{110m}Ag activated in the core appears to be a significant problem in some Soviet-made PWRs, and may lead to troubles in the continuous measurement of other contaminants in the primary side coolant. The laboratory model systems (such as the "foil" and "electrode lowering" methods, presented in Chapter 5) make it possible to study the kinetics and the mechanism of silver accumulation on the surfaces of e.g. austenitic stainless steel type 08X18H10T (GOST 5632-61) and carbon steel (both materials are frequently used in some Soviet VVER-type PWRs). In order to demonstrate the usefulness of the in-situ radiotracer studies in this field, this section covers some radiochemical research carried out in a model solution of the primary side coolant of PWRs, containing silver labeled with ^{110m}Ag. In addition, the sorption of silver and the possibility of measuring the corresponding radiation are studied in the presence of other radioactive species (samples from a nuclear power plant). Some data characterizing the steels, as well as the composition of solution samples originating from a nuclear power station are given in [37]. The composition of the model solution used is described in [14]. Other experimental conditions are as detailed in Section 3.1.

At the radiotracer experiments carried out in model solution of primary side coolant the silver concentration was set to 1×10^{-6} mol dm^{-3}. As demonstrated by the relevant literature data ([14, 68] and references cited therein), in an aqueous solution containing 1×10^{-6} mol dm^{-3} Ag(I) at around pH=7.7 silver occurs in +1 oxidation state. Ag^{+} ions are stable in neutral or slightly alkaline solution due to the very moderate hydrolysis, but it is also known that they tend to form Ag$_2$O. In addition, it is of special importance to note that at pH=7.7 (i.e. the pH of the model solution) the amount of hydrolysis products of silver (such as AgOH and Ag(OH)$_2^{-}$) is negligible [68]. Thus it can be concluded that under the above-mentioned experimental conditions the chemical form of Ag(I) in the solution phase may be ionic (Ag^+) and/or hardly soluble oxide (Ag_2O), from which the latter may exist as dissolved or colloid species. Upon polarization into positive direction the formation of various silver oxides with higher oxidation states (AgO, Ag_2O_3) cannot be excluded, as well as both the formation and deposition of atomic silver may take place in the lower potential region.

Figure 11 shows the time dependence of silver accumulation on a disk-shaped sample of austenitic stainless steel measured by the "electrode lowering" method in a model solution of primary side coolant. A highly time and potential dependent sorption of silver species takes place on the smooth electrode surface, and no quasi-equilibrium surface excess of deposited silver is reached even after a period of 18 h at a given potential value. The fact that a surface excess of $\Gamma = 2 \times 10^{-9}$ mol dm^{-3} corresponds to one monolayer coverage of Ag^+ ions [69] provides evidence that only a limited part of the real surface area of the steel sample (less than 25%) is occupied by silver ions. The silver deposition processes are essentially dependent upon the electrode potential of steel surface. This statement is highly supported by the Γ vs. E profiles presented in Fig. 12. Figure 12a shows the cyclic voltammetry of the stainless steel in a borate buffer solution with and without silver ions. This voltammogram demonstrates that the studied steel exhibits passive features over a wide potential range (between -0.10 and 1.10 V) both in the absence and in the presence of Ag^+ ions labeled with [110m]Ag. The Γ vs. E curves in Fig. 12b reveal a considerable decrease in Γ values at more anodic potentials than the open circuit corrosion potential (curve 1). On the other hand, a potential shift into cathodic direction results in a significant increase in the surface excess concentration of silver (curve 1' in Fig. 12b and Fig 12c). All these results indicate that the extent of the contamination caused by [110m]Ag^+ ions is strongly influenced by the potential values to be chosen within the passive region of the stainless steel surface.

As detailed above, the adsorption behavior of Ag(I) present in the solution at very low concentration ($c \leq 1 \times 10^{-6}$ mol dm^{-3}) is decisively dependent on its aqueous chemistry, as well as on the nature of the adsorbent. In neutral and slightly alkaline pH medium Ag^+ ions are very sensitive to oxidation by O_2 traces in the solution. Moreover, experiments performed with $Fe(OH)_3$ [70-71] in solution containing 1×10^{-7} mol dm^{-3} Ag(I) give an indication that hydrous iron oxide surfaces may also catalyze the formation of Ag(I) oxide. Assuming that corrosion products formed on carbon steel indeed increase the relative amount of silver(I) oxide (Ag_2O) in the solution, the time and potential dependences of Ag(I) accumulation on carbon steel are expected to vary significantly from that on austenitic stainless steel due not only to the differences in the nature and composition of steel surfaces, but also to the changes in the chemical state of the majority of Ag(I) species to be adsorbed.

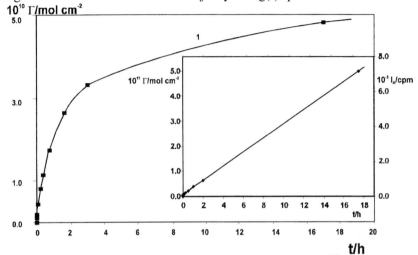

Fig. 11. Time dependence of the accumulation of silver labeled with [110m]Ag on the austenitic stainless steel in the model solution at open-circuit potential (E=0.75 V). For comparison, time dependence of the accumulation of silver labeled with [110m]Ag on the carbon steel in the model solution at open-circuit potential (E=0.39 V) is shown in the insert.

In accordance with the above expectations, comparative studies of silver contamination on carbon steel (see e.g. insert in Figs. 11) clearly demonstrate that the carbon steel exhibits markedly different sorption behaviors than the austenitic one under identical experimental conditions. As it has been revealed in [14], an active dissolution of carbon steel takes place in the potential region of -0.10 V to 1.10 V; therefore, one may expect enormous silver accumulation. In contrast, the extent of silver sorption in this case is one order of magnitude lower than on the austenitic stainless steel as reflected by the Γ vs. time curve in the insert in Fig. 11. It should, however, be noted that this curve does not show equilibrium feature, and in addition, the surface excess concentration values do not depend on the applied potential in the potential region of -0.10 V to 1.10 V. The surprising sorption feature of carbon steel is indicative of a distinctly different mode of the surface binding of silver species. At this stage in our studies, it is impossible to clarify the exact mechanism of the silver accumulation; however, it is probable that the various corrosion products (hydroxides and oxyhydroxides) dissolved from the arbon steel into the solution phase promotes the formation of silver(I) oxide (Ag_2O).

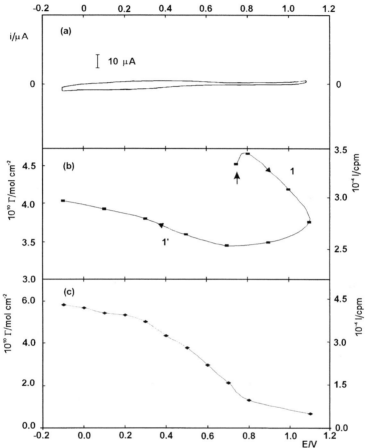

Fig. 12. (a) Cyclic voltammetry of disk-shaped austenitic stainless steel electrode in model solution. Scan rate 10 mV s^{-1}. (b) Potential dependence of adsorption of silver labeled with [110m]Ag on stainless steel in the model solution (curves 1 and 1', positive- and negative-direction plots, respectively). Arrow pointing upwards indicates the open-circuit potential. The experiment was carried out in continuous polarization mode, waiting for 30 min at each potential before obtaining intensity values. (c) Potential dependence of the adsorption of silver labeled with [110m]Ag on stainless steel in the model solution. The electrode was polarized into cathodic direction starting from 1.10 V. Other experimental conditions are as above.

Finally, Fig. 13 shows that the [110m]Ag accumulation on powdered austenitic stainless steel from solution samples, originating from nuclear power plant, can be studied by the measurement of the different radiations of the same isotope. Moreover, the b and c branches of curves 2 and 3 in Fig. 13, in accordance with the results presented in Fig. 12, reveal the effects of potential shifts on the sorption of ionic silver [37].

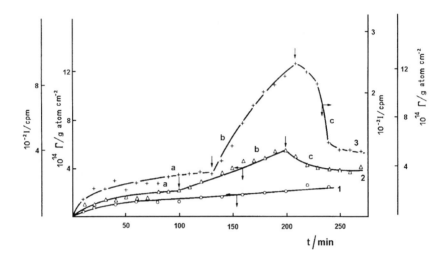

Fig. 13. Accumulation of silver species labeled with [110m]Ag from different solutions originating from NPP on austenitic stainless steel type 08X18H10T (GOST5632-61) under different experimental conditions: (1) β-radiation of [110m]Ag was measured under open-circuit condition, (2) K_α radiation of [110m]Ag was measured under: (a) open-circuit condition; (b) cathodic potential shift; (c) anodic potential shift. (3) K_α radiation of [110m]Ag was measured under: (a) open-circuit condition; (b) cathodic potential shift; (c) anodic potential shift. Other experimental conditions are as detailed in [37].

From these illustrative examples several implications arise, as follows:
(i) The substantial difference in the sorption behavior of Ag^+ an Ag_2O species, present in the solution phase, is expected to cause dissimilar accumulation at steel surfaces. The plausible contamination processes are schematically depicted in Fig. 14. It is probable that *on the stainless steel surface a deposition process owing to the cementation of Ag^+ ions rather than the accumulation of Ag_2O prevails over the contamination phenomena.* During the cementation the

dissolution of a metallic component of the steel sample is coupled with the reduction of silver ions to yield surface oxide-layer contaminated in depth with metallic silver.

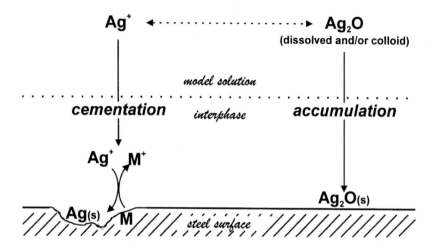

Fig. 14. Scheme of silver accumulation at steel surfaces.

(ii) Both the deposition (contamination) and the dissolution of deposited silver (decontamination and/or surface prevention) can be influenced by the electrochemical parameters of the system studied.

4. DECONTAMINATION TECHNOLOGIES USED IN NUCLEAR POWER PLANTS

4.1. General considerations

As a consequence of the radioactive contamination detailed in the previous sections, *decontamination technologies* have been elaborated in order to reduce the radiation field (dose rate) prior to in-situ inspection, maintenance or even dismantling during the normal operation (as well as the decommissioning) of nuclear power plants. The technology for decontamination of nuclear reactors and equipment is a fairly recent development (has a history of some 50 years), since the need for decontamination procedures did not extensively arise until some time after the first nuclear reactor was constructed, and radionuclides were produced and used in large quantities.

Decontamination is considered as a special cleaning; i.e. removal or decrease the amount of unwanted radionuclides from surfaces. Although both cleaning and decontamination require similar procedures, methods and materials, these two operations differ basically *by the type of substances removed and the degree of removal required.* Namely, in decontamination it may become necessary *to increase merely the efficiency of the removal of some radionuclides by a DF factor of 10, 100 or even more without detrimental changes of the treated surfaces.* Independently, both cleaning and decontamination require the same areas of knowledge such as chemistry of fouling, corrosion and corrosion inhibition, identification and control of reaction products, waste management, chemical engineering of various procedures etc. Therefore, it is not surprising that most of the procedures and chemicals utilized in decontaminating nuclear reactors were first developed and used for cleaning equipment in the non-nuclear industry. Novel techniques now being developed for e.g. chemical cleaning may be used later in decontamination.

In this section, we provide a simplified overview on decontamination techniques used in nuclear power plants. Special attention is paid to the description of methods suitable for the decontamination of industrial construction materials (metals, plastic, painted surfaces). It should, however, be noted that the very aim of this section is to give a classification upon some selection criteria rather than to outline all the advantages and disadvantages of the available techniques. The detailed information about the applied processes and procedures can be found in Ref. [6-9, 19].

In LWR type nuclear reactors, the procedures utilized for decontamination are primarily dependent upon two factors:

(i) *Type of contamination.* Specifically, if a fuel element has failed and has released a considerable amount of rupture debris (insoluble suspended materials (mostly metals), fuels), fission products and actinides, special reagents are required to dissolve and/or to remove these materials. On the other hand, if the contamination consists principally of activated corrosion products accumulated on the surfaces, one must use a procedure that removes the contaminated surface-layer down to the bare material. In this section, we will concentrate only the latter phenomenon.

(ii) *Type and surface condition of construction materials.* It is advisable to thoroughly review and list all materials (metals and non-metals) that will be decontaminated and that will be in future service. This review should cover important properties (such as type, chemical and phase compositions, thickness and structure of surface-layer (if any), porosity and surface roughness, mechanical properties, chemical stability and passivity, etc.) of the treated materials. This compilation should be used to make a preliminary engineering judgment of the effects of different commercially available decontamination procedures on radionuclide removal, corrosion and recontamination of surfaces, both during the decontamination operation and in subsequent service.

For instance, in LWRs there are at least five families of metals that should be characterized for decontamination evaluation. These are: austenitic stainless steels (e.g. type 304, 316, 321, 347, etc.); nickel base alloys (e.g. Alloy 600, 690, 800, etc.), chromium iron alloys; low alloy steels; and carbon steels. The austenitic stainless steels are of special importance owing to the fact that they are most widely used as construction material in LWRs (especially in PWRs) all over the world [4-11, 72]. On the surfaces of the stainless steel, which are in contact with the primary coolant, a coherent duplex oxide-layer of a thickness up to 10-12 μm is usually formed after various periods of normal operation. In case of reductive environments (as the case of PWR's water chemistry) the outer part of this duplex layer consists of mainly magnetite (Fe_3O_4) and occasionally some hematite (α-Fe_2O_3). The inner layer, next to the bulk metal, is a non-stoichiometric mixed oxide of spinell structure ($Cr_xNi_yFe_{3-x-y}O_4$, where $0 \leq x+y \leq 3$) [4-15, 17, 72]). The magnetite and the non-stoichiometric Ni-ferrite (iron partially substituted by nickel ($Ni_xFe_{3-x}O_4$, where $x \leq 1$, or $Co_yNi_xFe_{3-x-y}O_4$, where $y+x \leq 1$ and $y \ll x$)) are formed close to the surface and exhibit a measurable solubility in high-temperature coolant containing boric acid. In contrast, the Fe-chromites ($FeCr_2O_4$) and their substituted forms by Co and Ni ($Co_yNi_xFe_{1-x-y}Cr_2O_4$, where $x+y \leq 1$ and $y \ll x$) are generated in the inner region and extremely insoluble even at high temperature, as long as the reducing nature of the coolant is maintained. If the steel surface is treated with oxidative agents or oxygen is ingressed into the coolant, the chromites decompose to yield soluble chromates as a result of the Cr(III)$\rightarrow$Cr(VI) oxidation. The steel surfaces - initially covered with inactive oxide film - become contaminated mostly by the incorporation of corrosion product radionuclides (e.g. [60]Co, [58]Co, ^{59}Fe, ^{51}Cr, [54]Mn and [110m]Ag) [5-15]. As discussed earlier, the radioactive corrosion products in the cooling loop may create a serious radiation problem for the operational and maintenance personnel. Therefore, it is urgent to apply effective procedures for the surface decontamination, which avoid detrimental corrosion effects on the surfaces of operating equipment. To choose an appropriate decontamination method it is of special interest to know the contamination processes of the relevant radionuclides (such as ^{60}Co, [58]Co and [110m]Ag). In some earlier works (see e.g. [6-15] and references therein), the contamination phenomena caused by the above mentioned nuclides on austenitic stainless steel in various solutions of the primary side coolant were investigated. The experimental results reveal that (i) the ^{60}Co radionuclides are incorporated in both the outer magnetite (Ni-ferrite) and the inner Fe–chromite layers, and the radioactivity is mainly accumulated in the thin chromite layer; (ii) significant part of the ^{110m}Ag radionuclides is deposited onto (and into) the surface oxide-layer via some cementation processes (see previous section). Considering these metallographic and radioactive contamination findings, it can easily be understood that an effective decontamination should include the decomposition of the oxide-films, which formed on the steel surfaces, as well as, removal of loose and adherent radioactive corrosion layers. As a result of decontamination procedures, the corrosion rate of the surfaces may eventually be increased; therefore, to minimize the corrosion damages, the preparation of perfectly clean and passive surfaces in addition to decontamination is strongly recommended.

Further selection criteria that may be used upon classification of decontamination methods are as follows:

(iii)*Basic principles and methodology of procedures.* In this regards the range extends from the simple physical cleaning methods (vacuuming, grinding, washing, filter removal) to fairly sophisticated physical, chemical or electrochemical procedures (ultrasound, so-called "soft" and "hard" chemical technologies).

(iv)*Technical issues associated with the development and applicability of decontamination technologies.* These are as follows: decontamination efficiency and selectivity, corrosion and corrosion prevention, recontamination of treated surfaces, waste management and cost-benefit analysis.

The next part of this section provides a brief overview focused on the basic principles and methodology of decontamination techniques applied in nuclear power plant, covering in details the chemical decontamination technologies.

4.2. A brief overview of the applied decontamination technologies

4.2.1. Physical (mechanical) decontamination methods

Physical (mechanical) decontamination methods may be classified as *either surface cleaning* (scrubbing or high pressure water jetting) or *surface removal* (sandblasting, grinding) [19]. The surface cleaning techniques are used when the radioactive contamination is limited to the outermost surface region. These methods differ from surface removal procedure owing to the fact that their goal is rather the removal of the contaminant from the surface than the removal of the contaminated surface layer itself. Many surface cleaning techniques may be used as a secondary treatment following removal of the surface layer.

The physical (mechanical) procedures can be applied on any surface (metallic and non-metallic surfaces) and may achieve very acceptable results; for example, their application is a good choice on porous surfaces since other decontamination methods generally have lower effectiveness. Certainly, their efficiency depends decisively on the state and conditions of surface as well as on the physical–chemical properties of the contaminants. These methods may be used either as an alternative, simultaneously or in sequence with chemical decontamination technologies. Further advantages of their application: (i) The equipments needed are commercially available; (ii) The application costs are limited and dependent upon the actual conditions.

Beside the above advantages, there are some shortcomings, too. These methods require the essential surface area of the workplace to be accessible. Moreover, many procedures may produce airborne dust; therefore, respiratory protection (filter masks, in some cases tubes supplying oxygen from outside) is needed. Certain surface cleaning techniques generate substantial amounts of liquid waste that needs to be collected and treated.

All these physical treatments have been used for the decommissioning of NPPs and were tested after the Chernobyl accident [19, 73-74]. A compilation of the available physical (mechanical) methods and brief description of their characterizing features are given in Table 2.

4.2.2. Chemical decontamination technologies
4.2.2.1. Historical background and classification of the experimental methods

The commercial application of chemical decontamination technology has expanded significantly in recent 20 years at nuclear power plants. From initial applications to components and small subsystems to the current examination of full system decontamination, the entire technology has become much more sophisticated [6-9, 19].

Early water reactor decontaminations used existing chemical technologies. However, over the last three decades as the problem of corrosion products released into coolant systems was found to be generic in water reactors, novel decontamination techniques have been developed. Decontamination has emerged as an accepted tool to operators to mitigate the effects of activated corrosion products. In addition, many chemicals and techniques have been developed for routine decontamination during decommissioning of nuclear facilities.

Conventionally, the chemical decontamination reagents consist of various combinations of acids, detergents and complexing agents [6-8]. The decontamination processes of metal surfaces usually involve the following steps: oxidation and/or reduction, complexation (dissolution) and passivation (preparation of a corrosion-resistant, thermodynamically stable surface after removing of the contaminated surface-layer). The chemical decontamination can be carried out basically by two different ("soft" and "hard") techniques. "*Soft*" (mild) chemicals include non-corrosive reagents such as detergents, complexing agents, diluted acids or alkalis. These can be used when the system (or equipment) has to be treated without attacking the base material. "*Hard*" (aggressive) chemicals include concentrated strong acids or alkalis and other corrosive reagents. The dividing line between these two groups of decontamination methods is usually at about 1 m/m% concentration of the active reagent.

Many reactor circuits, their sub-systems and components were successfully decontaminated with "hard" processes based on citric, oxalic and phosphoric acids with and without pre-oxidative treatment involving potassium permanganate ([6-9] and references cited therein). Entire circuit of nuclear reactors were cleaned in the U.S.A., France and in the U.K. Following routine decontamination of the VVER-type PWR in the former East Germany with alkaline permanganate citric, oxalic acids (AP-CITROX) improved results were obtained with AP/citric acid/EDTA. Relatively strong reagents remain in use in some European countries (Slovakia, Hungary, Russian Federation) for the removal of loose and adherent active corrosion films from coolant circuits, more especially for sub-systems and/or components (e. g. steam generators, main pumps) [9, 66, 78, 82]. Although they are relatively corrosive, positive features of the "hard" technologies include short application times and good activity removal (i.e. basically high decontamination factors; DFs>10).

Developments in Canada during the early 1970s using diluted (< 1m/m%) chemical reagents led to the CAN-DECON (CANDU-DECONTAMINATION) process for use in PHWR coolant systems. Here the major corrosion product was magnetite (Fe_3O_4). Dilute chelating acids were found to be adequately successful in removing the active oxides present and achieving useful DF's between 3 and 6. Moreover, the corrosion penalty by using CAN-DECON was greatly reduced and the reagent could be continuously purified and regenerated by ion-exchange during the system decontamination. The process has been further developed and subsequently commercially applied with success in a number of North American water reactors [9]. CAN-DECON chemistry allows for effective dissolution of a range of corrosion products in BWRs (Fe_2O_3; Fe_3O_4; $NiFe_2O_4$) and, following a pre-oxidative stage, in PWRs ($FeCr_2O_4$; $NiCr_2O_4$; Cr_2O_3).

Table 2

A compilation of the available physical (mechanical) methods and their
characterizing features [19]

Decontamination techniques	Characterizing features
High pressure water jetting	It is suitable for decontamination bare and coated concrete, metal surfaces and ceramic tile. Expected efficiency is ca. 80 – 90 %. The temperature of the water needs to be higher than 5°C. During the treatment ca. 20 dm^3 m^{-2} of liquid waste may be generated.
Vacuum sweeping/brushing	It has high effectiveness on special industrial surfaces, for metals it is about 80 – 90 %. Higher degree of decontamination occurs if particles are small, bound in dust, and if surfaces are smooth and dry. During the process ca. 1 – 2 g m^{-2} of solid waste arises.
Grinding	Hand-held, power driven grinding equipment is used to abrade or remove thin layer of the surfaces. The expected efficiency is ca. 80 – 90 % in optimal case. The process generates ca. 50 – 100 g m^{-2} waste.
Sandblasting	Wet cleaning is used mainly to remove the smearable and fixed contamination from metal surfaces such as steel, scaffoldings, hand tools and machine parts. The wet system is a closed-loop, liquid abrasive decontamination technique, using a combination of water, abrasive material and compressed air. The dry-blasting technique is called commonly sandblasting or abrasive blasting jetting. This technique, which uses abrasive materials suspended in a medium (minerals, magnetite, and sand) that is projected onto the surface being treated, results in a uniform removal of surface contamination. It may provide high decontamination efficiency (75 – 97 %).
Ultrasound treatment	This process is based on the use of ultrasonic waves in a bath containing a chemical cleaning solution. The ultrasound is produced by a generator at a frequency between 20-50 kHz. Effective application (% reduction: 90 – 99) of the ultrasonic cleaning process requires optimization of a number of parameters.

In late 1984, laboratory evidence emerged to indicate that solutions containing oxalic acid under some conditions induce intergranular attack on sensitized type 304 stainless steel. A cautions period followed, when sufficient corrosion evaluation was conducted to understand and avoid conditions, which could potentially cause a problem. The LOMI (Low Oxidation Metal Ion) process, which does not contain oxalic acid, was first used at a BWR in 1984. Aside from the encouraging corrosion data, LOMI offered at least comparable DFs and required short application times. Meanwhile a new process, CAN-DEREM, which does not contain oxalic acid too, was developed from CAN-DECON, and now can be used on heat exchanger of the steam generators at PWRs. CAN-DEREM and LOMI are given equal prominence at this time for full system decontamination at LWRs.

In the primary circuits of PWRs, pre-oxidation of stainless steel surfaces is necessary to oxidise Cr^{3+} to soluble Cr^{6+}, leaving a chromium-depleted surface corrosion product which yields to dilute chelating acid. The original chrome-rich layer formed in coolant circuits has an unfortunately high affinity for ^{60}Co. In earlier "hard" processes strong alkaline permanganate was used for this pre-oxidation. Since 1977 research in the U.K. aimed at developing a new family of dilute reagents including alternative oxidizing systems for PWR-type oxides. Parallel work involved an alternative approach to the dissolution of LWR oxides in general. The LOMI reagents emerged in which dissolution of magnetite based films and pre-oxidized ferrites from PWRs are achieved reductively rather than by proton (acid) attack of the oxides. Since 1980 the developing technology of LOMI reagents has been quickly applied in loops and water reactor systems. The main advantages were lower system corrosion rates and some improvement in activity removal efficiency for very short reagent application times. Some developments in the U.K. include the use of dilute nitric acid/potassium permanganate (NP) reagent for chrome-rich situations; this forming the first stage of a PWR oxidative treatment process (POD). In 1984 the NP/LOMI process was introduced commercially in the U.S.A. Over 560 m^2 of reactor coolant loop pipework and the heat exchanger in the coolant purification system were successfully decontaminated at a BWR. Local radiation levels were reduced by factors as high as 60 with an overall average DF of 23 [9].

All full system decontaminations of LWR type reactors performed in Europe in early 1990s have used the CORD-UV process [79-80]. A feature of CORD-UV, which appeals to utilities, is the relatively low levels of waste generated (it uses permanganic acid instead of potassium permanganate). Another attractive feature is the use of ultra violet light for in-situ decomposition of the decontamination chemicals, essentially to carbon dioxide and water. Schematic of the Loviisa 2 coolant system during decontamination by CORD-UV process is shown in Fig. 15 [80].

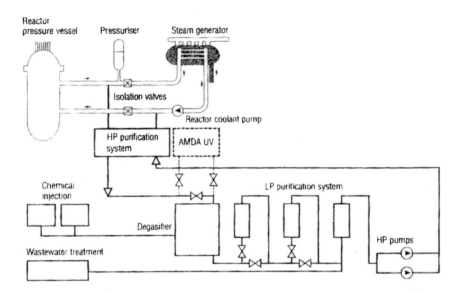

Fig. 15. Schematic of the Loviisa 2 (VVER-440) primary coolant system during decontamination by CORD-UV technology [80].

In the light of the above historical background, it can be stated that chemical decontamination is an evolving technology for control of radioactive corrosion products. While it is being used more frequently for radiation level control in LWRs sub-systems, entire coolant circuit decontamination including fuel is not widely practiced. The general problem of minor contamination of components, tools and equipment used in nuclear power plants is often associated with corrosion product migration and retention at surfaces and within crevices on a wide variety of materials.

Owing to the fact that many chemicals and procedures have been developed for routine decontamination during normal operation and/or decommissioning of nuclear facilities, a simplified classification of the major (well-known) decontamination technologies is worthy to be presented. A collection of the most important methods utilized for chemical decontamination is shown in Table 3. In case of the decontamination of metal surfaces a distinction has been made between the procedures used in *closed systems* (e.g. full system decontamination of the primary circuit of LWR-type reactors or the partial decontamination of closed loops) and the processes used in *open* tanks (e.g. decontamination of dismantled equipment).

4.2.2.2. Corrosion peculiarities

Corrosion has always been recognized as an area of potential concern to decontamination users. It is essential to find out the corrosion effects of decontamination chemicals on the materials used in nuclear power plants and to understand the observed phenomena.

Within the frame of a comprehensive program to qualify currently-available decontamination processes (LOMI, CAN-DECON and CITROX), extensive testing was carried out ([7-8] and references therein). These tests include general corrosion measurements, constant extension rate tensile test, pipe crack test and crack growth rate measurements. Generally, austenitic stainless steels, nickel base alloys, low alloy steels, carbon steels and non-metallic materials specimens were used for the examination of various chemical decontamination processes. Although these studies were carried out under BWR conditions in almost every case, the results of examinations are generally applicable to PWR full system decontamination, too.

Discussion of the available results and evaluating the decontamination solvents has been successful in avoiding significant corrosion damage during the decontamination. The effect of corrosion on a metallic surface can take many forms. Specifically, these forms are general, pitting, crevice, galvanic corrosion, intergranular attack (IGA), stress corrosion cracking (SCC) and intergranular stress corrosion cracking (IGSCC) [7-8, 10, 17]. Although the majority of alloys testing were performed on the 300 series austenitic stainless steels, there has been sufficient data for nickel-base alloys and some carbon and low alloy steels. Initial testing was limited to material loss (weight loss), but sophisticated conditions have been applied to perform crack growth rate measurements under cyclic loads, constant extension rate tensile tests (CERT) and corrosion potential measurements. Standard geometry test specimens were used extensively, and large diameter pipe (with welds) was selectively tested to determine crack initiation and crack propagation under cyclic loads. Types of the corrosion attacks caused by the decontamination processes (LOMI, CAN-DECON and CITROX) on selected structural materials are summarized in Table 4 [7-8].

The LOMI (NP-LOMI, AP-LOMI) tests data indicate that the general corrosion rates during decontamination processes are low for austenitic, low alloy and carbon steels (< 1 μm/h). Some isolated shallow pitting but no IGA or IGSCC has been observed for stainless steels. The data available for stress corrosion performance of stainless steels are essentially acceptable. Early isolated instances of shallow IGA and IGSCC are noted for the nickel base alloys. No effects were observed from the residues of LOMI reagents left in contact with carbon steels. Constant extension rate tensile test, crack growth rate and pipe tests conducted in a BWR environment using decontaminated specimens generally indicate no acceleration in crack growth rates or predisposition towards accelerated crack initiation as a result of LOMI exposure. For some nickel base and low alloy steels, significant crack propagation acceleration was identified in solutions. However, this crack propagation is still typical for non-decontaminated alloys.

General corrosion rates during CAN-DECON (AP/CAN-DECON) decontamination processes of carbon and low alloy steels were observed to be moderate to high (1 to 8 μm/h). Corrosion rates for austenitic materials were low. Indication of IGA and IGSCC on austenitic stainless steels and accelerated corrosion in low alloy steels has been observed in case of CAN-DECON process. IGA of nickel base alloys was also noted. Shallow pitting occurred for the AP/CAN-DECON on austenitic specimens with a metal crevice. Additional evidence of a deleterious effect was observed in CAN-DECON CERT test using sensitized austenitic stainless steel. Accelerated crack growth rate was observed with nickel base alloy and carbon steel specimens exposed to CAN-DECON.

The comprehensive test of CITROX (AP-CITROX) decontamination processes show that the general corrosion rates for carbon and low alloy steels were generally less than 1 μm/h. IGA has been observed for austenitic stainless steels and nickel base alloys.

In summary, some adverse effects (IGA, pitting, general attack) have been detected with organic acid reagents. Laboratory studies suggest that the presence of oxalic acid is necessary for IGA to occur. Process restrictions to minimize corrosion have been defined, and there is ample data available now for utilities to select a process for particular application (see [7-8] and references cited therein).

Table 3

Chemical decontamination method utilized in nuclear facilities

Decontamination method	Closed system	Open system	Short description
Oxidation processes			
Cerium/Sulfuric acid		x	These methods are applicable for cleaning of cooling system and steam generators of nuclear reactor. The process is based on the use of a strong oxidizing agent (Ce^{4+}) dissolved in low pH nitric acid or sulfuric acid [19, 75-76].
Cerium/Nitric acid		x	
MEDOC	x	x	MEDOC (Metal Decontamination by Oxidation with Cerium) is based on the use of Ce^{4+} as strong oxidant in sulfuric acid with continuous regeneration with ozone. It is primarily applied for decommissioning [62].
Oxidation-reduction processes			
APACE	x	x	It is a two-step process of alkali permanganate as an oxidizing pre-treatment followed by ammonium citrate to remove the oxide layer, and use of EDTA for chelating iron oxide in solution [6, 19, 73-74]
APOX AP-CITROX	x	x	(APOX: Alkaline permanganate and oxalic acid; AP-CITROX: Alkaline permanganate, citric acid and oxalic acid) These methods are used for removing crud/oxides from metal (mostly from austenitic stainless steel, low alloyed steel and carbon steel) surfaces. In case of carbon steel, oxalic acid reacts with steel to form a highly insoluble ferrous oxalate film, which requires treatment with nitric acid and sulfuric acids to remove [6, 19, 74]. Formation of the undesired structure of oxide-layer may occur on treated surfaces of stainless steels, too [77].
AP-NHN		x	(AP-NHN: Alkaline permanganate and a mixture of an acid, complexing and reduction agents). The procedure is developed for decontamination of VVER type nuclear reactors in order to eliminate a potentially hazardous reagent (oxalic acid) [78, 82].
CAN-DECON, CAN-DEREM	x	x	They were developed for use in CANDU reactors in order to remove the crud and oxide from the surfaces, involving application of dilute reagents (citric acid, oxalic acid) and EDTA for chelating iron oxide in solution to prevent redeposition onto the cleaned surface. The CAN-DEREM method does not contain oxalic acid [7-8, 19, 73-74].
LOMI	x		(LOMI: Low Oxidation-states Metal Ions) This method uses vanadium picolinate ion to reduce steel corrosion layers (Fe^{+3}) to soluble state (Fe^{+2}) with V^{+2}, which is oxidized to V^{+3} [7-8].
CORD	x	x	Usable mainly for full system decontamination of BWR and PWR reactors. The first step is the surface pre-oxidation with permanganic acid, it follows reduction and decontamination steps with oxalic acid [19, 79-80].

Table 4
Corrosion attacks caused by the decontamination processes (LOMI, CAN-DECON and CITROX) on selected structural materials [7-8]

Process	Austenitic stainless steel	Nickel base alloy	Low alloy steel	Carbon steel
	Corrosion attack			
LOMI AP-LOMI and NP-LOMI	Pitting	IGA IGSCC	No data available	No attack
CAN-DECON and AP/CAN-DECON	IGA IGSCC Pitting	IGA	Galvanic IGA IGSCC	No data available
CITROX and AP-CITROX	IGA	No data available	IGA	No data available

On evaluating the detrimental corrosion and metallographic effects of decontamination reagents containing oxalic acid, it is of special importance to highlight the *formation of a surface-layer that exhibits unwanted structure and chemical composition.* Specifically, oxalic acid reacts with iron on surfaces of carbon steel and 400 series stainless steels to form a highly insoluble ferrous oxalate film [6, 19, 74]. Recent systematic studies of austenitic stainless steel type 08X18H10T (GOST 5632-61), which corresponds to AISI 321 and DIN 1.4541, have revealed that chemical decontamination of the SGs by the AP-CITROX procedure does exert an undesired transformation effect on the stable constituents of surface layer (magnetite, hematite, spinell and even the bulk austenite), leading to the formation of a "hybrid" structure of the amorphous and crystalline phases [77]. The latter oxide-layer (i) exhibits great mobility in the primary coolant, and (ii) influences significantly the extent of the radioactive contamination and the amount of the corrosion products in the primary circuit.

In order to provide a compelling evidence of the formation of "hybrid" structure layer grown on steel tube surface, the metallographic cross sections of samples were prepared and analyzed by scanning electron microscopy (SEM) [77]. As seen in Fig. 16, the SEM micrographs obtained by the detection of the backscattered electrons justify the undesired changes; i.e. the formation of a "hybrid" structure film with a thickness up to 5 μm.

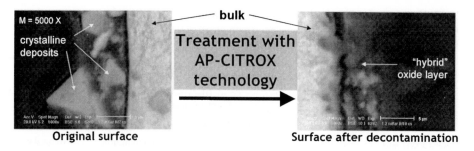

Original surface   **Surface after decontamination**

Fig. 16. Effects of the AP-CITROX process on the surface structure and morphology of austenitic stainless steel tubes of SGs [77].

The above detrimental effects of the chemical decontamination on the corrosion state of the inner surface of the heat exchanger tubes may be interpreted as follows:

During the application of the AP-CITROX method in plant environment the chemical treatment of the stainless steel surface is inhomogeneous, and the queasy-equilibration dissolution of the iron content from the surface oxide layer during the oxalic-acid – citric-acid treatment cannot assure. In the knowledge of the technological parameters (0.4 cm^3/cm^2 decontamination solution to treated steel surface area ratio, temperature not more than 95 °C) and considering the solubility of the Fe(II)-oxalate in hot water (0.026g/100 cm^3 water [81], which corresponds to 1.6 mg/dm^3 Fe^{2+} uptake), it is probable that during the chemical treatment a considerable amount of Fe(II)-oxalate precipitates on the inner surface of the heat exchanger tubes. The Fe(II)-oxalate deposits can not be eliminated even during the cleaning and passivation steps of the AP-CITROX technology. It is to be noted that similar observations were published in [82]. After the decontamination procedure the SGs are open to the air for a longer period of time; consequently, the inner surfaces are covered with aqueous solutions (in some cases with a mixture of steam and water) saturated in dissolved O$_2$. During this time period, a considerable amount of amorphous Fe(III)-hydroxides may be formed from the Fe(II)-species in the surface region of the protective oxide-layer, especially if oxidative agents (e.g. MnO$_2$) remain on the surface. Therefore, part of the iron originally bounded in the form of stable oxides (magnetite, spinell, hematite) could be transformed into amorphous Fe(III)-hydroxides and remained on the surface oxide layer as an undesired result of the decontamination technology. The mechanism described above, and the formation of *a hybrid structure of surface oxides,* are supported by the results of some surface spectroscopic and diffraction (CEMS, XRD and XPS) studies [77].

4.2.3. Electrochemical decontamination

Electrochemical decontamination (electropolishing) in principle can be considered to be chemical decontamination of conductive surfaces (mostly metals and alloys) assisted by an electrical field. It is the electrolytic removal of the metal layer in highly ionic solution by making use of an appropriate electrical current density. Almost all the metal (ferrous or non-ferrous) surfaces can be electropolished. A compilation of the metals and alloys that can be electropolished is as follows: aluminum-non silicon, beryllium-copper alloys, brass, bronze, copper, hastelloy, gold, nickel, nickel-silver, steels, stainless steels, titanium [19, 73].

The electrochemical decontamination procedures are generally effective and provide high decontamination factors. Their effectiveness is influenced by the presence of adhering materials, such as oil, grease, oxides and paint or other coating; therefore, these should be removed before decontamination. This technique produces a combination of surface properties, which can be achieved by no other methods of surface finishing.

Electrochemical decontamination could be applied by immersion of the contaminated item in an electrolyte bath or by passing a pad over the surface to be decontaminated. The electrolyte is continuously regenerated by recirculation.

The main disadvantage of electrochemical decontamination is the generation of liquid (ca. $5 - 15$ dm^3 m^{-2}) waste. The treatment and conditioning of this waste require appropriate processes to be considered when selecting the decontamination option. If organic acids are used, decomposition of the organic acids yields non-acidic waste.

More detailed description of these procedures is given in [6, 19, 73].

5. SUMMARY

The very aim of this chapter was to highlight the significance and potentialities of the application of radioactive-tracer technique elaborated by G. Hevesy in the studies of contamination-decontamination problems of nuclear reactors. Bearing in mind that most of the prospective readers are not very familiar with sources, and significance of the radioactive contamination as well as the decontamination strategy at nuclear reactors, an attempt was made to discuss the theoretical and technical details of these issues. Significant part of the present article is focused on the methodology and applicability of the radiotracer methods used for the investigation of radioactive contamination-decontamination processes on construction materials of industrial importance. Owing to the fact that results of the radiotracer research in several areas have been reviewed in the previous chapters, the radioactive labeling methods applied exclusively for studying contamination-decontamination processes at solid/liquid interfaces were taken into consideration.

To demonstrate the usefulness of the upgraded versions of two in-situ radiotracer methods (so-called "foil" and "electrode lowering"), selected experimental situations in the field of contamination caused by corrosion product radionuclides (^{60}Co and ^{110m}Ag deposition) have been described in some details.

Finally, the last section of this chapter deals with the selection criteria and classification of decontamination techniques utilized in nuclear power plants. Special attention has been paid to the historical background and applicability of chemical decontamination, covering in details the detrimental corrosion effects of some well-known procedures.

ACKNOWLEDGMENTS

This work was supported by the Paks NPP Co. Ltd. (Paks, Hungary), and the Hungarian Science Foundation (OTKA Grant No. T 031971/2000).

LIST OF ABBREVATIONS

APACE	Alkaline Permanganate and Ammonium Citrate + EDTA (chemical decontamination method)(
AP-CITROX	Alkaline Permanganate and Citric acid + Oxalic acid chemical decontamination method
APOX	Alkaline Permanganate and Oxalic acid (chemical decontamination method
BR	Breeder Reactor
BWR	Boiling Water Reactor
CA-DEREM	Chemical decontamination technology using diluted reagents (citric acid, EDTA)
CAN-DECON	Chemical decontamination procedure using diluted reagents (citric acid, oxalic acid, EDTA)
CANDU	Canadian Deuterium Uranium Reactor
CEMS	Conversion Electron Mössbauer-Spectroscopy
CERT	Constant Extension Rate Tensile test
CORD	Chemical decontamination method using dilute reagents

	(permanganic acid, oxalic acid)
DF	Decontamination Factor
Dx(%)	Residual surface activity in percentage
GCR	Gas-Cooled Reactor
GFBR	Gas-cooled Fast Breeder Reactor
GW(e)	Gigawatt (electric)
HTGCR	High-Temperature Gas-Cooled Reactor
HWR	Heavy Water Reactor
IGA	Intergranular Attack
IGSCC	Intergranular Stress Corrosion Cracking
LMFBR	Liquid Metal Fast Breeder Reactor
LOMI	Low Oxidation-states Metal Ions (chemical decontamination method)
LWR	Light Water Reactor
m/m%	Mass percentage
MEDOC	Metal Decontamination by Oxidation with Cerium (chemical decontamination (decommissioning) technology)
MR	Magnox Reactor
MSBR	Molten-Salt Breeder Reactor
NPP	Nuclear Power Plant
PHWR	Pressurized Heavy Water Reactor
PWR	Pressurized Water Reactor
RBMK	Soviet-made reactor (Pressure tube boiling water reactor)
SCC	Stress Corrosion Cracking
SEM	Scanning Electron Microscopy
SG	Steam Generator
SGHWR	Steam Generating Heavy Water Reactor
VVER	Soviet-made PWR-type reactor (water-water energetic reactor)
WR	Water-cooled Reactor
XPS	X-ray Photoelectron Spectroscopy
XRD	X-Ray Diffraction

REFERENCES

[1] K. Varga, G. Hirschberg, P. Baradlai and M. Nagy, In: E. Matijevic (ed.), Surface and Colloid Science, Kluwer Academic/Plenum, New York, 2001. pp.341-393.

[2] The Nuclear World 2002. Annual Report of the IAEA 2002. (on-line available at: www.iaea.org)

[3] Nuclear Power in the World Today. IAEA Nuclear Engineering International Handbook, 2003. (on-line available at: www.iaea.org)

[4] A. Bodansky, Nuclear Energy, AIP Press, New York, 1996.

[5] G. Choppin, J Rydberg, and J.O. Liljenzin, Radiochemistry and Nuclear Chemistry. Butterworth-Heinemann Ltd, Oxford (1995).

[6] J.A. Ayres (ed.) Decontamination of Nuclear Reactors and Equipment, The Ronald Press, New York, 1970.

[7] C. J. Wood, Prog. in Nucl. Energ., 23 (1990) 35.

[8] C. J. Wood and C.N. Spalaris, Sourcebook for Chemical Decontamination of Nuclear Power Plants, EPRI Report, NP-6433, Palo Alto, California, 1989.

[9] G.C.W. Comley, Prog. in Nucl. Energ. 16 (1985) 41.

[10] (a) D.H. Lister, Water Chem. of Nucl. React. Systems. 6. BNES, London, 1992.
(b) D.H Lister, Some aspects of corrosion in cooling water systems and their effects on corrosion product transport.
EUROCORR 2003, Budapest, Hungary, 28 September-2 October 2003, Proceedings (on CD-ROM).

[11] A.P. Murray, Nucl. Technology, 74 (1986) 324.

[12] G. Romeo, Characterization of corrosion products on recirculation and bypass lines at Millstone-1. Research Project 819-1, NP-949, Interim Report, 1978.

[13] 3rd International Seminar on Primary and Secondary Side Water Chemistry of Nuclear Power Plants. Balatonfüred, Hungary, 16-20 September, 1997, Proceedings.

[14] G. Hirschberg, P. Baradlai, K. Varga, G. Myburg, J. Schunk, P. Tilky and P. Stoddart, J. Nucl. Mater., 265 (1999) 273.

[15] K. Varga, G. Hirschberg, Z. Németh, G. Myburg, J. Schunk and P. Tilky, J. Nuclear Mater., 298 (2001) 231.

[16] J. Mc Keown, IAEA Bulletin, 45 (2003) 24.

[17] F. P. Ford and P. L. Andresen In: P. Marcus, J. Oudar (eds.), Corrosion Mechanism in Theory and Practice, Marcel Dekker, New York, 1995. p. 501.

[18] K. Varga, Ph.D. thesis, Veszprém, 1990.

[19] K. Eged, Z. Kis, G. Voigt, K. G. Andersson, J. Roed and K. Varga, Guidelines for planning interventions against external exposure in industrial area after a nuclear accident. Part 1, GSF-Berich 01/03, ISSN 0721-1694, 2003.

[20] G. Reinhard, Isotopenpraxis, 18 (1982) 41; 18 (1982) 157.

[21] K. Varga, Kémiai Közlemények, 83 (1996) 77.

[22] M. Simnad and R. Ruder, J. Electrochem. Soc., 98 (1951) 301.

[23] J. Alexa, Jad. Energ., 29 (1983) 396.

[24] M. Kurbatov, G.B. Wood and J. Kurbatov, J. Phys. Chem., 55 (1951) 1170.

[25] H. Tamura and E. Matijevic, J. Colloid Interface Sci., 90 (1982) 100.

[26] H. Tamura, E. Matijevic and E. Meites, J. Colloid Interface Sci., 92 (1983) 303.

[27] P. Tewari, A. Campbell and W. Lee, Canad. J. Chem., 50 (1972) 1642.
[28] P. Tewari and W. Lee, J. Colloid Interface Sci., 52 (1975) 77.
[29] C. Wu, M. Yang and C. Lin, Radiochim. Acta, 33 (1983) 57.
[30] M. Blesa, R. Larotonda, A. Maroto and A Regazzoni, Colloid Surf., 5 (1982) 197.
[31] R. Borggaard, J. Soil. Sci., 38 (1987) 229.
[32] H. Karasawa, Y. Asakura, H. Sakagami and S. Uchida, J. Nucl. Sci. Technol., 23 (1986) 926.
[33] S. Ardizone and L. Formaro, Annali di Chimica, 77 (1987) 463.
[34] C. Tseng, M. Yang and C. Lin, J. Radioanal. Nucl. Chem., 85 (1984) 253.
[35] C. Mellish, J. Payne and G. Worrall, Radiochim. Acta, 2 (1964) 204.
[36] P. Benes, and J. Kucera, Collect. Czech. Chem. Commun., 36 (1971) 1913.
[37] K. Varga, E. Maleczki and G. Horányi, Electrochim. Acta, 35 (1990) 817.
[38] A. Kolics, E. Maleczki and G. Horányi, J. Radioanal. Nucl. Chem., 170 (1993) 443.
[39] A. Kolics and K. Varga, J. Colloid Interf. Sci., 168 (1994) 451.
[40] A. Kolics and K. Varga, Electrochim. Acta, 40 (1995) 1835.
[41] A. Kolics and G. Horányi, Appl. Radiat. Isot., 47 (1996) 551.
[42] G. Aniansson, J. Phys. Chem., 55 (1951) 1286.
[43] V. E. Kazarinov and V. N. Andreev, In: E. Yeager, J. O'M. Bockris, B. Conway, S. Sarangapani (eds.), Comprehensive Treatise of Electrochemistry Vol. 9, Plenum Press, New York, 1984, p. 393.
[44] A. Wieckowski, In: R. E. White, J. O'M. Bockris, B. Conway (eds.), Modern Aspects of Electrochemistry, Vol. 21, Plenum Press, New York, 1990, p. 65.
[45] E. K. Krauskopf and A. Wieckowski, In: J. Lipkowski, R. P. Ross (eds.), Frontiers of Electrochemistry, Vol. 1, VCH Publishers, New York, 1992, p. 119.
[46] P. Zelenay and A. Wieckowski, In: H. D. Abruna (ed.), Modern Techniques for In Situ Surface Characterization, VCH Publishers, New York, 1991, p. 479.
[47] G. Horányi, Electrochim. Acta, 25 (1980) 43.
[48] G. Horányi, B. Electrochem., 5 (1989) 235.
[49] G. Horányi, Rev. Anal. Chem., 14 (1995) 1.
[50] G. Horányi, In: P. Somasundaran, A. Hubbard (eds.), Encyclopedia of Surface and Colloid Science , Marcel Dekker, New York, 2002, p. 1.
[51] G. Horányi, In: I.T. Horváth (ed.), Encyclopedia of Catalysis, Vol. 3, Wiley-Interscience, New Jersey, 2003, p. 115.
[52] G. Horányi, In: E. Gileadi, M. Urbakh (eds.), Encyclopedia of Electrochemistry, Vol. 1, Wiley-VCH Verlag GmbH, Weinheim, 2002, p. 349.
[53] D. Lister, D. Charlesworth and B. Bowen, Corrosion, 27 (1971) 281.
[54] D. Lister, Nucl. Sci. Eng., 61 (1976) 107.
[55] D. Lister, Corrosion, 35 (1979) 89.
[56] S. Uchida, Y. Ozawa, E. Ibe and Y. Meguro, Nucl. Techn., 59 (1982) 498.
[57] F. Kawamura, K. Funabashi, M. Kikuchi and K. Ohsumi, Nucl. Techn., 65 (1984) 332.
[58] H. Rommel, Kernenergie, 21 (1978) 49.
[59] H. Rommel, G. Sachse, H. Schlenkrich and I. Mittag, J. Radioanal. Nucl. Chem., 58 (1980) 153.
[60] H. Rommel, Kernenergie, 26 (1983) 23.
[61] J. Bosholm, H-J. Hoffmann, B. Schiller and K. Winkelmann, Kernenergie 24 (1981) 382.

[62] M. Ponnet, M. Klein, M. Vincent, H. Davain and G. Aleton, Thorough chemical
      decontamination with the MEDOC process. EPRI International Decomissioning and
      Radioactive Waste Workshop, Dounreay Site-Thurso, Scotland, Sepember 17-19, 2002,
      Proceedings.

[63] CER Technology-In situ Techniques for Surface Characterization, CORMET Testing
      Systems (on-line available at: www.cormet.fi).

[64] H. Kanbe, T.Inoue, T. Tomizawa, H. Koyama and H. Itami, Nucl. Techn., 60 (1983) 367.

[65] T. Honda, K. Ohashi, Y. Furutani and A. Minato, Corrosion (Houston), 43 (1987) 564.

[66] K. Varga, Z. Németh, J. Somlai, I. Varga, R. Szánthó, J. Borszéki, P. Halmos, J. Schunk and
      P. Tilky, J. Radioanal. Nucl. Chem., 254 (2002) 589.

[67] R. O. James and T. W. Healy, J. Colloid Interface Sci., 47 (1972) 65.

[68] C.F. Baes and R.E. Mesmer, The Hydrolysis of Cations, John Wiley & Sons, New York,
      1976.

[69] D. M. Kolb, in: H. Gerischer, C. W. Tobias (eds.), Advances in Electrochemistry and
      Electrochemical Engineering, Vol. 11, Wiley & Sons, New York, 1978, p. 127.

[70] S. Music and M. Ristic, J. Radioanal. Nucl. Chem., 120 (1988) 289.

[71] S. Music, Isotopenpraxis, 21 (1985) 143.

[72] K. Varga, P. Baradlai, G. Hirschberg, Z. Nemeth, D. Oravetz, J. Schunk and P. Tilky,
      Electrochim. Acta, 46 (2001) 3783.

[73] Nuclear Energy Agency, 1999. Decontamination Techniques Used in Decommissioning
      Activities. Report by the NEA Task Group on Decontamination, NEA Report-1707. (On-line
      available at: http://www.nea.fr/html/rwm/reports/1999/decontec.pdf.)

[74] U.S. Department of Energy, 1994. Decommissioning Technology Descriptions:
      Decontamination. DoE, Office of Environmental Management, (On-line available at:
      http://www.em.doe.gov/define/techs/decon.html.)

[75] Murray, A.P. et al., Method of decontaminating metal surfaces. European Patent Application,
      Westinghouse Electric Corporation (US), No. 0164937 A1, 1985.

[76] Murray, A.P. et al., Method of decontaminating metal surfaces. European Patent
      Specification, Westinghouse Electric Corporation (US), No. 0164988 B1, 1989.

[77] (a) K. Varga, Z. Németh, Z. Homonnay, A. Szabó, K. Radó, D. Oravetz, J. Schunk, P. Tilky
      and F. Oszvald, J. Nucl. Mater. (submitted).   (b) Z. Homonnay, E. Kuzmann, S.
      Stichleutner, K. Varga, Z. Németh, A. Szabó, K. Radó, K. É. Makó, L. Kövér, I. Cserny, D.
      Varga, J. Tóth, J. Schunk, P. Tilky, G. Patek and F. Oszvald, J. Nucl. Mater. (submitted)

[78] D. Mayersky and M. Solcanyi, Corrosion Aspects or decontamination of VVER type
      primary circuit parts, Chemistry of Water Reactors, NICE, 1994.

[79] H. Willie and H. O. Bertholdt, System decontamination with CORD and decontamination for
      unrestricted release, IAEA, Wien, Nuclear power Performance and Safety, 1988.

[80] U. Linden, Full system decon. The Loviisa 2 experience, Nuclear Eng. Int., 40 (1995) 41.

[81] D. R. Lide (ed.), CRC Handbook of Chemistry and Physics, CRC Press, London, 1994.

[82] M. Prazska, J. Retbarik, M. Solcanyi, and R. Trtilek, Czech. J. Phys., 53 (2003) A687.

Radiotracer Studies of Interfaces
G. Horányi (editor)

Chapter 10.2

Environmental problems

Á. Veres

Institute of Isotope and Surface Chemistry, Research Center for Chemistry, Hungarian Academy of Sciences, H-1525 Budapest, P. O. Box 77, Hungary

1. INTRODUCTION

Certain important environmental issues are closely related to both radioisotopes and processes taking place at interfaces. More specifically, the migration of radioisotopes in geological media is of great importance, since the final disposal of nuclear wastes can reasonably be performed storing them in geological media. The geological media is expected to ensure the long-term isolation, which should exceed several times the half-time of isotopes present in the spent fuel. During the long time storage the isotopes may spread in the geological media. The spreading is expected to occur primarily in aqueous media in most cases, thus the migration of isotopes is, primarily, controlled by processes taking place at the liquid-solid interfaces of the groundwater and the stone. Thus in this particular case the "radiotracer" is used in large amounts, and the main efforts are focused to delay the migration and the appearance of the radioisotopes in the biosphere. To predict the large-scale processes laboratory and in situ experiments should be performed at smaller scale and in shorter time periods where radioisotopes can advantageously be used.

In spent fuels long lived fission product (LLFP), minor actinides (MA) and Pu are present. They belong to different segments of the periodic table, thus their chemical properties are distinctly different. For the final disposal of them three main options have been considered recently:

i)   direct disposal,
ii)   disposal after (chemical) reprocessing,
iii)   disposal after exposing the spent fuel to further nuclear treatments, e.g. transmutation of LLFP (^{99}Tc; 2.1×10^5 yr, ^{135}Cs; 3×10^6 yr, ^{129}I;

1.67×10^7 yr), MA (^{237}Np; 2.1×10^6 yr, ^{243}Am; 7.4×10^3 yr, ^{245}Cm; 8.5×10^3 yr) and ^{239}Pu; 2.44×10^4 yr.

In order to give some insight into the main technological steps of nuclear waste treatment in Fig. 1 a generalized and simplified flow diagram of the treatment is shown.

The problem of waste disposal is a central issue in the activity of many national and international organizations. This situation is clearly reflected by the great number of meetings, conferences and workshops convened by various organizations during the last forty years. Only some of them could be mentioned here.

In 1962, the International Atomic Energy Agency convened an ad hoc panel meeting on radioactive waste disposal into the ground. Since that time the Agency has periodically convened panels and/or other meetings to update knowledge concerning this field and efforts were made to disseminate new information by publishing Technical *Report* Series [1-5], TECDOC [6-8] and Proceedings [9-10].

An important series of conferences, MIGRATION conferences was started in 1987 (Munich, Germany). Since that time biennale conferences were held at different places, and venue of the eighth conference was in Bregenz, Austria, 2001.

In the frame of the so-called Radioactive Waste Management Project five workshops were organized for the discussion of the following topics:

I)   Field tracer transport experiments;

II)   Basis for modeling the effects of spatial variability of radionuclide migration;

III)   characterization of water-conduction features and their representation in models of radionuclide migration;

IV)   confidence in models of radionuclide transport for site-specific performance assessment;

V)   geological evidence and theoretical bases for radionuclide-retention process in heterogeneous media.

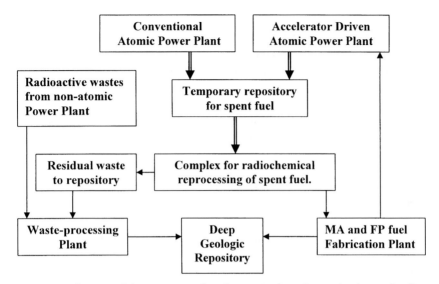

Fig. 1. Flow diagram of the treatment of nuclear waste from the production to the disposal

The fifth workshop in these series was held in Oskarshamn, Sweden, 7-9 May, 2000.

While the aim of the previous chapter was to discuss problems connected with contamination and decontamination of nuclear reactors the present chapter will be devoted to environmental problems connected with the radioactive wastes. A great stress will be laid on the clarification of the role of processes occurring at the interfaces (solid/liquid, solid/gas, liquid/gas inter-phases), involved equally in the surface decontamination experiments and radioactive waste-management.

One of the most important problems of the long-term disposal of radioactive waste is how to predict on the basis of short-term experiments and observations the long-term behavior of the systems. Different approaches were elaborated for this prediction using the term "extrapolation" for this procedure. An attempt will be made to present the various approaches to extrapolation considering migration, sorption and corrosion processes which could occur during the long-term disposal.

As sorption processes play a central role under the conditions of disposal, a wide variety of the experimental techniques is available for the study of disposal of radioactive wastes via sorption phenomena. Therefore, we give a short survey of publications, in which the authors applied mainly radiochemical methods to solve the problem connected with the long-term disposal of radioactive waste.

At the end of this chapter a relatively new approach, the transmutation method, for nuclear waste treatment will be presented.

The very principle of transmutation method is as follows:

High intensity proton beams, provided by accelerator, can produce very intense neutron flux through spallation reactions. These neutrons are useable to drive a subcritical nuclear reactor in which long-lived isotopes of nuclear waste transmute into stable or short-lived isotopes. These plants generate electricity and their additional advantages are their versatility and safety. It is the most important that the extremely hazard reactor run-away accidents can be avoided, because the sub-critical reactor can be stalled immediately by switching off the accelerator.

2. GENERAL CONSIDERATIONS FOR DEPOSITION OF NUCLEAR WASTES

In recent years we have seen a considerable world-wide activity to find a solution to the problems of storage and/or disposal of a category of hazardous wastes arising mainly from the operation of nuclear power plants and to lesser extent from the application of radioisotopes in industry and medicine. Although various versions of technologies for the management and disposal of radioactive wastes have been investigated, the disposal option appearing to be the most viable is their underground isolation in suitable geological environments.

The development and design of a repository system faces several requirements, of which the selection of solutions providing long-term performance is of outmost importance. For assessing the long-term performance of waste isolation, the effects of various reactions and the interaction occurring with the engineered barrier system, such as corrosion of canister, alteration of buffer materials and chemical interactions of the constituents released from waste products should be fully understood.

Generally, a multibarrier system should be devised for the long-term performance of waste isolation [11].

Safety assessment of a repository should take into account all barriers built or selected to protect environment from possible impact of the disposed waste or spent fuel.

A primary barrier preventing release of radioactive contaminants is the waste form, which is either spent fuel or solidified waste (glass, cement and other solid waste). One of the major tasks of studies is the description of degradation processes and quantification of contaminant releases to determine the source term of safety analysis.

The radioactive waste is placed in some kind of container (for instance, from stainless steel) forming the next barrier. The corrosion and sorption behavior of the containers and their interaction with the radioactive waste are important factors in the long-term performance of the repository system.

Finally the geological barriers of different levels ensure the safety of waste isolation.

In the following a short survey on these barriers will be given in the next subsections.

2.1. Investigation of the role of waste forms and containers

Some illustrative examples will be shown to visualize the character of studies and considerations connected with the investigation of the behavior of waste forms and containers.

Wattal et al. [12] focused their studies on the degradation of borosilicate high-level waste glasses in granite environments. For the studies of degradation of waste form, a borosilicate glass powder of – 16+25 mesh size particles was used and boiling distilled water was applied as the leaching agent. Since the vitrified waste product would be ultimately buried in a deep geological repository, they investigated the groundwater from the probable repository site in order to consider the possible role of the composition of groundwater in the leaching process.

In the studies by Kim et al. [13-14] a similar investigation was carried out using boron as an indicator of leaching of borosilicate glass. The leach rate of boron was related to the ratio of the surface area to the volume of leachate passed through the specimens. The concentration of silicon in the leaches with the crushed glass samples was in the range of 30 ppm to 60 ppm. The cumulative release of boron from the crushed glass samples decreased with the increase of leaching time. The concentrations of U, Nd, La and Zr in the leaches were below 1 mg/m^3, and the leach rate of uranium is too low, below 4×10^{-7} g m^{-2}-day.

Lee et al [15] investigated the corrosion behavior of stainless steel and vitrified waste product under geological repository conditions by radiotracer method. They investigated the pitting corrosion of stainless steel and they observed a well-defined crevice site after the exposure of sensitized stainless steel sample even to 100 g/m^3 Cl$^-$ solution at 80 °C with Ca-bentonite. The U and actinides can be retarded by a compacted bentonite. There was no difference in the components, except the depletion of sulfur, and properties of bentonite before and after leaching up to about 2 years. Sridar and Dunn [16] reported a crevice corrosion in SS316 not being initiated within 40 days of its exposure in 1000 g/m^3 Cl$^-$ solution, while Lee et al. pointed out that the well defined crevice site could be initiated within three month exposure of sensitized SS316L even in 100 g/m^3 Cl$^-$ solution at 80 °C.

Guasp et al. [17] tested binary lead alloys with tin, antimony and bismuth in synthetic groundwater and seawater at 60 °C and 75 °C by electrochemical and long-term weight tests. It was found that lead-tin alloys with tin content of 3.5 % or higher had a better corrosion resistance than commercial lead and than

Pb-Sb and Pb-Bi alloys. In a galvanic corrosion, the evaluation in that commercial lead has higher corrosion rates than high purity lead, (see Refs. 18 and 19).

Electrochemical behavior of low alloy steel in chloride containing saline waters is similar to that of high purity iron. Semino et al. [20] studied the electrochemical and corrosion behavior of steels in chloride containing saline waters. In their polarization measurements they observed three zones in the anodic curve: an active zone, a narrow passivity zone and, at higher over-potentials, a big increase in the current associated to a pitting phenomenon.

According to the literature, two distinct corrosion periods for canisters placed in a deep vault can be defined [21]: i) a hot oxidizing period, where a localized corrosion process is expected; ii) a cool non-oxidizing period, where stifling and repassivation of localized corrosion sites should occur, followed by a general corrosion process. The fastest corrosion process should occur at the initial oxidizing period of the vault, where oxygen saturated groundwater in the buffer pores reaches the metal canister surface. There will be a time delay for groundwater to wet the entire thickness of the surrounding ("buffer") material before reaching the metal surface. This time delay will depend on the buffer compaction and the hydraulic pressure, among other factors [22]. On the other hand, the effective oxygen diffusion coefficient in the buffer material will depend on its saturation degree [23].

In the Canadian model [24] oxygen concentration in the buffer is supposed to be equal to oxygen solubility in water at room temperature and at 25 °C (2.54×10^{-7} mol cm^{-3}). In the British UKAEA model [25] it is considered that all the oxygen trapped in the black-fill material is completely solubilized in the groundwater at 25 °C at normal pressure. This assumption gives higher oxygen concentration than in the former model (8.16×10^{-6} cm m^{-3}).

Grambow et al. [11] studied the long-term stability of spent nuclear fuel waste packages in Gorleben salt repository. The authors investigated the corrosion stability of the container material under repository conditions. In a repository, the waste container acts as a barrier against the mobilization of radionuclides in two ways, namely as a physical barrier and as a chemical barrier.

2.2. Retention processes in geological media

The existence and importance of retention process is recognized not only in the field of radioactive disposal, but also in other areas of non-radioactive hazardous wastes. There are differences in the factors that affect the operation of retention processes at different types of sites. Both retardation and immobilization processes are important in all cases that will be mentioned in this chapter, although only the retardation processes of sorption and matrix diffusion are widely incorporated in performance assessment (PA) models. Short definitions of some notions are as follows:

1) *Sorption*: A radionuclide that is sorbed is attached to a solid surface by some kind of mechanism, including ion exchange. Irreversible sorption is the situation where the time-scale of desorption is very slow.

2) *Matrix diffusion*: The diffusion of radionuclides between regions of flowing water, such as fractures, and connected regions of stagnant water, such as pore spaces in the fracture wall rock; matrix diffusion may be increased by surface diffusion, or restricted by anion exclusion.

3) *Colloid-facilitated radionuclide transport*: The association of radionuclides with either natural groundwater colloids or colloid arising from the presence of a repository, and the subsequent transport of radionuclide bearing colloids through the geosphere. The distance over which transport occurs depends, in part, on colloid stability and filtration.

4) *Surface precipitation*: The growth of a solid phase exhibiting a primitive molecular structure that repeats itself in three dimensions on the surface of a different contiguous solid phase.

5) *Performance-assessment (PA) modelling*: In sorption processes use of conservatively selected distribution coefficients (K_d) is widespread and likely to continue. Some PAs use nonlinear sorption isotherms. Others use semi-empirical functions to represent dependence of K_ds on geochemical conditions (pH, Eh).

   Matrix heterogeneity is often treated in a simply manner PA models due to uncertainties in the characterisation of the matrix. Some programmes have developed and applied PA models that allow explicit modelling of different matrix zones.

   Immobilisation processes have not so far been taken into account in PA in a quantitative manner. This omission is generally the result of the absence of relevant site-specific data and uncertainties associated with the slowness and complexity of the processes.

   Sorption is one of the many processes that can contribute to the retention of radionuclides in the geosphere. The processes of adsorption and ion exchange are widely observed in laboratory and field experiments, as well as in natural systems.

   Mainly three types of models have been applied to describe radionuclide sorption at mineral surfaces under equilibrium conditions between aqueous and sorbed phases.

   i)   The first one is based on empirical partitioning relationships, these are widely used in the PA. The simplest approach uses the concept of as single distribution coefficient, K_d.

ii)   The second is the mechanistic model that describes the details of the
       formation   of   chemical   species   at   mineral   surfaces   using
       thermodynamic formalisms and the surface complexation concept.
iii)   The semi-empirical site-binding models utilize the concepts of the
       former two models.

The surface species postulated in mechanistic models are supported by
evidence from high-resolution techniques. Thus, mechanistic models are well
developed for simple mineral/water interfaces, but appropriate interpretations
for the geological situations are more complex. The semi-empirical models
represent a less rigorous approach to sorption modeling than the fully
mechanistic model. The semi-empirical model has not been widely used to date,
but may represent a good compromise between the data collection requirements
and the need to predict radionuclide distribution under various chemical
conditions.

The practicable partitioning of total radionuclide mass between mobile
solution species and immobilized species associated with geosphere solid phases
is considered in all performance assessment (PA) programs.

The matrix diffusion processes occur in fractured rock systems. Several
methods exist, that have been suggested, to characterize matrix pores. The
breakthrough curves generated by these experiments are sometimes indicative of
retention by diffusion into a matrix with homogeneous characteristics, at least
within the portion of the matrix accessed in the course of the experiment. A
further problem with using field transport experiments to provide evidence of
matrix diffusion for performance assessment purposes is the long timescales of
relevance to performance assessment.

Colloid-facilitated radionuclide transport is recognized as an important
factor in recent PA and has been the subject of considerable experimental and
theoretical study in order to develop a fundamental understanding of the relevant
processes. The mechanisms of colloid generation, deposition, radionuclide
uptake, and transport were discussed in many publications.

The immobilization of radionuclides in the bulk of a solid has the potential
to provide a highly effective retention mechanism. Incorporation of
radionuclides in a solid may occur during the co-precipitation as a new phase
from a supersaturated solution, during the re-crystallization process that all
crystal surfaces undergo, or by solid-state diffusion, especially where density
defect is high. Short-term laboratory measurements on bulk systems often do not
allow to observe slow immobilization processes beside the usually much more
rapid   sorption   process.   Thus,   fundamental   understanding   of   these
immobilization processes is, in many cases, poor. However, high-resolution
techniques allow the observation of subtle processes on the nanometer scale and
offer the prospect of improved understanding of this area. As to specific
observations, dynamic equilibrium constantly rearranges material allowing

incorporation of radionuclides by re-precipitation, and radionuclides adsorbed on surface can migrate into bulk material by solid-state diffusion, where they are less available for release. These slow processes are, nevertheless, important in the context of immobilization over the long timescales of interest in performance assessment.

2.2.1. Sorption from aqueous media

These processes are controlled, on one hand, by the geochemical conditions characteristic for the surrounding geological media and by the state, composition of groundwater, on the other hand. In the following, we try to illustrate the approach to problems emerging during the experimental and theoretical study of processes occurring in the course of sorption from groundwater into geological media.

Davis [26] investigated the distribution coefficients for U(VI) adsorption as a function of pH on pure ferrihydrite and quartz surfaces (carbon dioxide gas, on hematite, etc.). An alluvial aquifer sediment total U(VI) concentration of 10^{-6} M was derived. The empirical modeling approaches have the advantages that the collection of experimental data is relatively simple and straightforward. The K_d distribution coefficients are defined as:

$$K_d = RN_{ads}/RN_{aq} \tag{1}$$

where RN_{ads} refers to the quantity of radionuclide sorbed and RN_{aq} refers to the aqueous concentration of the same radionuclide. The approach followed in [27] is a typical representation of empirical modeling. However, the parameters of all empirical partitioning relationships are conditional, and should not be utilized to describe radionuclide partitioning beyond the experimental conditions. This a major drawback to the empirical modeling approaches. For example, it was shown that the distribution coefficient for adsorption of uranium(VI) on sediment and pure mineral phase might vary by more than eight orders of magnitude over a wide range of chemical conditions, see Ref. 27.

The materials are seldom pure chemical species and thus a number of different reactions may occur in the testing or using a single material. Even, depending on the requirements the same material can be used in different functions. For instance, the principal techniques that are considered at various installations using clay minerals on a plant scale can be broadly categorized in two groups:

i) The use of clay minerals as ion exchanger in both the batch and column type of contacting devices.

ii) The use of clay minerals as additives and product conditioners as well as barriers in disposal pits and trenches.

The physicochemical characterization of natural ion exchange materials with respect to application in waste disposal of radioactive solutions is relatively easy for single ion exchange components. This becomes difficult, however, for complex mixture of constituents reacting according to a mechanism other than ion exchange. Unfortunately, this often happens to be true with soils or natural minerals including those used in local waste disposal. Numerous special procedures have been developed for the determination of capacities (ion exchange, sorption) and for the preparation of minerals or soils.

The application of radiotracer is one of these methods. For instance, in the case of a clay material the study of cation exchange using ^{110}Ag and ^{90}Sr/^{90}Y model systems [28] could be presented here.

In contrast to conventional methods, no elution of the exchanged ions is necessary here. Conversion to the desired ionic forms is performed with 1N silver nitrate labeled with ^{110}Ag, using 1N strontium chloride labeled with ^{90}Sr/^{90}Y. In the case of an exchange with Ag^{+} ions, bright daylight has to be avoided. After centrifuging, the samples were washed with water, filtered, dried and counted. The activity of the exchanger loaded with radioactive ions is measured directly to obtain the capacity. Ion exchange capacities thus obtained agree well with values obtained with conventional methods. Since many different factors can influence the capacity of clay materials, one cannot expect the different methods to give identical values for the same material.

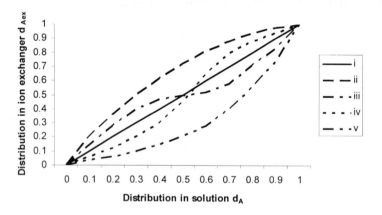

Fig. 2. Isotherms of ion exchange. The curves were obtained under the following conditions: In the solid line (i) there is no preference for ion *A* with respect to *B*. The line (ii) has ion exchanger preferring ion *A* to ion *B*. The line (iii) ion exchanger prefers ion *B* to ion *A*. The line (iv): ion exchanger prefers ion *A* at low equivalent fractions of ion *A*, and ion *B* at high equivalent fractions of ion *A*, and the line (v): ion exchanger prefers ion *B* at low equivalent ionic fraction of ion *A*, and ion *A* at high equivalent ionic fraction of ion *A*.

If an ion exchanger is placed in a solution containing ions different from those of the exchanger, reversible ion exchange reactions occur according to the following equations:

$$A^+_{ex} + B^+ \leftrightarrow B^+_{ex} + A^+ \tag{2}$$

in the case of a cation exchanger and

$$A^-_{ex} + B^- \leftrightarrow B^-_{ex} + A^- \tag{3}$$

in the case of an anion exchanger, where the *ex* subscript refers to the ions of the solid exchanger.

Activity equilibrium distribution of ions A and B in the solution will be established. Convenient quantities to characterize this distribution are the equivalent ionic fractions d_A and d_B of the ions A and B respectively in the solution, and the corresponding values d_{Aex} and d_{Bex} in the solid ion exchanger.

However, this method requires that the local equilibrium should be attained in the exchanger during elution of the column. This can be achieved if the rate of ion exchange is sufficiently high. Since this condition cannot always be satisfied with mineral ion exchangers, serious errors may arise. Only carefully pre-treated ion exchangers in well-defined ionic forms can be expected to give reproducible results. The ion exchange isotherms are obtained by plotting d_{Aex} against d_A, as shown in Fig 2.

When determining the ion exchange capacity of a clay mineral, it may be advantageous to denote the exchangeable base forming cations, e.g. $Ca^{2+}/100$ g dry weight as the S-value. The total cation exchange capacity of clay, including ions such as H^+ and Al^{3+}, is then called the T-value. Sometimes the fraction S/T is given as a percentage.

Brush et al. studied the chemical retardation of dissolved actinides [29]. The performance assessment calculations demonstrated that, in the absence of human intrusion, no radionuclides would escape from the repository to the accessible environment during the 10 000 year regulatory period. Considering their potential effects on the long-term performance, the order of these elements is Pu > Am >> U > Th >> Np.

Triay et al. [30] carried out an empirical study of the sorption of Th(IV), U(VI), and Np(V), Pu(V), and Am(III) by samples of dolomite-rich rock of Culebra. Because the detection limit of this technique is about 1%, the samples that yielded the K_d for PA contained a lower concentration of clay minerals than the Culebra as a whole (about 1 to 5%). They also carried out a few experiment with dolomite from the H-19 Hydropad and clay-rich rock from the lower, unnamed member of the Rustler. They also studied the effect of the concentration of dissolved actinide on sorption.

Brady and his colleagues [31] carried out a mechanistic study of the sorption of Nd(III), Th(IV), U(VI), Np(V), Pu(V) and Am(III) from 0.05, 0.5 and 5 M NaCl solutions by samples of well characterized, pure dolomite from Norway. They used a limited residence-time (1 min) reaction vessel to minimize the extent of dolomite dissolution, to avoid actinide precipitation. This allowed them to study the effects of pH (from about 3 or 4 to 9 or 10 in most cases), of the CO_2 concentration and the concentration of various dissolved species on sorption.

Lucero [32] and his group investigated the column-transport with intact, 14.5 cm diameter cores obtained from the Culebra. The flow rates are at or close to the upper limit of the range of in-situ fluid velocities. Therefore, they were only able to calculate minimum values of K_d. These minimum's are within the ranges obtained from the empirical and mechanistic sorption studies. They did not observe breakthrough of Th(IV), Pu(V), or Am(III) during the column-transport study. Therefore, it was only possible to determine minimum values of K_d for these actinides.

Nakata et al. [33] studied the sorption and reduction processes of neptunium(V) on the surface of magnetite, and the following conclusions were drawn: i) the amounts of sorbed Np on magnetite under anaerobic conditions are about 2 or 3 times greater than that under aerobic conditions. The dominant sorption behavior under anaerobic condition was quite different from that under aerobic conditions; ii) The dominant oxidation state of Np sorbed under anaerobic conditions is Np(IV), while that under aerobic conditions is Np(V); iii) The amounts of sorbed Np and sorption behavior under anaerobic conditions was not noticeably different from that under aerobic conditions because Fe(III) does not take part in the reduction of Np(V); iv) The reduction of Np(V) does not take place in the liquid phase by Fe(II) ion released from magnetite but by Fe(II) on the surface of magnetite, and only in the case when Np(V) ion is in contact with magnetite.

Honeyman [34] studied the role of colloids in radionuclide retention and transport through geologic media. Inorganic colloids are generated by two primary mechanisms:

i) in situ formation of colloids

ii) mobilization of existing colloids.

He summarized that most systems of interest as sites for nuclear waste repositories will be "symmetrical", at least in the far field (transport models provide the description of processes within the geological environment). Near-field systems (models, which cover all processes and features which affect the migration of radioactive/contamination from the waste forms through the engineered barriers to the host forms) are likely to become "asymmetrical" with time. Colloid-facilitated radionuclide transport in symmetrical systems will likely be minimal.

The current uncertainty over the importance of colloid-facilitated radionuclide transport is associated with a substantial lack of knowledge of system heterogeneity. More data sets are needed for developing and testing models. Methodologies are needed for up scaling. Potentially asymmetrical systems need to be identified. Research is needed for the greater understanding of the role of chemical and physical heterogeneity on colloid-facilitated radionuclide transport. Also research is needed for the deeper understanding of the role of the sorption of radionuclides by colloidal- and immobile-phase. Greater emphasis should be to be placed on organic- and bio-colloids.

Ochs et al. [35] investigated the uptake of oxo-anions by cements through solid-solution formation. They determined the K_d values for fresh and leached hydrated Portland and high-alumina cements, and in addition with selenite and selenate on synthetic ettringite ($Ca_6Al_2(SO_4)_3(OH)_{12} \cdot 26OH_2$). A comparison of the results of K_d measurements on different fresh and leached hydrated cements samples and on pure ettringite strongly suggests that selenate is taken up by cements through solid solute formation with ettringite. Uptake kinetics shows opposite trends for primary and secondary ettringite, and can be viewed as end-members of the same process. It is pointed out that very long equilibration times were required to observe these effects.

The authors developed a solid-solution model for Se(VI) uptake by ettringite, which allowed defining a simple relation between K_d of Se(VI) and dissolved/solid sulfate concentrations. This model was able to estimate uptake of different oxo-anions on a variety of hydrated cements under different conditions.

Fanghänel et al. [36] discussed the thermodynamical constants of Np(V) in concentrated salt solutions. They used many experimental data from the literature for the determination of the desired model parameters. In order to demonstrate how predictions can be made based on their model, the solubility of Np(V) is calculated for the different hydroxide and carbonate solid phases as a function of H^+ or CO_2^{2-} at different $NaClO_4$ and $NaCl$ concentrations. In carbonate solution two solid phases are formed, either $NaNpO_2CO_3 \cdot nH_2O$ or $Na_3NpO_2(CO_3)_2 \cdot nH_2O$, depending on the carbonate concentration and ionic strength. The calculation of the solubility depends, to a certain extent, on whether the crystal water of solid phase is considered or not, because the water activity considerably decreases at high electrolyte concentrations. The conclusions of the authors are: Using the Specific Ion Theory (SIT) approach [37], with the ion interaction parameters evaluated in their work, the activity coefficients of the NpO_2^+ ion, hydrolysis species and carbonate complexes and thus equilibrium constants can be accurately determined from dilute solutions to high $NaClO_4$, and $NaCl$ concentrations. The solubility of different Np(V) solid phases are predicted as a function of pH carbonate concentration and ionic strength within the maximum uncertainty of ± 0.3 logarithmic units in both $NaClO_4$ and $NaCl$ solutions.

2.2.2. Solid solution and secondary phase formation

Radionuclides can be mobilized during the corrosion of high level waste under the attack of highly concentrated salt brines in nuclear waste repository located in geological formations. The leached radionuclides can be re-immobilized *via* the formation of sparingly soluble secondary mineral phases. This can occur either by sorption or by co-precipitation/solid solution formation with secondary formed minerals.

In order to assess the radiological hazards arising from the deep disposal of nuclear waste, the chemical processes leading to radioisotope release and retention must be quantified. The interface between the waste matrix and intruding groundwater has a particular role. Co-precipitation may be significant process in controlling radionuclide release during spent fuel dissolution in a breached geological disposal site by water.

Rousseau et al [38] applied two experimental procedures to study co-precipitation of radionuclides with UO_{2+x} (s) as host phase. Their conclusions were that for co-precipitation in the ThO_2/UO_{2+x} system two-precipitation domains could be distinguished: homogenous solid solution formation at pH < 4 and pure phase formation at pH > 4. Yet, it will be necessary to characterize solid phase for pH > 4 to compare to solution analysis. For understanding of the formation of this solution, it is necessary to know the precipitation mechanism of UO_2. They proposed homogeneous precipitation to be a result of the fast growth process of UO_2.

Zimmer et al [39] investigated the formation of secondary phases after long-term corrosion of simulated high-level radioactive waste glass in brine solution at 190 °C. They studied borosilicate glass containing 25 wt. percentages of model high-level waste oxides. The glass crushed in a tungsten carbide ball mill and the resulting powders were extracted to two different grain size fractions (70-100 μm and 200-280 μm). They observed that the chemical change of the aqueous solution strongly depends on the dissolution behavior of the glass and the formation of secondary phases. The complex solution/glass interactions are strongly related to the pH of the solution. The formation and stability of secondary phases depend on various parameters, such as pH, solution composition and reaction time. The rare earth elements such as La, Ce, Nd, and Gd showed to be associated with some of these phases. The abundance of these elements in the secondary phase also depends on their concentrations in the unaltered glass (Gd > Nd > Ce > La). Due to the complexity of these processes, there is a need for a detailed understanding of the mechanism of glass dissolution, the formation of secondary phases and their stability.

In alteration layer of high-level waste glasses corroded in aqueous solutions less soluble elements are retained by sorption or by formation of secondary alteration products controlling the solution concentrations. Contact of glass with an aqueous solution results in diffusion of water into glass surface and ion exchange of the alkali ions against H_3O^+. The dissolution of the glass occurs by hydrolysis of the silicate network. The reverse reactions are referred to as condensation reactions.

Luckscheiter and Nesovic [40] demonstrated that the sorption behavior of Eu also is a representative for homologue trivalent actinides. They also studied the sorption behavior of Am in de-ionized water and in brines. The results of these sorption studies with Eu and Am in water and in brines using pre-corroded glass as substrate show clearly that the sorption behavior of Eu is representative for trivalent actinides. In general the sorption behavior of Eu and Am is very similar. The desorption coefficients determined from the Ce and Nd concentration leached out from the glass in water and brines correspond quite well to the value for Am and Eu. Consequently, the retention of rare earth elements and trivalent actinides found in long-term glass corrosion experiments can be explained by sorption onto the alteration layers. The weak sorption of rare earth elements and Am onto natural as well as synthetic, poor- crystalline smectite clay below pH 6, allows to draw the conclusion that clay mineral in the gel layer during glass corrosion is not responsible for the strong sorption of Me(III) cations on the gel.

Calcium silicate hydrates (CSH) are the main components of cement paste. These nano-porous materials consist of aggregates of nano-crystalline particles of high specific area (200-300 m^2g^{-1}). Their layered structure consists of three sheets (a calcium oxide plane sandwiched between two tetrahedral silicate chains). Viallis-Terisse et al. [41] studied the specific interaction of caesium with the surface of CSH. The sorption of caesium at CSH surface was investigated, both through sorption isotherm data and by solid-state NMR experiments. The results reveal that CSH with low Ca/Si ratios present strong sorption properties towards caesium with K_d reaching a value close to 400 ml/g. However, the sorption capacities decrease when the caesium concentration increases in the solution, suggesting that a progressive saturation of the surface sites may occur. On the contrary, CSH with high Ca/Si ratios present a low distribution coefficient (10-20 ml g^{-1}), which, however, remains constant over the whole concentration range, indicating the absence of site saturation. Sorption isotherms obtained for caesium retention on the CSH surface reveal the great ability of low Ca/Si ratio CSH to adsorb caesium cations, probably through the partially ionized silanol sites. Concerning chloride, only a little amount seems to sorb at the CSH surface, even at high Ca/Si molar ratio, where the surface charge of CSH is positive.

2.2.3. Thermal and radiation effects on water-rock systems

For geologic repository of long-lived waste a great number of requirements should be met concerning the sort and nature of rocks, where storage is planned.

One of these requirements is the thermo-radiative stability of minerals and rocks making up the bulk of the storage as the engineering construction. First, it concerns the stability of physical and chemical properties of minerals and rocks during the time of their contact with radioactive wastes. The various types of radiation effects in minerals have been investigated in detail. However, no methods of prediction can be found in the literature, therefore at present time we are not able to predict the possible changes in the physical and chemical properties of minerals.

Litovchenko et al. [42] used an extrapolation method, which based on the analysis of radiation defects, which appear in mineral structures under γ-irradiation. They analyzed the following radiation defects:

i)   electron-hole paramagnetic centers;

ii)   structural OH-groups destruction;

iii)   changes in charged state of ions in mineral structures (for example, transition Fe^{2+}, Fe^{3+}).

They presented extrapolated data of thermal and radiation effects in minerals (quartz, feldspar, apatite amphiboles, biotite and kaolinite).

An analytical approach for the long-term extrapolation was tested using the variation of radiation defects concentration. They demonstrated that rocks in which minerals do not contain appreciable quantities of Fe^{2+} ions and OH-groups could be a prospective place for long-lived radioactive waste dumping. First, the variations of radiation defects concentration must be checked immediately on exposure to γ-irradiation and heat. Pushkaryova et al. [43] published that in the system "clay mineral - heavy water" there was exchange between heavy isotopes of hydrogen and protons of structural OH-groups under γ-irradiation. However, it should be noted that the extrapolation methodology used by them was not applied satisfactorily. It was found that extrapolation to long periods is a complex problem, and in order to obtain satisfactory results further contributions from geological, chemical, physical and biological studies are required.

Merli and Fuger [44] investigated the thermo-chemical behavior of Np and Nd oxides, and presented a few results on the thermo-chemistry of $Np_2O_5(cr)$, $NpO_2OH \cdot 2.5H_2O(am)$, $Nd(OH)_3 \cdot H_2O(am)$ and $Nd(OH)_3(Cr)$. The Np_2O_5 was prepared by calcinations in air at 573 K of $NpO_3 \cdot H_2O$ which was obtained from oxidation of neptunium(V) hydroxide suspended in water at 363 K. The actinide and lanthanide oxides and hydroxides are important compounds in the problematic high-level waste disposal since such and related compounds may appear as solid phase because of the interaction of waste with the natural water.

Neck et al. [45] investigated the hydrolysis behavior of the NpO_2^+ ion at 25 °C by solubility experiments in the pH range 7 to 14 in 0.1 M, 1.0 M, and 3.0 M $NaClO_4$ solutions under pure Ar atmosphere. They measured the time dependent concentrations of Np(V), and the chemical state of Np(V) hydroxide precipitates in each solution was confirmed. They observed that in 0.1 M $NaClO_4$ the precipitate remained amorphous over several months, while in 1.0 M $NaClO_4$ the precipitate changed from an amorphous to a more stable aged state within a relative short time. In 3.0 M $NaClO_4$ the aged modification of NpO_2OH is formed from the beginning. Applying the ion interaction theory (SIT) they determined the thermodynamic constants: $logK_{sp}$ = -8.76±0.05 for amorphous $NpO_2OH(s)$ and $logK_{sp}$ = -9.44±0.10 for aged $NpO_2OH(s)$.

Technetium (as ^{99}Tc) is one of the redox-sensitive elements (including selenium, uranium and neptunium) with radionuclides of importance that can be used in assessments of the performance of high-level radioactive repositories. Several investigations have been carried out on technetium(IV) solubility. These provide a good basis for the estimation of the pH dependence of Tc(IV) solubility in the pH range of interest(pH 6-10).

Therefore Baston et al. [46] studied the solubility and migration behavior of technetium in Boom Clay. They prepared Te(IV) hydrous oxide by the reduction of pertechnetate with zinc in hydrochloric Tc-acid. The Te(IV) also electrodeposited onto Pt foil from a solution of pertechnetate in hydrochloric acid. The solubility of Tc(IV) was measured from under-saturation above precipitated Tc(IV) hydrous oxide and above electrodeposited Tc(IV) hydrous oxide in Boom Clay water. They investigated two kinds (A and B) of synthetic Boom Clay water samples containing inorganic solutes only. Further solubility measurements were also made in pre-filtered Boom Clay water samples. The results for hydrous oxide in waters were in the range 8×10^{-8} to 4×10^{-7} Tc/mol dm^{-3}, and for electrodeposited oxide in water were in the range 3×10^{-9} to 4×10^{-9} mol dm^{-3}. Solubility measured in the presence of Boom Clay is at lower limits of the ranges measured in the absence of Boom Clay.

The principle of percolation experiment with an undisturbed Boom Clay is shown in Fig. 3. The radiation source was a few μl of ammonium pertechnetate solution (about 115-150 kBq of ^{99}Tc) dried onto a filter paper. This was then sandwiched between the two Boom Clay cores of about 3.5 cm lengths. Everything was then confined in stainless steel and sealed to prevent oxygen ingress in order to reproduce the anticipated in situ chemical conditions imposed by the clay.

Boom Clay interstitial water in equilibrium with a 0.4% CO_2 atmosphere was percolated at constant flow rate through the core. Water samples were collected at regular intervals at the outlet of the cells and their ^{99}Tc concentration was measured by proportional detector or liquid scintillation counting. At the end of the percolation experiment, the clay cores were progressively removed from the permeameter and cut into thin slices to allow the determination of the migration profile of technetium in the solid clay.

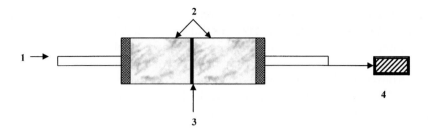

Fig. 3. Schematic representation of a percolation experiment with a Tc source sandwiched
between two Boom Clay cores. (1) shows the Clay water inlet and outlet, (2)
represents percolation, (3) is the radionuclide source and (4) represents the counting
instrument.

The data obtained from four percolation experiments are presented in Table
1. From the four experiments the Cerberus in situ test (γ) demonstrated an
experiment with γ-irradiation and heating simulating Cogema canister after 50
years of cooling time. Yet the Cerberus in situ ref. test serves as reference for
the Cerberus test with the fresh interstitial water [47].

The results of the combined solubility, modeling and migration study
illustrate how the complementary techniques contribute to the understanding of
the behavior of Tc in Boom Clay experiments. The technetium concentration in
the in situ Cerberus test is four times higher than in the corresponding in situ
reference experiment. This may indicate a slightly raised solubility of Tc(IV) at
elevated temperature or persistence of a trace of Tc(VII) under the influence of
radiolysis. From the fitting of the technetium concentrations as a function of
time, the value of the retardation factor of technetium in Boom Clay has been
estimated between 5 and 10. The consistency between laboratory solubility
calculations suggests that technetium is indeed present as Tc(IV) under the
Boom Clay chemical conditions.

Exposure of the system to γ-radiation dose rates of several hundred Gy h^{-1}
resulted only minor differences in the behavior and the local conditions around a
high level waste repository.

Maes et al. [48] used the electromigration as a qualitative technique to
study the migration behavior and speciation of U in Boom Clay. Two clay cores
of 5 cm length and a diameter of 3.8 cm were prepared, like it is on Fig. 3. The
initial activity added was 4.1×10^4 Bq of ^{233}U. The investigation was started
using $UO_2(CO_3)_3^{4-}$ which was considered to migrate as non-solubility controlled,
non-retarded species. It was shown that these species are not stable under the
Boom Clay conditions and species are formed with another speciation. Since the
newly formed species move towards the cathode they are either neutral or

positively charged. The electro migration experiments indicate that the dominant U-species in the solution is probably the solubility limited $U(OH)_4$.

Table 1.
Results obtained from four percolation experiments performed with ^{99}Tc in situ and in the surface laboratory on Boom Clay.

In situ and Laboratory Migration experiment	Hydraulic Conductivity	Fraction Tc released	Tc Concentration	Retardation Factor	K_d
(lab code)	$(m\ s^{-1})$	(%)	$(mol\ dm^{-3})$	(-)	$(cm^3\ g^{-1})$
Cerberus in situ test (γ)	$2.1\pm0.20\times10^{-12}$	0.05	$3.2\pm0.4\times10^{-8}$	5	0.8
Cerberus in situ test ref.	$1.8\pm0.20\times10^{-12}$	0.12	$7.9\pm0.6\times10^{-9}$	10	1.8
Tc Lab. A. experiment	$1.5\pm0.05\times10^{-12}$	0.17	$9.0\pm2.0\times10^{-9}$	10	1.8
Tc Lab. B. experiment	$1.4\pm0.06\times10^{-12}$	0.23	$1.2\pm0.2\times10^{-8}$	5	0.8

Sekine et al. [49] discussed radiolytic formation of Tc(IV) oxide colloids in pertechnetate solutions irradiated with bremsstrahlung of 30 MeV from a linear accelerator. They measured the fractions of TcO_4^- in the irradiated solution (from 2.6 kGy to 407.6 kGy) under different conditions. The colloid produced at 17 ± 3 °C consisted of tiny particles of 2 nm in diameter, whereas round-shape particles were formed by the irradiation at 40 °C. The reduction of TcO_4^- occurred mainly through the processes involving a bimolecular reaction of TcO_4^- with e_{aq}^- followed by successive disproportionate irradiated of Tc(VI) and Tc(V). According to the author the colloid produced in the Ar-saturated solution was very stable against further coagulation. The colloid in neutral solution had a negative charge that would stabilize in the dispersed solution. The occurrence of tiny particles of $TcO_2 \cdot nH_2O$ of 2 nm diameters could be a clue to understanding their generation mechanism. Such tiny particles could be stabilized in solution immediately after the generation of nuclei of colloids.

2.2.4. Diffusion and migration processes.

In general, migration of contaminants can be characterized as a *far field* issue; nevertheless, its *near field* part should not be neglected. The far field and near field transport models may be identified as follows:

- *Far field transport* for which models provide the description of processes within the geological environment up to the biosphere.
- *Near field transport* for which models cover all processes and features which affect the migration of radionuclides/contaminants from the waste

forms through the engineered barriers to the host formation (mainly aqueous and gas facilitated transport).

As migration is directly related to the presence of groundwater, both saturated and unsaturated conditions in rock/soil have been considered. Migration involves sorption and desorption processes which are complicated phenomena depending on the characteristics of the fluid, the contaminant, and the materials through which they penetrate. In migration studies of radionuclide, deterministic and stochastic approaches are also used.

Susan L. S. Stipp discussed immobilization of radioactive contaminants [50]. It was the aim to describe some of the chemical processes responsible for uptake and release, to provide mechanistic evidence from high-resolution analytical techniques. Some minerals are more resistant to remobilization processes and storage within the bulk of a mineral is more effective than sorption at surfaces where desorption is possible. It was observed that the effectiveness of the contaminant uptake depended strongly on the relative rates of competing processes. For example, adsorbed ions can be considered immobilized provided the rate of desorption is slow in comparison to the time-scale of decay of the component. Ions trapped within a solid solution can be mobilized again if the solid exsolves to make a mixture of two or more separate phases. Transformation of one mineral phase to some more stable phases often releases material back to the solution that is adsorbed on or incorporated within a solid solution. Diffusion through the solid is generally slow, but for minerals where defect density is high, diffusion can hide adsorbed materials or it can bring buried materials back to the surface. To determine uptake capacity by chemical methods measuring low concentrations with very high spatial resolution are necessary. The author is of the opinion that techniques exist, but if funding for instruments for basic research remains scarce, future prospect for filling in currently unknown information look bleak.

Lázár [51] within the Hungarian project studied the migration and sorption of radionuclides in Boda mudstone. This formation is considered as a prospective host rock. The spreading of radionuclides in a geological medium was described with a simple approximation. The isotopes are transported primarily by the hydraulic flow of groundwater; however, the isotope transport is retarded by sorption-desorption processes in comparison to the flow rate of groundwater. In [51] the following approximation was used:

$$\frac{v_{rad}}{v_{gw}} = \frac{1}{1 + K_d \rho \left(\frac{1-\varepsilon}{\varepsilon}\right)} \tag{4}$$

where v_{rad} and v_{gw} are the transport rates of isotopes and groundwater respectively, ρ is the density, and ε is the porosity of the rock.

Table 2
Sorption of isotopes on claystone

Isotope	K_d Maximum	K_d Minimum	K_d Mean
^{137}Cs	4900	526	1554
^{60}Co	5700	1240	3273
^{65}Sr	142	63	91
^{125}I	2.2	017	1.2

As it is seen, the extent of retention is primarily determined by the K_d values. Distribution coefficient values (K_d) were determined at 1:10 solid to liquid ratio by the author, see Table 2. The investigated Boda Claystone is a typical metamorphosed sedimentary rock. The main mineral constituents are: quartz (~ 10 %), albite (20-40%), illite-muscovite (20-40%), chlorite (~10%), calcite (<10%) and hematite (<10%). The influences of micro-racks and macroscopic fractures on the migration of isotopes were also investigated in separate experiments.

Experiments under ambient and simulated in situ conditions were extrapolated using diffusion-based approaches. Effective diffusion coefficients were determined for the most mobile iodine isotope and those were found of 10^{-9} m^2 s^{-1}, and 10^{-11} m^2 s^{-1} in aqueous and dry experiments, respectively. The result shows that under steady conditions the migration distance for the most mobile radionuclide is only of a few hundred meters during the first 10 thousand years of the life of the repository.

Molera and Eriksen [52] carried out a careful diffusion study of cations Na$^+$, Sr^{2+} and Cs$^+$, which are predominantly sorbed by electrostatic interaction. Furthermore they investigated the diffusion of Co^{2+}, for which the dominant sorption mechanism at pH>7 is inner-sphere complexation and/or surface precipitation. The experiments were performed at different densities of clay compaction (1.8, 1.6, 1.2, 0.8, 0.4 g cm^{-3}) using a through diffusion technique. They estimated the mobile fraction of radionuclide as:

$$D_a = \frac{\varepsilon D_p}{\varepsilon + K_d \rho} \qquad (5)$$

where D_a is the apparent diffusion coefficient (cm^2 s^{-1}), ε is the porosity, D_p is the pore water diffusion coefficient (cm^2 s^{-1}), K_d is the distribution coefficient (cm^2 g^{-1}) and ρ is the dry bulk density (g cm^{-3}) of the porous medium. According to Eq. (5) the value of apparent diffusion coefficient should change nearly linearly with the reciprocal value of K_d. For the surface related diffusion the Eq. (5) can be modified

$$D_a = \frac{\varepsilon D_p}{\varepsilon + K_d \rho} + \frac{\rho K_d D_s}{\varepsilon + K_d \rho} \qquad (6)$$

where D_s is a surface related diffusion coefficient. Assuming that a fraction of sorbed cations is mobile and denoting its value in time average by f, and assuming that the tortuousity is constant within pore space, Eq. (6) can be written as

$$D_a = \frac{D_w \delta \left(\varepsilon + f K_d \rho \right)}{\tau^2 \left(\varepsilon + K_d \rho \right)} \tag{7}$$

where D_w is the diffusion coefficient of the cation in free water, f (mobile fraction) is a factor that is directly to the excess mobility of the cation in the clay, δ is the constrictivity and τ^2 is the tortuousity.

The matrix diffusion is one of the primary retention mechanisms for radio-nuclides in existing and proposed geologic repositories. Haggerty [57] studied within the frame of matrix diffusion the so called heavy-tailed residence time distributions and their influence on radionuclide retention. The behavior of a solute (tracer, radionuclide, or other pollutant) within the matrix may be characterized by its residence time distribution $g^*(t)$. This is a probability density function that gives the time of the release of a molecule from the matrix following its entrance into the matrix at $t = 0$. In conventional matrix diffusion, the matrix residence time distribution is given by:

$$g^*(t) = 2\upsilon \frac{D_a}{a^2 \sum_{j=1}^{\infty} \exp\left(-A_j \frac{D_a}{a^2} t \right)} \tag{8}$$

where υ and A_j are geometry-dependent coefficients. The matrix block size (a) is typically assumed to be uniform.

The only possible heavy-tailed residence time for conventional matrix diffusion is the case of an infinitely large matrix block, where:

$$g^* \sim \left(\frac{D_a}{t} \right)^2 \tag{9}$$

In Ref. [53] the author examined more general cases where $g^*(t)$ is defined by a power-law, i.e.:

$$g^* \sim t^{1-k} \tag{10}$$

where k is a positive number.

The relationship between the matrix residence time distribution and the concentration is:

$$c = t_{ad}\beta_{tot} \frac{M}{Q} \frac{\partial g^*}{\partial t} \tag{11}$$

where c is the concentration; t_{ad} is the advection time between the injection and measurement points; M is the mass of the pulse injected into the stream-tube that is measured; Q is the flow rate of the water in which the solute is measured; and β_{tot} is the capacity coefficient, which is the ratio of the mass in the advecting water to mass in the matrix equilibrium.

It should also be noted that for a non-sorbing species, $K_d = 0$ and the apparent diffusion coefficient is equivalent to the pore diffusion coefficient (D_p). Broadly speaking, there are two sources of heavy-tailed behavior in matrix diffusion. The first is the power-law distribution of D_a/a, and the second is the fractal pore geometry. The author discussed both and drew the following conclusions:

i)   A single pore diffusion coefficient and matrix block size will not be able to present the diffusion process at all time or space-scales.

ii) The scale-dependent diffusion rate coefficient used decreases with time and space-scales. This is caused by the effective value of D_a/a^2 decreasing with the time-scale of observation. This has the potential to cause an over-estimated of the value of D_a/a^2 to be ascribed within the PA calculation.

iii) The retention capacity of host rock may be smaller than apparent in laboratory and field tests because the whole of the pore space is not accessible via diffusion over the transport time.

iv) Certain experimental designs and/or parameter estimation techniques may be particularly susceptible to non-conservative parameter estimates, and should be avoided. Problems with small-scale tracer tests may be compounded by the use of single, uniform values of matrix diffusion parameters and by not collecting or adequately using late-time/low-concentration data.

Nagao et al. [54] discussed the migration behavior of Eu(III) in sandy soil in presence of dissolved organic material. At the laboratory column experiments, the maximum value of non-equilibrium sorption was attained more rapidly at higher input concentrations of dissolved organic materials. This effect can be explained by the assumption that variation of molecular size of dissolved organic materials by the addition of Eu is small. The differences in the mobility of Eu in the presence of dissolved organic material with different characteristics are not completely understood. However, it is noted that the complexation properties of individual dissolved organic material might be very different and thus the mechanism too. Therefore, characteristics of dissolved organic materials are important factors controlling the migration behavior of the Eu/dissolved organic material complexes in sand layer.

One of the near field barrier techniques used in geo-technical studies is to determine data needed for the extrapolation of the cyclical impact of different processes. Slovák and Pacovský [55] investigated the influence of the discontinuity interfaces on physical and chemical properties of backfill material. An extrapolation method was applied on findings of experimental research. (This is a cyclic loading method for determination of the rheological stability parameters, which are widely used in the geotechnical evaluations). The authors examined the basic function of bentonite. The project aimed at the development of a method for the evaluation of the role of discontinuities in buffer, backfill and sealing system. Extrapolation relied on analytical tools, which consisted of cyclical application of experimental conditions (temperature and moisture) on bentonite blocks. Formation of discontinuity faults in the blocks induces irreversible changes of the geo-technical properties of the bentonite.

2.2.5. Unification of the database for surface complexes of radioisotopes

As we can see from the above-mentioned examples, most of the aqueous/surface speciation methods are based on the quantitative treatment of adsorption reactions. The related intrinsic adsorption constant (K_{int}) may not be directly compared between different surfaces of different solid sorbents because the K_{int} values appear to depend on the total site density Γ_T (related to surface) and/or maximum Γ_{max} density (related to surface species). The values of K_{int} may also depend on the choice of the model of solid-water interface. It was recognized that fitting at the same value of Γ_T for the same model is essential for obtaining a consistent set of K_{int} for different metal sorbents. However, it remains unclear how the K_{int} values fitted at different Γ_T parameters can be consistently compared between different surfaces of a mineral particle and even among different minerals, as it was attempted for triple layer model by Sahai and Sverjensky [56].

The different algorithms require the separate mass-balance constraints of the form $N_{t,s} = A_T\Gamma_T$, (were A_T is the total surface area in m^2, and Γ_T is density in mol/m^2) to be set for each surface/site type. Uptake of radioisotopes may also take place in solid-solution aqueous-solution (SSAS) systems for which the application of the law-of-mass action (LMA) algorithms becomes cumbersome.

Kulik, see Ref. [57] and references therein, demonstrated that the high-order multiphase non-ideal SSAS equilibria can also be formulated using LMA algorithms, if the standard molar properties of the solid-solution end-members are known or can be retrieved from the experimental or natural solubility data.

These studies also show that such thermodynamic sorption modeling is possible indeed if unambiguous definitions of standards and reference states for surface species, surface activity terms and chemical elemental stoichiometry of surface functional groups and complexes can be provided. Then the standard partial molal properties of surface species can be determined and collected into a uniform thermodynamic database, and, if necessary, converted into thermodynamic equilibrium constants K and intrinsic adsorption constants K_{int} for any site density parameter $\Gamma_T \neq \Gamma_0$ for usage in the LMA speciation codes. Furthermore the LMA codes can be modified by introducing the calculation of the surface activity terms together with the aqueous activity coefficients, and in this way circumvent introduction of the additional $N_{t,s}$ mass balance constraints.

Far for reference state, three complications occur that require non-thermodynamic correction of surface concentration relative to that dictated by the chemical potential.

 i)   Te density of reactive sites available for the monolayer binding of a sorbate may be limited by crystallographic or physical constraints, and this limit may be approached sometimes with help of other competing surface species.

 ii)   Adjacent surface-bound species on real surfaces may interact laterally in poorly understood physical phenomena, in principle, accountable for through activity coefficients.

 iii) Amphoteric oxide surfaces in water develop negative or positive surface charge in response to addition of base or acid to the bulk electrolyte.

Another important (and intensively debated) issue is how and to what extent the inherent heterogeneity of mineral surfaces must be reflected in the surface complexation models, and what are the activity coefficients of surface-bound species. In this context, Kulik believe that further development of surface activity terms concept may eventually lead to much better description of the irreversible processes in the time-dependent "process extent" modeling on the basis of the partial equilibrium principle.

Further development of the surface activity terms concept may eventually lead to a much better description of irreversible processes in the time-dependent modeling on the basis of the partial equilibrium principle. Overall, using the GEM approach the "sorption continuum" involving radioisotopes on mineral-water interfaces can be modeled only in chemical stoichiometry, without additional balance constraints for surface sites.

This opens a feasible way towards future development of a uniform database for radionuclide sorption for aqueous species, gases and minerals. In the future, application of the radionuclide adsorption thermodynamic database would facilitate evaluation of the "smart K_d" ranges relevant in the performance assessment studies, and in interpretation of new geochemical or experimental data, correlation and prediction of thermodynamic adsorption constants.

3.   DECONTAMINATION OF DIFFERENT COMPONENTS OF NUCLEAR FACILITIES DURING DECOMMISSIONING

The term "decommissioning", as used in the nuclear industry, means the actions taken at the end of a facilities useful life and the retirement of the facility from service in a manner that provides adequate protection for the health and safety of the decommissioning worker and environment. The techniques and equipment used to decontaminate radioactive surfaces and dismantle nuclear facilities are similar to those used in non-nuclear industry but have been modified and improved, in many cases, to suit the special nuclear or decommissioning application. Moreover, some techniques and equipment have been developed specifically for decommissioning purposes, for example spalling to remove the surface layer of concrete. See, for example Table 3:

Table 3.
Summary of surface decontamination methods with some characteristic properties

Surfaces	Cleaning methods	Advantages	Disadvantages
Dry contaminated surfaces	Conventional vacuum technique with filter only.	Good dry porous surfaces. Avoids water reactions.	All dusts must be filtered. Machine is contaminated.
Nonporous surfaces (metal. paint, plastic, etc.)	Employing water shot from high pressure hoses.	Contamination may be reduced by 50 %	For oiled surfaces cannot be used.
Nonporous surfaces (painted or oiled surfaces)	Steam cleaning from top to bottom by using detergent.	Reduce contamination by approximately 90 % on painted surfaces.	Spray hazard makes wearing of waterproof outfits necessary.
Nonporous surfaces (industrial film)	Rub surface and wipe with dry rag. Use clean rag for each application	Dissolves industrial film which holds contamination. It may be reduced by 90 %	The method not efficient on long-standing contamination.
Nonporous (e.g. greasy or waxed) surfaces paint or plastic finishes	Organic solvents. Also may be applied by standard wiping procedures.	Quick dissolving action. Recovery of solvent possible by distillation.	Requires good ventilation and fire precautions. Toxic to personnel.
Metal surfaces especially with porous deposits.	Inorganic acid. Dip-bath techniques advisable for movable items.	Corrosive action. It may be moderated by addition of corrosion inhibitors to solution.	Good ventilation required. Acid mixtures should not be heated
Painted surfaces (horizontal)	NaOH, Ca(OH)$_2$. Lye paint removal mixture.	Minimum contact with contaminated surfaces. Easily stored	Danger to personnel.
Painted surfaces (vertical, overhead)	10 % Tri Sodium Phosphate applied by wiping technique	Reduces activity to tolerance in one or two applications.	Destructive effect on paint. Not to be used on Al or Mg.
Surface removal	Abrasion technique. Keep surface damp to avoid dust hazard.	Activity may be reduced to as low a level as desired.	Impracticable for porous surfaces.

3.1. Decontamination of concrete and metal structures

The objectives of decontamination during decommissioning are to reduce occupational exposure, to permit reuse of the item being decontaminated or to facilitate waste management. Radioactive contamination occurs on concrete and metal surfaces during the operation of all types of nuclear facilities. The sources of radioactive contamination can be divided into two categories: i) radioactivity induced by neutron activation of certain elements in reactor components, such as the reactor vessel and adjacent structures; ii) the radioactive material deposited on the internal and external surfaces of various out-of-core systems as surface contamination. Before effective decontamination and dismantling work can commence, the location and characteristics of the contamination must be identified. These data are required to choose the decontamination process and the requirements of shielding, etc.

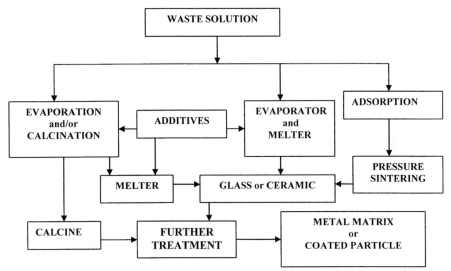

Fig. 4. Flow diagram of the basic solidification processes.

The primary aim is to convert the high-level waste solution into a mixture of oxides, which can then be further treated, by the addition of glass forming materials and reacting at high temperature, to form glass or ceramic. Fig. 4 illustrates the basic routes. It is clear that many variants are possible and have, in fact, been the subject of investigation and development. During the past years, attempts have been made to improve the properties of the product by further treatment. One of the simplest steps is evaporation and denitration (when nitric acid was used for decontamination, see Table 3.) of the waste solution to form dried products, which can either, be in granular or solid form. In the first developed process the waste was incorporated in a glass as distinct from being converted to a calcined oxide. The waste solution was mixed in a ceramic crucible with mixture of 85% nepheline syenite (a naturally occurring silicate mineral) and 15% lime to form a gel.

The IAEA Technical Report Series No. 286 (1988) presented techniques and equipments used or developed for decontamination or dismantling work during the decommissioning of all type of nuclear facilities except mining and milling. Methods described in the document do not cover a complete list of available procedures. In the report four major areas of interest are summarized as:

1. Pre-treatment and chemical decontamination processes for metal surfaces.
2. Decontamination processes and equipment for both concrete and metal.
3. Cutting and demolishing techniques and equipment for concrete.
4. Segmenting techniques and equipment for metal components.

The characteristic features of contamination layers can be summarized as follows:

In primary reactor systems, the radioactive contamination on the internal surfaces is a result of the deposition from the reactor coolant of neutron-activated particles, dissolved elements of fission product and transuranic elements released when there is a failure of fuel cladding. These deposits become part of the oxide layer, which forms on the inside wall of the piping. This layer has a complex structure that depends on a variety of parameters such as coolant chemistry, temperature of formation, system materials and operating time. It is usually desirable to characterize these layers to identify the best decontamination methods to remove the surface oxides.

In fuel reprocessing facilities, the acidic treatment inhibits the formation of an internal oxide layer and thus deposition of radionuclides is limited. In other types of nuclear facilities, such as hot cells and mixed oxide fuel fabrication plants, high levels of contamination can exist in process vessels, cells, etc., as a result of normal operations. In nuclear facilities, many external surfaces are contaminated because of leakage and spills from the process systems and from demolition, maintenance and waste management activities.

Of particular concern is the potential contamination of concrete surfaces by water-borne contamination. Unless the surface of the concrete is sealed, water-soluble radionuclides can penetrate deeply into the concrete.

Chemical decontamination processes can be divided into two groups:

i) low concentration (usually being about 1% concentration of the active reagent) and

ii) high concentration methods. The reagents used in these processes can also be incorporated into foams, gels, pastes, which are applied to the metal or concrete surface to be decontaminated.

High concentration chemical decontamination processes can involve one or more stages using different chemical solutions with intermediate rinses. The fields of application include reactor coolant circuits, items that can be immersed in to a tank and transport containers. Electrochemical decontamination process is used to remove contamination on or in metal surfaces by the controlled removal of thin layer of surface metal.

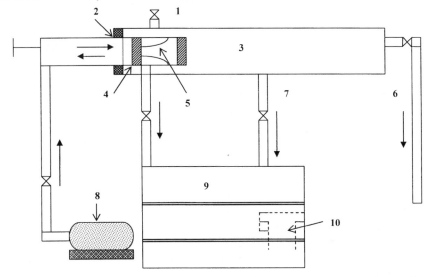

Fig. 5. Schematic diagram of movable electro polishing equipment in a corrosion test loop. (1) N_2 purge, (2) seal, (3) contaminated pipe, (4) insulators, (5) movable stainless steal cathode, (6) rinse water drain, (7) electrolyte drain, (8) electrolyte pump, (9) electrolyte reservoir, (10) heaters.

For example, the electro-polishing is a special case of electrochemical decontamination, which traditionally uses phosphoric acid as an electrolyte and a high current density to produce a smooth surface, which is desirable if the item is not to be reused [58]. If the item is not to be reused, electrolytes such as nitric acid can be used at low current density. Figs. 5 and 6 shows a schematic diagram of movable electro-polishing equipment in a corrosion test loop [59-60].

At the Gundremmingen BWR power station in Germany, several tonnes of contaminated metals from the turbine building were treated by electro polishing to bring the activity on the metal to levels suitable for unrestricted release [61].

388   Á. Veres

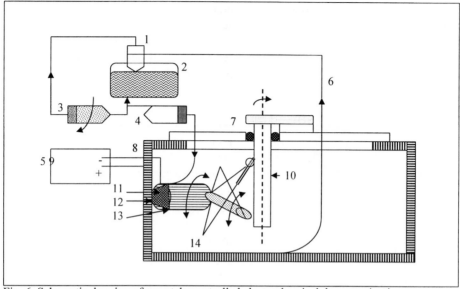

Fig. 6. Schematic drawing of remotely controlled electrochemical decontamination
    equipment. The numbers indicate: (1) water jet pump, (2) electrolyte tank, (3)
    circulating pump, (4) feed pump, (5) direct current supply, (6) recirculation unity, (7)
    rotating gear, (8) main circulating pump case, (9) joins, (10) bearing axle, (11) sponge,
    (12) cathode, (13) pressure spring, (14) moving swinging arms.

Electrochemical decontamination process was used to remove
contamination on or in metal surfaces by the controlled removal of a thin layer
of surface metal, including corrosion films that would otherwise be difficult to
remove.

The process is labor intensive; exposure can be a problem with highly
active material unless the equipment is adapted for remote operation, and non-
conducting surface layers must be removed before the process can be applied.
With reusable items, close control must be exercised to retain tolerances.

The scarifier technique is best suited for the removal of thin layers of
contaminated surface. The scarifier tool is recommended for applications where
the concrete surface is to be used after decontamination. Studies are being done
on the practicality of modifying high production commercial machines for use in
removing tightly adhering contamination from large horizontal surfaces [62].
The scarified surface is generally level. Machines are available to give a coarse
or a fine finish.

The three related cleaning techniques of brushing, washing and scrubbing
are widely and frequently used at nuclear facilities since they are simple and
inexpensive and can result in relatively high decontamination factors. Detergents
and solvents can be used in the solutions if the loose contamination associate is

of grease or oil. Abrasive powders or pads can be used if the contamination is associated with the rust or embedded near surface. Since there is a variety of solvents, chemical agents, etc. available, the technique used can be adapted to suit the degree of difficulty of the job to be done. Special care should be taken if any of the solvents give rise to inflammable or hazardous vapors.

High-pressure steam lances are also available. If steam is used as a cleaning medium, it should be saturated or wet and have a high enough velocity to ensure wetting of the entire surface to be cleaned. Soluble abrasives, e.g. boron oxide, can be added to help to remove tightly adhering surface coatings such as paint and rut. Vacuum cleaning is probably the most widely and frequently used decontamination method to clean metal and concrete surfaces and to collect dust, etc., resulting from decontamination and demolition processes. The method is simple and efficient and can be used for loose particles on both wet and dry surfaces.

Several processes are used experimentally to remove the top layer of concrete by spalling. In a flame scarifying process the top layer of concrete is heated to cause differential expansion and spalling. Pieces of up to several square centimeters erupt from the surfaces. Electrical resistance or induction heating of steel reinforcing bars causes the bars to expand more than the surrounding concrete. This makes the concrete spall from the bars. Yasunaka et al. [63] heated the water in the concrete layer to form high-pressure steam by high frequency (2540 MHz) microwaves devices, which causes the layer to break apart. The unit consists of a microwave generator, power motor, guide tubes and irradiation port. An isolator is provided to protect the equipment from reflected microwaves. The concrete debris and dust generated during the process are sucked up by a vacuum device attached to the irradiation port. The depth of spalling can be controlled by adjusting the scanning speed of the irradiation port.

These processes are in the development stage.

3.2. Testing and monitoring of gas cleanup systems.

Most of the solidification processes shown on Fig. 4 involve evaporation (and if required, denitration) of highly active waste solution. In addition to steam and oxides of nitrogen, the off-gas contains a variable carry-over of radioisotopes; either as fine particulates or as violent elements. The purpose of the off-gas system is to trap the carry-over, which could cause operation and maintenance problems and ensure that the gaseous discharge from the process to the atmosphere is within acceptable limits.

Different type of filter may be employed, depending on the concentration and particle size of the effluent aerosol, the removal efficiency required, and the temperature of the effluent and its chemical form (corrosiveness). High-Efficiency Particulate Air (HEPA) filters are normally used for effective particulate removal. Individually HEPA filters should have an efficiency of at least 99.95-99.97% for a standard test aerosol. Where HEPA filters are employed, it is usual also to install pre-filters in order to protect HEPA filters. It is important to test the effectiveness of filters.

The most common method now in use for testing the effectiveness of installed filters is the DOP aerosol method. The aerosol employed consists of liquid droplets of di-octyl-phthalate (DOP). Mainly two methods are applied. One of the pneumatic methods employs compressed air to atomize liquid DOP. In this case the equipment may become a little unwieldy, especially for flow rates in excess of 5000 m^3 h^{-1} when large volumes of air are compressed. A spray of DOP droplets is passed through a heater in which the DOP is vaporized. The vapor is then re-condensed in a cooler gas stream. Sorption devices are required to satisfactorily remove iodine in various forms it is released, one of the dominant and penetrating forms being methyl iodide in gas phase [64]. Techniques and equipment for in situ testing of filters under harsh or severe environmental conditions are currently being examined.

Friedrich and Lux [65] used a bimolecular model to describe the adsorption process in continuous flow columns filled with solid adsorbent. The analytical solution of the model for low gas concentrations and a cascade-type numerical method for higher gas concentrations were developed. An airflow apparatus using activated carbon as adsorbent and methyl-iodide labeled with ^{125}I as adsorbate was constructed for measuring breakthrough and activation curves, see Fig.7.

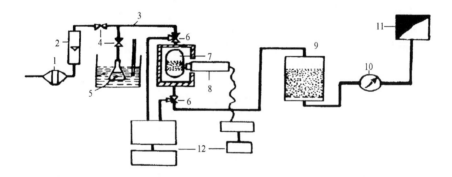

Fig. 7. The laboratory apparatus. (1) Air filters, (2) rota meter, (3) mixing tube, (4) valves adjusting for CH_3I vapor and air flow, (5) liquid methyl iodide in an ultra thermostat, (6) gas samplers, (7) adsorption column, (8) scintillation detector, (9) safety adsorber (10) air pump, (11) central ventilation system and (12) gas chromatograph and scalar recorders.

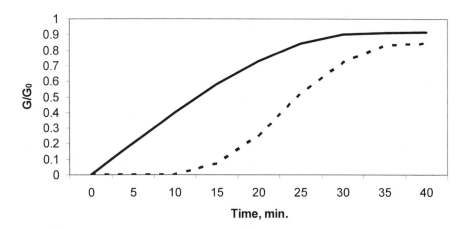

The vapor of CH_3I labeled with ^{125}I was the product from liquid CH_3I. The concentration of gaseous CH_3I (in range of 0.01-0.05 mg l^{-1}) was adjusted by the temperature of a thermostat and the airflow. The experiments were carried out at ambient temperature and 25-30 % relative humidity. Figure 8 shows a pair of simultaneously measured accumulation and breakthrough curves. Difficult environmental conditions, which are possibly more common, are encountered for instance in fuel and waste processing plants. The difficulties here are usually due to humidity, high temperature or corrosive off-gas streams and sometimes all three together. Some of the commonly used in situ methods are suitable for testing in non-ambient conditions but rarely in situations where these three factors are combined. Regarding the testing of filtration systems under accident conditions, more information is needed concerning the range of values for a number of parameters. An increasing number of test rigs are available which can better simulate accident conditions. So the factors (temperature, humidity, mechanical vibrations, irradiation, etc.) and the testing of cleanup systems in these conditions can be investigated.

In connection with the trend towards the use of more effective and reliable filters, for example in order to obtain much higher decontamination factors, it would be desirable to increase the sensitivity of filter methods. It may be that the criteria are set for accident conditions, but the test methods currently employed are not capable of demonstrating compliance with the criteria under these conditions.

4. TRANSMUTATION OF LONG-LIVED NUCLEAR WASTES BY SPALLATION REACTION AND ITS RADIOTRACER CONSEQUENCES

Spallation is a nuclear process, known from 1947 [66], in which a high-energy particle (proton, deuteron, alpha), in GeV energy range, impinging on a heavy target (tungsten and above) produces a large number of neutrons and some other particles (see Fig. 9). Some of the nucleons ejected during the first step have still sufficient energy to induce other spallation reactions with neighboring nuclei, leading to a multiplication of the emitted neutrons. Typically a 1 GeV proton on a thick heavy metal (W, Pb, U) produces around 25-45 neutrons.

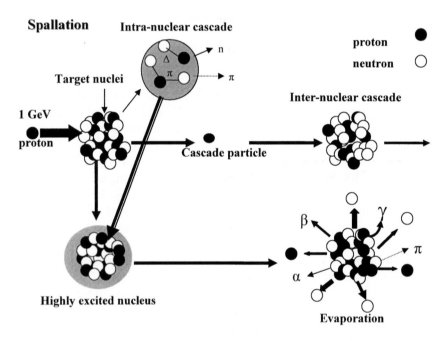

Fig. 9. In a spallation process neutrons and other particles are released by bombarding a heavy metal target with high energy protons from a high-power accelerator. The highly excited target nuclei evaporate up to 55 fast neutrons. After cooling them down by so-called moderators they can be used to drive, for example, a subcritical reactor.

To transmute long-lived radioisotopes into stable or short-lived elements we can use mainly two nuclear reactions: the capture of one neutron, or neutron induced fission (for actinides). Both processes occur in a reactor neutron flux. The spallation reaction produces an intense neutron flux, induced by a high intensity proton beam (10 to 250 mA) with particle energy in the GeV range in a heavy target. The extra neutrons provided by the accelerator allow maintaining the chain reaction while burning the long-lived nuclear waste.

Studies on the transmutation capabilities are being conducted in several countries. The US Accelerator Driven Transmutation of Waste (ADTW) program is focused on the back end of the once-through fuel cycle light water reactor [67]. A major program is proposed for the European Community to develop the MEGAPIE experiment to demonstrate high-power spallation target technology necessary for ADTW [68]. Japan in the OMEGA program has made a large financial commitment for research and development for the partition of transuranic from high-level waste and for transmuting TRU (transuranic elements) in either reactor or accelerators-driven transmuter [69].

A block diagram of accelerator configuration

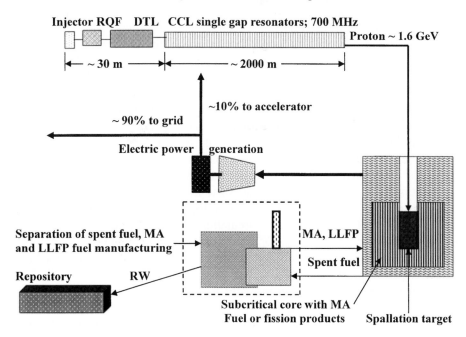

Fig. 10. Concept of an accelerator driven transmutation system (ADTS). The accelerator configuration consists of about 2 km long, 700 MHz, coupled-cavity linac (CCL), injected protons through radiofrequency quadrupoles (RFQ) and the drift tube (DTL). Other marks are: Long-lived fission product (LLFP); minor actinide MA; Radioactive Waste (RW). In the following we only deal with the separation methods of the spent fuel and nuclear waste which is indicated with dropped square.

Russian research institutes collaborated on the study of an ADT option for protection and management of plutonium, including both dismantling of nuclear weapons and nuclear power activities [70]. In such systems (Fig. 10), spallation reactions induced by high intensive beam (10 to 250 mA) of GeV protons on a heavy target produce an intensive neutron flux. These neutrons are used to drive a subcritical blanket.

Despite innovation in nuclear power plant technology, nuclear energy is unable to gain complete public acceptance due to long-term radiological toxicity of nuclear waste arising out of the back-end of fuel cycle. As any nuclear procedure involves a radiation burden to the operators, it is appropriate to apply conditions for partitioning and transmutation (P&T) techniques, which minimize environmental impacts apart from other technical considerations. One of this is the subcritical assembly, which is a reactor-like system, but unlike a conventional nuclear reactor in which fission goes on in a chain reaction without supply of external neutrons, a particle accelerator beam produces these neutrons when it is incident on a suitable target located within the assembly, see Fig. 12. The most important advantage of the accelerator driven system (ADS) is that if the accelerator is switched off the subcritical reactor comes to a standstill, thus the very hazardous runaway accidents can be avoided and isolation of radioactive material from man and the environment can be ensured.

Accelerator driven transmutation implies and depends very strongly on the chemical processes. Two primary options exist for a reference technology: the comparatively well-known aqueous separations and the pyrochemical separations that provide some specific advantages useful for accelerator transmutation of waste. Concerning the waste of ADS, the method of aqueous separations can be used, which is a well-developed technology. For the second stage of ADS fuel cycle, – the recirculation of the fuel in ADS, aqueous chemistry cannot be used, because one would like to recirculate this fuel without too long cooling time. Chemical reprocessing technology not sensitive for high radioactivity must be applied.

Pyrometallurgical separation provides these capabilities and is considered to be more proliferation resistant. Pyro-processes withstand the high heat and radiation anticipated during the processing of fuel that has been irradiated in the ADS transmuter.

Both aqueous and pyro-based separation facilities can be modularized and constructed at the plan sites. Thus, off-site transportation will include only spent fuel from current nuclear power reactors and waste forms from accelerator transmutation of waste plants; separated transuranic will never leave a secure site, which has significant advantages for public acceptance, safety, and proliferation resistance.

The radiolytical condition in the operation of the system is one of the important subjects, which determine to what extent the proposed technologies fulfil requirements of safety, reliability and technical standards. All the problems connected to radiation chemical processes during plant operation (water radiolysis, radiation induced corrosion of metallic structure elements and chemical transformations) are important and the ways of optimal and safe operation, which can be rationally chosen have to be investigated.

4.1. Aqueous separation and transmutation of nuclear waste

The proposed chemical engineering facility related to a sub critical heavy-water nuclear reactor operating in the continuous reprocessing regime can be subdivided into an in-reactor section and a unit for continuous discharge and reloading of fuel into and from a radiochemical facility. The reprocessing complex is used for successive extraction of U, Pu and Np, extraction and group partioning of other actinides and of lanthanides, vitrification of residual short-lived radionuclides. Then the components return to the reactor loop that has to be recirculated for transmutation. The continuous purification of a side stream from the reactor loop results in a constant concentration in the loop of components to be transmuted and of fission products. The question of maximum permissible concentration of the materials to be disposed as well as other technological and ecological problems of a transmutation cycle based on aqueous technologies also needs to be solved. In the frame of OMEGA program, a partitioning process was developed for separating elements from high-level waste (HLW) into four groups, namely trans-uranium elements, Sr-Cs, the Tc-Pt group metals (PGM), and others, see Fig. 11. A preliminary assessment study has been carried out in laboratory-scale experiments. The recovery might be more than 99.97% for Am and Cm, 99.85% for Np and 99.9% for U. About 1% of lanthanides might be contained in the Am and Cm product. The disposal of the other group directly into a deep underground repository without any long term cooling may be possible.

Various reagents have been developed for actinide extraction (see, for instance the scheme of a Japan four-group partitioning process in Fig. 11).

In a proposed concept, an integral power plant for transmutation of nuclear waste is based on aqueous technologies for nuclear reactors and chemical separation. The plant runs in a closed cycle of liquid nuclear waste transmutation and separation. Kulikov et al. [71] investigated:

- The aqueous separation and transmutation of nuclear waste and the breakage of irradiated suspended particles.

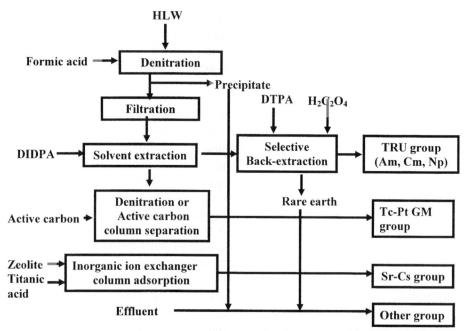

Fig 11. Four group partitioning process of high-level long-lived waste (HLLW) [80].

The intermediate and low activity waste in fluid form need to be collected, concentrated, and returned to the main partitioning cycle. It would be important to ensure that long-lived radionuclides are not released to the environment. It would be also important to solve the recirculation of the unburied actinides. A practical method to meet these requirements is not yet available. The question of maximum permissible concentration of the materials to be disposed as well as other technological and ecological problems also needs to be solved.

• Suppression of radiolytic decomposition of the heavy water.

Kinetic studies of radiolytic gas production in the presence of some acceptor were made in order to determine the volume of radiolysis gas formed from heavy-water solutions of salts of Th and trans-uranium nuclides. Irradiation experiments were made using metallic ampoules. The best results were obtained by the high-voltage discharge between Pt electrodes. In this case, the concentration of unburned D_2 was found to be ≤ 0.1 %.

• Breakage of suspended particles during the irradiation and its effect on corrosion.

It was shown that equilibrium size particles failed to contain recoiling fission products. Particles of 40-45 μm diameters were found to decrease the energy deposited in the tube walls by a factor of 10. Namely, the protective effect depends on the particle size in the suspension. The results show also that the radiolytic and erosion-corrosion behavior of heavy-water based salt solution and oxide suspension was similar. They also observed that the catalytic recombination of deuterium and oxygen into heavy water blocked the radiolytic heavy-water decomposition for the irradiated solution and suspension.

4.2. Non-aqueous partitioning processes for metal spent fuel

There are various activities of developing non-aqueous processes for metal fuel, oxide fuel and nitride fuel. Non-aqueous process uses liquid metals or molten-salts, which are more resistant to radiation degradation and provide a possibility of more compactness.

Whereas several methods of pyrochemical reprocessing are known, only two methods based on electrochemical separation have been investigated extensively and developed up to electric power plant scale. One is the molten salt electro-refining method for metallic fuels developed by the Argonne National Laboratory (USA) and the other is the pyrochemical method for oxide fuels developed by the Research Institute for Atomic Reactors (Russia). Japan has two propositions under consideration. One is to separate minor actinides (Np, Am, Cm) and long-lived fission products from irradiated oxide fuels by electro-refining after reducing them to metals and transmute them in subcritical reactors. The other is to use the nitride fuels in fast reactors and reprocess them either with molten salt electro-refining or denitriding them in molten cadmium. France has proposed to use pyrochemical processes for separation of minor actinides and fission products. European Commission is presently funding a project (PYROREP) for partitioning and transmutation in which electro-refining in molten chloride or fluoride medium is the main process. The Czech national "Pyrochemical Partitioning" program is directed primarily on the development of the front-end part of the fuel cycle technology for the molten salt reactor systems with liquid based fluoride melts [72]. The dominant purpose of the pyrochemical methods development should be associated with the development of liquid fuel based reactor transmutation systems. The problem in separation of long-lived individual minor actinides and some fission products, which could be solved effectively only by using rather complicated technologies are particularly concerned. Some of these drawbacks could be eliminated by using pyrometallurgical processes.

High radiation resistance of inorganic reagents allowing faster fuel recycling, considerable process compactness and production of a small amount of waste could be expected. The application of pyrometallurgical processes calls for tackling extraordinary demands on the construction as these processes are proceeding at temperatures above 500 °C with the application of very aggressive chemical reagents.

In the spent fuel treatment system, uranium and a majority of fission products are separated from the transuranic elements and the most hazardous long-lived fission products. The electrochemical extraction employed in the treatment plant produces a stream of un-separated plutonium and higher actinides that is mixed with the selected fission products.

One of the requirements is in principle the separation of enough uranium (99%) so that no significant new plutonium or other fissile actinides are produced in the transmutation process. Following electrochemical extraction of the uranium, the remaining fission product stream is encapsulated in solid metal fuel elements (transmutation assembles), to be introduced in the subcritical core, for irradiation. Most of the radioactivity contained in the discharged waste would decay within a shorter time, with only weak residual activity of negligible environmental impact remaining afterwards.

The pyro-metallurgical separation provides some specific advantages useful for accelerator-transmuted wastes. This alternative path for processing the spent fuel may offer advantages including greater proliferations resistance, but requires some technology development. However, such processes are at early stage of development compared with hydrometallurgical processes already in operation.

In the spent fuel treatment unit (Fig. 12), uranium and a majority of the fission products are separated from the transuranic elements and the most hazardous fission products [73-74]. The electrochemical extraction employed in this method produces un-separated plutonium and higher actinides that are mixed with selected fission products. The uranium is sent out and stored for possible recycle and further use. Another stream containing the spent fuel cladding metal, the majority of the fission products from the spent fuel, and the remaining fission products from the transmuted waste is prepared for permanent disposal. Following electrochemical extraction of the uranium, the remaining stream is encapsulated in solid metal to be introduced in the subcritical burner for irradiation. In the waste cleanup process, eventually all the fission products contained in the irradiated waste are portioned into three forms: active metals, noble metals and lanthanides.

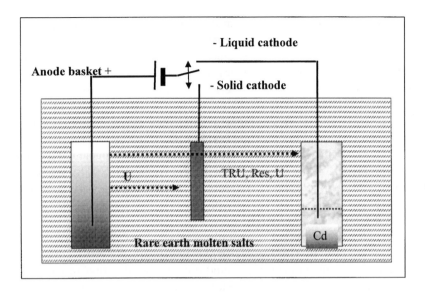

Fig. 12. Schema of an electrochemical unit proposed for separation of spent fuel, in rare earth molten salt. Uranium and a majority of the fission products are separated from the transuranic elements and the most hazardous long-lived fission products.

Before 1990 many pyro-processing techniques were used to facilitate the disposal of a fuel that could not otherwise be sent directly to a geologic repository. The uranium metal fuel is dissolved in LiCl+KCl molten bath, the U is deposited on a solid cathode, while the stainless steel cladding and noble metal fission products remain in the anode and are consolidated by melting to form durable metallic waste form. The transuranics and fission products in salt are then incorporated into a zeolite matrix, which is hot pressed into a ceramic composite waste.

The PYRO-A process was developed by the Argonne National Laboratory (US). It is a pyrochemical process for the separation of transuranic elements and fission products containing in the oxide powder resulting from de-nitration of the UREX raffinate. The nitrates in the residual raffinate acid solution are converted to oxides, which are then reduced electrochemically in a LiCl-Li$_2$O molten salt base. The chemically more active fission products (Sr, Cs) are not reduced and remain in the salt. The metallic product is electro-refined in the same bath to separate the transuranic elements on a solid cathode from the rest of the fission products. They also developed the PYRO-B process that separates the transuranic, technetium and iodine from the transmuter fuel. A typical transmuter fuel is free of uranium and contains recovered transuranic in an inert matrix such as metallic zirconium. In this process an electro-refining step is used to separate the residual transuranic elements from the fission products and recycle the transuranics to the reactor fissioning.

However, after the initial separation of uranium, all further accelerator-transmuted waste (ATW) separations and processing steps are based on pyro-metallurgical processing for two reasons. First, bulk separations provide greater proliferation resistance, and second, the pyro-process is more tolerant of the decay heat and radiation environments anticipated during the processing of fuel that has been irradiated in the target/blanket.

In the future it will be important to develop pyrochemical technologies and introduced them for the management of ATW as well as to applications in nuclear subcritical power plant systems.

5. CONCLUDING REMARKS

In conclusion, retention processes in the geosphere are well understood in a qualitative sense, understanding of sorption, immobilization processes, matrix diffusion, as well as, colloid-facilitated radionuclide transport, are well supported by laboratory and field experiments.

In many cases, our understanding is insufficient to support realistic modeling of the processes in the geologic media, and over the scales of space and time that are of interest in performance assessment. The uncertainties often lead to the adoption of simplifying assumptions in performance assessment models, including the omission of immobilization processes.

Due to the uncertainties associated with the complexity of geological systems and the long timescales of interest, conservatively selected empirical partitioning relationships are widely used to present sorption in performance assessment transport models. Immobilization processes have so far not been taken into account in performance assessments in a quantitative manner.

The characteristics of the site and repository concept, and the degree to which these are understood, determine how much performance can be attributed to different retention processes and how much needs to be attributed in order to meet acceptance criteria. Of key importance is the interaction between regulators, performance assessors and technical specialist in various scientific disciplines, in order to focus the efforts of scientific specialists on the most critical issues.

The pyrochemical technology is one of the potential methods for the separation of the radioactive components in the ADT nuclear fuel cycle. Not only economic advantage but also environmental safety and strong resistance proliferation are required for fuel cycle. In order to satisfy the requirements, minor actinides recycling applicable to different type of reactor cycles by pyro-process has been developed during more than ten years. The main technology is electro-refining for U and Pu separation and reductive-extraction for TRU separation, which can be applied on oxide and metal fuels through reduction process. The application of this method on separation of transuranic elements and long-lived fission products could contribute to the safe geological disposal and hence improvement of public acceptance.

We believe that progress in this field gives a new possibility to get rid of the biggest part of the long-lived hazardous nuclear wastes. The results of radiotracer studies also may successfully contribute to this progress.

LIST OF ABBREVIATIONS

ADS   Accelerator Driven System
ADT   Accelerator Driven Transmutation
ADTW Accelerator Driven Transmutation of Waste
ATW   Accelerator Transmuted Waste
BWR   Breeder Water Reactor
CCL   Coupled-Cavity Linac
CSH   Calcium Silicate Hydrates
DOP   Di-Octyl-Phthalate
DTL   Drift Tube Linac
GEM   Gibbs Energy Minimization
HEPA High-Efficiency Particulate Air
HLW   High-Level Waste
LLFP   Long Lived Fission Products
LMA   Law-of-Mass Action
MA   Minor Actinides
P&T   Partitioning and Transmutation
PA   Performance Assessment
PGM   Tc-Pt Group Metals
RW   Radioactive Waste
RFQ   Radio-Frequency Quadruples
SIT   Specific Ion Theory
SSAS Solid-Solution Aqueous-Solution
TRU   Transuranics

REFERENCES

[1] International Atomic Energy Agency, *Technology of Radioactive Waste Management Avoiding Environmental Disposal*, Technical Report Series No. 27 (1964)
[2] International Atomic Energy Agency, *Use of Local Minerals in the Treatment of Radioactive Waste*, Technical Report Series No. 136 (1972)
[3] International Atomic Energy Agency, *Techniques for the Solidification of High-Level Wastes*, Technical Report Series No. 176 (1976)
[4] International Atomic Energy Agency, *Decontamination and Demolition of Concrete and Metal Structures During the Decommissioning of Nuclear Facilities*, Technical Report Series No. 286 (1988)
[5] International Atomic Energy Agency, *Performance of Engineered Barriers in Deep Geological Repositories*, Technical Report Series No. 342 (1992)
[6] International Atomic Energy, *In Situ Experiments for Disposal of Radioactive Wastes in Deep Geological Formations*, IAEA-TECDOC-446 (1987)
[7] International Atomic Energy Agency, *Experience in Selection and Characterisation of sites for Geological Disposal Radioactive Waste*, IAEA-TECDOC-991 (1997)
[8] International Atomic Energy Agency, *Extrapolation of short term observation to time periods relevant to the isolation of long lived radioactive waste*, IAEA-TECDOC-1177 (2000)
[9] International Atomic Energy Agency, *Treatment and Storage of High-Level Radioactive Wastes*, (Proc. Symp. Vienna, 1962) IAEA (1963)
[10] USDOE, *International Decommissioning Symposium*, Proc. Int. Symp. Pittsburgh (1987)
[11] R. Grambow, A. Loida and E. Smailos, Nuclear Technology 121 (1998) 174
[12] P. K. Wattal, R. K. Mathur, P. K. De, V. N. Sastry, J. N. Mathur IAEA-TECDOC-1177 (2000) 91
[13] S. S. Kim, G. H. Lee, K. S. Chun, Radiochim. Acta 79 (1997) 199
[14] S. S. Kim, G. H. Lee, J. G. Lee, K.Y. Jee, K. S. Chun, Anal. Sci. 13 (1997) 361
[15] W. J. Lee, S. I. Pyun, J. W. Yeon, K. S. Chun, I. K. Choi, Materials Sci. Forum 289-292 (1997) 915
[16] N. Sridar and S. S Dunn, J. Corrosion 50 (1994) 857
[17] R. Guasp, L. Lanzani, P. Bruzzoni, W. Cufre, C. J. Semino, IAEA-TECDOC-1177 (2000) 27
[18] R. O. Cassiba and S. Fernandez, J. Nucl. Mater. 161 (1989) 95
[19] A. A. Abdul Azim, V. K. Gouda, L. A. Shalaby, S. E. Afifi, Br. Corros, J. Corrosion 8 (1973) 76
[20] C. J. Semino, A. L. Burkart, M. E. Garcia, R. Cassiba, J. Nucl. Mater. 238 (1996) 198
[21] D.K. R. Trethewey and P. R. Roberge, (eds.) *Modelling Aqueous Corrosion*, NATO ASI Series, Boston (1994) 201
[22] C. J. Semino and J. R. Galvele, Corrosion Sci. 16 (1976) 297
[23] D. A. Dixon et al. *Water uptake and stress development in bentonite and bentonite-sand buffer materials*, AECL-11591 (1966)
[24] F. King, M. Kolar, D. W. Shoesmith, Proc. Corrosion/96 Research Topical Symp. (Denver, Colorado), ed.: NACE (1996) 380
[25] G. P. Marsh et al. *Evaluating the general and localised corrosion of carbon steel containers for nuclear waste disposal*, IAEA 641-T1-TC-703/4
[26] J. A. Davis, Proc. *Workshop on Radionuclide Retention in Geological Media*, (Oskarshamn, Sweden 7-9 May 2001) OECD (2002) 71

[27] U.S. *Environmental Protection Agency, Understanding variation in partition coefficient K_d values*, EPA 402-R-99-004A, Washington D.C. 1999

[28] H. P. Boehm and K. H. Leiser, Z. anorg. Chem. 304 (1960) 207

[29] L. H. Brush, C. R. Bryan, L. C. Meigs, H. W. Papenguth and P. Vaughn, Proc. Workshop on Radionuclide Retention in Geological Media, (Oskarshamn, Sweden 7-9 May 2001), OECD (2002) 179 and 187

[30] I. R. Triay et al. *Summary and Synthesis Report on Radionuclide Retardation for the Yucca Mountain Site Characterisation Project*, Milestone 3784, LANL, Los Alamos, New Mexico (1997)

[31] P. W. Brady, H. W. Papenguth and J. W. Kelly, Applied Geochemistry 14 (1999) 569

[32] D. A. Lucero, G. O. Brown, and C. E. Heath, *Laboratory column experiments for radionuclide adsorption studies of the Culebra dolomite member of the Rustler formation*, Albuquerque, SAND97-1763 (1998)

[33] K. Nakata, S. Nagasaki, S. Tanaka, Y. Sakamato, T. Tanaka and H. Ogawa, Radiochim. Acta 90 (2002) 665

[34] R. D. Honeyman, Proc. Workshop on Radionuclide Retention in Geological Media, (Oskarshamn, Sweden 7-9 May 2001) OECD (2002) 91

[35] M. Ochs, L. Lothenbach and E. Giffaut, Radiochim. Acta 90 (2002) 639

[36] Th. Fanghänel, V. Neck and J. I. Kim, Radiochim. Acta 69 (1995) 169

[37] L. Ciavatta, Ann. Chim. (1980) 551

[38] G. Rousseau, M. Fattahi, B. Grambow, F. Boucher and G. Ouward, Radiochim. Acta 90 (2002) 523

[39] P. Zimmer, E. Bohnert, D. Boshbach, J. I. Kim and E. Althaus, Radiochim. Acta 90 (2002) 529

[40] B. Luckscheiter and M. Nesovic, Radiochim. Acta 90 (2002) 537

[41] H. Viallis-Terisse, A. Nonat, J.-C. Petit, C. Landesman and C. Richet, Radiochim. Acta 90 (2002) 699

[42] A. Litovchenko, E. Kalinichenko, V. Mazykin, B. Zlobenko, Mineral Journ. 18 (1996) 7

[43] R. A. Pushkaryova, A. S. Lytovchenko, M. A. Plastinina, Mineral Journ. 18 (1999) 85

[44] L. Merli and Fuger, Radiochim. Acta 66/67 (1994) 109

[45] V. Neck, J. I. Kim and B. Kanellakopulos, Radiochim. Acta 56 (1992) 25

[46] G. M. N. Baston, P. R. De Cannière, D. J. Ilett, M. M. Cowper, N. J. Pilkington, C. J. Tweed, L. Wang, S. J. Williams, Radiochim. Acta 90 (2002) 735

[47] L. Noynaert, G. Volckaert, P. R. De Cannière, P. Meinendockx, S. Labat, R. Beaufays, M. Put, L. Wang, M. Aertsens, A. Fonteyne, F. Vandervoort, The Cerberus Project: *Demonstration test to study the near-field effects of a HLW canister in argillaceous formation*. European Commission Nuclear Science and Technology Report. EUR 18151 EN (1998)

[48] N. Maes, H. Moors, L. Wang, G. Delècaut, P. De Cannière and M. Put, Radiochimica Acta 90 (2002) 741

[49] T. Sekine, H. Narushima, Y. Kino, H. Kudo, M. Lin and Y. Katsumura, Radiochimica Acta 90 (2002) 611

[50] S. L. S. Stipp, Proc. Workshop on Radionuclide Retention in Geological Media, (Oskarshamn, Sweden 7-9 May 2001) OECD (2002) 101

[51] K. Lázár, IAEA-TECDOC-1177 (2000) 57

[52] M. Molera, T. Eriksen, Radiochimica Acta 90 (2002) 753

[53] R. Haggerty, Proc. Workshop on Radionuclide Retention in Geological Media, (Oskarshamn, Sweden 7-9 May 2001) OECD (2002) 81

[54] S. Nagao, R. R. Rao, R. W. D. Killey, and J. L. Jung, Radiochim. Acta 88 (1998) 205

[55] J. Slovák, J. Pacovsky, IAEA-TECDOC-1177 (2000) 45

[56] N. Sahai and D. A. Sverjensky, Geochim. Cosmochim. Acta 61 (1997) 2827

[57] D. A. Kulik, Radiochimica Acta 90 (2002) 815

[58] R. P. Allen *Decontamination technology – a US perspective*, Proc. Symp. Pittsburgh, Vol. 2. USDOE, Washington, DC (1987) IV-75

[59] M. E. Pick and M. G. Segal, Nucl. Energy 22 (1983) 433

[60] O. Pavlik, *Decontamination of Nuclear Facilities by Electrochemical Methods*, Report of the Institute of Isotopes, Budapest to the IAEA, Research Contract 3667, (1985)

[61] N. Eickelplasch, and M. Lash, Proc. Int. Conf. Bournemouth, 1983, British Nucl. Soc., London (1983) 379

[62] M. M. Barbier and C. V. Chester, *Decontamination on Large Horizontal Concrete Surfaces Outdoors*, CONF-800542-2, Oak Ridge Nat. Lab., TN (1980)

[63] H. Yasunaka, M. Shibamoto, T. Sukegawa, T. Yamate and M. Tanaka, Proc. Int. Symp. Pittsburgh, USDOE Washington, DC (1987) IV-109

[64] H. Till, Proc. Symp. Vienna, IAEA (1980) 123

[65] V. Friedrich and I. Lux, J. Radioanal. Nucl. Chem. Letters 93(1985) 309

[66] R. Serber, Phys. Rev. 72 (1947) 1114

[67] D. E. Beller, J. Van Tuyle, D Bennett, G. Lawrence, K. Thomas, K. Pasamehmetoglu, Ning Li, D. Hill, J. Laidler, P. Fink, Nucl. Instr. Meth. A463 (2001) 468

[68] M. Salvatore, I. Slessarev, G. Ritter, F. Fougeras, A. Tchistiakov, G. Youinou, A. Zaetta, Nucl. Instr. Meth. A414 (1998) 5

[69] S. Matsuura, Nucl. Phys. A654 (1999) 417c

[70] V. D. Kazaritsky, P. P. Blagovolin, M. L. Okhlopkov, V. R. Mladov, V. F. Batyaev, V. V. Seliverstov, E. B. Strakhov, V. I. Volk, A. Y. Vakhrushin, M. I. Zavadsky, Z. G. Ilina, S. V. Petrin, E. F. Fomukshkin, Nucl. Instr. Meth. A414 (1998) 21

[71] L. A. Kulikov, Nucl. Instr. Meth. A414 (1998) 36

[72] J. Uhlir, *Pyrochemical reprocessing technology and molten salt transmutation reactor systems*, OECD/ENA Workshop on Pyrochemical Separations, Avignon, France (2000)

[73] J. C. Browne, F. Venneri, Ning Li and M. A. Williamson, APH N.S. Heavy Ion Phys.7 (1998) 249

[74] F. Venneri, M. Williamson, Ning Li, M. Houts, R. Morley, D. Beller, W. Sailor, G. Lawrence, *Disposition of nuclear wastes using subcritical accelerator-driven systems*, The Uranium Institute Twenty Fourth Annual Symposium 2000

Strontium: 100,101,104-107,111,125,126,327,368,379,395,396,399

Structural relaxation: 204

Sub-boundaries: 187,188,191

Supporting electrolyte: 43,47,55-59,67,69,75,83,85,90

Suppression of radiolytic decomposition: 396

Surface:
  accumulation, 99,109,115,117,391,
    method, 198,199
      of radionuclides, 321,322,329,332,334-339
  complexation, 343
    model (SCM), 102,105,108,111,113,118
  concentration calculation, 295,297,298,307
  diffusion, 167,198,365
  excess, 60,61-65,67,79,135,142,144,156,296,323,330,335,336
    relative, 61-63,66
  oxide, 89-91
    layer, 90,318,322,329,335,339,341,342,350-353,386
  sites, 47,73,85,89-91,106,108,111,133,373,382,383
    acid, 16,17,25,31
    active, 9, 13,15,19
    characterization of, 13,18
    coordinatively unsaturated (CUS), 17
    protonated, 76,77,88,90

Suspension: 104,108-114,321,322,397

Thermal and radiation effects:
  on water-rock systems, 373-377

Thin gap technique: 48,49,50,69,283,286,289,291-293,295,300-302,304,305,
                308,323

Thin Layer Activation (TLA): 81

Thiophene: 17,18,20